A SEPARATE CREATION

A Separate Creation

THE SEARCH FOR THE BIOLOGICAL ORIGINS OF SEXUAL ORIENTATION

Chandler Burr

NEW YORK

Illustrations on pp. 65 and 129 by Patricia Wynne.

Charts on pp. 154, 155, 156 courtesy of the "Journal of NIH Research"

Illustration on p. 29 copyright © 1991 by the American Association for the Advancement of Science. Courtesy of Simon LeVay.

Illustration on p. 138 courtesy of Jane Gitschier.

LIBRARY OF CONGRESS CATALOGING-IN-PUBLICATION DATA

Burr, Chandler, 1963–
A separate creation : the search for the biological origins of sexual orientation / Chandler Burr.—1st ed.
p. cm.
Includes bibliographical references and index.
ISBN 0-7868-6081-2
1. Sexual orientation—Physiological aspects. I. Title.
QP81.6.B87 1996
155.7—dc20 95-50776
CIP

Designed by M. J. DiMassi

FIRST EDITION

10 9 8 7 6 5 4 3 2 1

This book is, with good reason, for Mrs. Flynn, whom I love.

ACKNOWLEDGMENTS

Writing acknowledgments is both an elating and a daunting task, particularly daunting in the face of the disparate personalities, politics, fears, brute power, and fascinating and frustrating subtleties of the art known as the biological sciences.

Thanks go to Laura Allen (for endless cheerful, patient, detailed clarifications), Christopher Alexander, Michael Bailey, Evan Balaban (for his ideas and, equally, his good spirit), Mike Baum, Jon Beckwith, Mark Bloom, David Botstein, Nancy Brackett, Robert Cabaj, Norman Carlin, Francis Collins, Rob Collins, David Cox, Anne-christine d'Adesky, Wayne Drevets, Jim Eaton, Anka Erhardt, Steve Fodor, Doug Futuyma, Laurence Frank (for hyenas and more), Lisa Geller, Jane Gitschier, Leonard Glanz, Steve Glickman, John Golin, Roger Gorski and his nuclei, Richard Green, David Hackos, David Haig, Jeffery Hall, Janet Halley, Greg Herek, Nan Hunter, Julianne Imperato-McGinley, Richard Isay, Thane Kriener, Eric Lander, Fran Lewitter, Simon LeVay (who gave me such a broad and solid basis), Arthur Levine, Richard Lewontin, Paul Licht, Charles Link, Bruce McEwen, Terry McGuire, Heino Meyer-Bahlberg, Walter Miller, John Money, Klaus Müller, Jeremy Nathans, Kate O'Hanlan, Phil Riley, Jasper Rine, Steve Rosenthal, David Singer, Maxine Singer, Cassandra Smith, Ladd Spiegel, Robert Spitzer, Stuart Tobet, Tim Tully, Tom Wehr, Rick Weiss and Natalie Angier (for hours of conversation and a pull toward the other side), Scott Wersinger, and Larry Wright. Given their help, the clichéd caveat that all mistakes are the author's is, I hate to admit, particularly applicable in my case.

I have relied on Amy Bloom's brilliant "The Body Lies," which originally appeared in *The New Yorker;* Dennis Overbye's *Lonely Hearts of the Cosmos;* Katherine Dunn's *Geek Love;* Russ Rymer's crystalline *Genie;* Carl Sagan and Anne Druyan's *Shadows of Forgotten Ancestors;* Thomas Kuhn's *The Copernican Revolution;* John D'Emilio's *Sexual Politics, Sexual Communities;* and my favorite oncologist, Dr. Christopher Buckley (and his colleague Richard Snow) and *Thank You for Smoking.* Alan Bérubé's book *Coming Out Under Fire* and James Harrison's documentary *Changing Our Minds* were both trenchant and indispensable. Paul Hoffman produced the single best piece of science writing it's ever been my pleasure to come across, "The Man Who Loves Only Numbers" in the November 1987 *Atlantic Monthly,* and it is reflected everywhere here.

I. C. McManus did crucial trans-Atlantic duty. Steve Washington (and now Jennifer Washington) housed this book in New York during its production. Peter Ballinger and James Glucksman survived it and sustained it and me. Joe Tomkowitz lent support, moral and immoral. And Jan Witkowski gave me access to the Banbury Center's wonderful programs and to him, his great good humor, and his interest in thinking through thorny explications of genetics. Hastings Wyman and Stuart Brodsky found the title.

If this book has a godfather it is Cullen Murphy, managing editor of *The Atlantic Monthly.* It was born in the *Atlantic*'s March 1993 cover, which I wrote and Cullen edited, called "Homosexuality and Biology." Navigating the corpus callosum and Richard Isay quotes, he expertly combed out the Gordian Knots I faxed him and handed me the end of the string. Thanks also to Bill Whitworth, the *Atlantic*'s editor; his single piece of stern guidance was essential. Michael Crichton supplied early encouragement. I'm more grateful than they know, and perhaps even can know, to my editor, Rick Kot, and my agent, Eric Simonoff, for, quite simply, bringing this book into existence, with an awesome determination, despite everything. And to Debby Manette for copyediting it, and David Cashion and Edie Klemm for holding it all together. My unofficial editor was my sister, Devon Burr, who devoted hours of work and made invaluable comments.

Finally, I would like to thank five people in particular. Bill Byne, Dean Hamer, Angela Pattatucci, and Richard Pillard each gave me months of their time, their best ideas, their patience and support (and opposition), and the heart of their work. (Richard also played host and cooked wonderful things, for which he will be canonized.) They let me into their labs, their arguments, and their moods and biases, which were always enlightening, as well as into their heads. And James Fallows, Washington editor of the *Atlantic,* among a hundred other things, was the first professional journalist to believe in me as a writer and is my friend, two things that make me feel very fortunate. All five of them trusted me, as did Rick and Eric, and there is nothing, in the end, more valuable than that.

Personally, I think that if we only engage in projects we think will be noncontroversial, using as our yardstick whether what we do can be misused, then we're going to wind up doing nothing because nothing of scientific worth will pass that test. And we'll always be afraid of what we find. —*Dr. Angela Pattatucci*

Many of the views which have been advanced [here] are highly speculative, and some no doubt will prove erroneous; but I have in every case given the reasons which have led me to one view rather than another. . . . The great principle of evolution stands up clear and firm, when these groups of facts are considered in connection with others, such as the mutual affinities of the members of the same group, their geographical distribution in past and present times, and their geological succession. It is incredible that all these facts should speak falsely. He who is not content to look, like a savage, at the phenomena of nature as disconnected cannot any longer believe that man is the work of a separate act of creation. . . . We thus learn that man is descended from a hairy quadruped, furnished with a tail and pointed ears, probably arboreal in its habits. . . .

The main conclusion arrived at in this work . . . will, I regret to think, be highly distasteful to many persons. . . . But we are not here concerned with hopes and fears, only with the truth as far as our reason allows us to discover it. I have given the evidence to the best of my ability. —*Charles Darwin,* Chapter 21
THE DESCENT OF MAN (1871)

"How far we are!" he sighed.

"From what?"

"From ourselves," said the Bishop. . . . "The very idea that they have already slept tonight in Spain fills me with terror."

"We cannot intervene in the rotation of the earth," said Delaura.

"But we could be unaware of it so that it does not cause us grief," said the Bishop. "More than faith, what Galileo lacked was a heart." —*Gabriel García Márquez,*
OF LOVE AND OTHER DEMONS

CONTENTS

PART I

WHAT WE KNOW:
THE BLACK BOX

Chapter One

THE BLACK BOX

THE HISTORY of science has been anything but tranquil. In 1859, when Charles Darwin proposed to the world the theory of evolution, the world was repelled. It was revolted by the idea that man was descended from a hairy quadruped, furnished with a tail and pointed ears, probably arboreal in its habits. It was aghast that a former Christian cleric would propose it. Most of the preeminent scientists of the day renounced both what Darwin had observed and the conclusions he had drawn from it. Adam Sedgwick, Darwin's old geology teacher and friend, denounced evolution and the biological mechanism of natural selection. Thomas Carlyle called it "a Gospel of dirt." Along with his personal integrity and his motives, Darwin's science was attacked.

And yet Darwin's science, as science, was all but a sideshow. People rejected Darwin's theory not because the evidence was against him. It was not, and most who assailed him understood nothing of his arguments. What sickened the world was a science that dragged human beings down from their position of primacy to the level of the other animal species. What sickened the world was the diminishment of man, who was no longer and would never be again, as Darwin would one day rather brutally put it, a separate act of creation.

The first major biological investigation of sexual orientation was published in 1991, a neuroanatomical search that jumped from the pages of the periodical *Science* to the *New York Times* and *Time*, then to CNN and *Nightline*, and from there to the dinner tables and offices of the country. Within a relatively short time, it was followed by several genetics forays and heightened interest in hormonal evidence. The various scientists pursuing this biological mystery—neuroanatomists Simon LeVay and Laura Allen, human geneticists Richard Pillard and Michael Bailey, molecular geneticists Dean Hamer and Angela Pattatucci, and endocrinologists Heino Meyer-Bahlburg and Anka Erhardt—were proposing that homosexuals were, in fact, the work of a separate act of creation, in this case biological creation. Gay people, they were arguing, were biologically distinct human beings.

A storm of protest erupted, some of it criticism (again) aimed ostensibly at the validity of the science, but the science was notably beside the point. The turmoil was caused by the impact of the research on humankind's perception of itself. Where evolution had threatened to bring human beings down from a separate pinnacle to the level of the animals, this science threatened to raise homosexuals up from the level of the subhuman or aberrant and place them on the human plane. If its implications were important—what, it was asked apprehensively, would it mean to the Church's view of sexuality and reproduction if there were a subgroup of human beings genetically directed to have nonreproductive sex?—the implications derived their importance from a profound equalizing effect, a leveling of political, social, and even theological hierarchies. Because if this research showed that homosexuals were biologically different from other people in tiny variations in genetic helices and patterns of microscopic neurons deep inside them, it also said that in the larger, important ways—in their basic humanity, in their capacity for feeling and thought, in the aspects of day-to-day life—heterosexuals and homosexuals were the same.

Thus the explosion, which was immediate and furious. It was also utterly misdirected, a fury based on confusion.

Something about the research of sexual orientation has gone almost universally unnoticed: We have assigned it an awesome significance that it does not possess. We have, by failing to note a simple, elementary principle governing the most basic process of biological research, grossly overinterpreted what finding a gay gene, neuron, or hormone actually means.

The research does not mean what we think it does.

∽

This book is about a biological investigation into a black box.

"Black box" is a term of art used by biologists to refer to any human trait that we observe, study, and measure in each other but whose origins—which genes formed it, which hormones made it that way—remain mysterious. A black box is every biological facet of ourselves that we identify but do not know how it operates, what makes it tick.

When the media report on the biological research of the black box called homosexuality, they do so under the ubiquitous headline: "Homosexuality: Genes or Choice?" The headline is short and appears moderate. It says, "What this story means is if we find a gene for homosexuality, then people don't choose to be gay, but if we don't find a gene, people do choose to be gay. We need to find a gene to know things about the trait." This interpretation made by the press is entirely incorrect.

Nature has kept her plan, DNA, hidden for millions of years. Today we are reconstructing it slowly, painstakingly, bit by tiny bit. From looking at the human animal, we see an almost unending list of traits we want to understand, all the mysteries that make us up waiting to be solved biologically: eye color, height, cystic fibrosis, cancer, intelligence, Tay-Sachs disease, baldness, athletic ability, resistance to some viruses and susceptibility to others, skin tone and muscle mass, allergies, and sexual orientation. Some traits can be defined simply by looking at the person, such as hair color or height. Some, such as cancer or blood type (A, B, or O), cannot. Some human traits are behaviorally expressed such as manual dexterity, sexual orientation, hand-eye coordination, and schizophrenia, and some are not, such as blood type (A, B, or O), race, or the hardness of tooth enamel. Some are disease traits: hemophilia, schizophrenia, cancer, color blindness. Some are politically and religiously charged. Race is one of these, a hugely complex "common inheritance" made up of myriad traits controlled by thousands of different genes (though as yet we have no idea how many genes or where they are). Others are not politically charged: the hardness of the enamel on your teeth, which is controlled by a single gene, whose location is known, and whose functioning we understand—"Tooth enamel hardness" is a trait that is no longer a black box; we have opened this trait and now understand, inside, what makes it tick.

For biologists, each of these traits is today, or was at one time, a black box. We observed them, defined them, measured them, evaluated and assessed them, and watched them trickling through generations, but we did not know what created them.

Today, most human traits are still black boxes. One black box has been the object of decades of empirical observation, and researchers have amassed a fairly complete external description of it in the scientific literature, what is sometimes called a "trait profile." This is the first stage of biological research. (What generates this black box, what makes it what it is are the questions biologists are just beginning to pursue.) We have measured the external dimensions of this black box, and on that external level we know it well. This is what we know.

1. Biologists refer to the trait as a stable dimorphism, expressed behaviorally.

2. It exists in the form of two basic internal, invisible orientations. Over 90 percent of the population accounts for the majority orientation and under 10 percent (one reliable study puts the figure at 7.89 percent) for the minority orientation, although there is still debate about the exact percentages.

3. Only a very small number of people are truly equally oriented both ways.

4. Evidence from art history suggests the incidence of the two different orientations has been constant for five millennia.

5. A person's orientation cannot be identified simply by looking at him or her; those with the minority orientation are just as diverse in appearance, race, religion, and all other characteristics as those with the majority orientation.

6. Since the trait itself is internal and invisible, the only way to identify a person's orientation is to observe the person's behavior or reflexes that express it. However:

7. The trait itself is not a "behavior." It is the neurological orientation expressed, at times, behaviorally. A person with the minority orientation can engage, usually due to coercion or social pressure, in behavior that seems to express the majority orientation—several decades ago, those with the minority orientation frequently were forced to behave as if they had the majority orientation—but internally the orientation remains the same. As social pressures have lifted, the minority orientation has become more commonly and openly expressed in society.

8. Neither orientation is a disease or mental illness. Neither is pathological.

9. Neither orientation is chosen.

10. Signs of one's orientation are detectable very early in children—often, researchers have established, by age two or three—and one's orientation probably is defined, at the latest, by age two, and quite possibly before birth.

These first intriguing observations began to catch the attention of researchers. The trait looked biological in origin. They began to press ahead systematically with their inspection, fleshing out the answer to the first question biologists always ask of a trait: "What is it?" This question must be answered before a scientist can pursue the second, quite different question: "Where does it come from?" The data that began flowing back indicated that the trait might well have a genetic source.

11. Adoption studies show that the orientation of adopted children is unrelated to the orientation of their adoptive parents, demonstrating that the trait is not environmentally rooted.

12. Twin studies show that pairs of identical (monozygotic) twins, who have identical genes, have a higher-than-average chance of sharing the same orientation compared to pairs of randomly selected individuals; the average (or "background") rate of the trait in any given population is just under 8 percent, while the twin rate is just over 12 percent, over 50 percent higher.

But the most startling and intriguing clues came from studies that began to reveal the faint outlines of the genetic plans that underlie the trait.

13. The incidence of the minority orientation is strikingly higher in the male population—about 27 percent higher—than it is in the female population, a bit of data hinting to the biologist how the gene or genes responsible might be operating.

14. Like the trait eye color, familial studies show no direct parent-offspring correlation for the two orientations of the trait, but the minority orientation clearly "runs in families," handed

down from parent to child in a loose but genetically characteristic pattern.

15. This pattern shows a "maternal effect," a classic telltale sign
of a genetically loaded trait. The minority orientation, as expressed in men appears to be passed down through the
mother.

The trait profiled above, of course, is handedness, a stable, behavioral bimodal polymorphism with the majority orientation, right-
handedness, expressed in over 90 percent of the population and the
minority orientation, left-handedness, in around 8 percent. There are
very few truly ambidextrous people, and the art history evidence suggests these ratios of right-, left-, and ambidexterity have been constant
for five millennia. Handedness is interesting in relation to the trait
sexual orientation because of the striking similarities between the two.
Those who know the literature would know immediately that the trait
profile above is not for sexual orientation, which differs from handedness in several ways: the population ratios for each trait's two orientations vary somewhat (while left-handed people comprise 8 percent of the population, the current estimated figure for homosexuals
is between 2 and 6 percent), and identical twin (MZ) concordance
figures are radically different. Twin concordance for left-handedness
is 12 percent against a background rate of 8 percent whereas for homosexuality, MZ concordance is 50 percent against a background of
only around 5 percent, indicating that homosexuality has a much
higher purely genetic component that left-handedness. (Also, and
more subtly, the telltale "maternal effects" that both traits display are
expressed somewhat differently.)

But these are the exceptions highlighting the fact that the trait
profiles of handedness and sexual orientation are extraordinarily
alike, and virtually everything we know about the one, we know about
the other. Neither left- and right-handedness nor hetero- and homosexual orientation can be identified simply by looking at a person.
Since both are internal orientations, the only way to identify them is
by the respective behaviors that express them, motor reflex and sexual response. Handedness shows up in children starting at age two or
before, and John Money of Johns Hopkins University puts the age of
the first signs of sexual orientation at the same age. Neither left-handedness nor homosexuality correlates with any disease or mental illness
(although there are studies showing a higher correlation between left-

handedness and, for example, schizophrenia). The grammar school coercion of left-handed children to use their right hands was ended years ago.

Sexual orientation and handedness also function well as working analogies. If you are right-handed, take a pen in your left hand and try to write your name. With some effort, you can probably get it down semilegibly, but the fact that you have engaged in left-handed behavior does not make you a left-handed person. Behavior is irrelevant; your orientation is what counts. And you are just as right-handed sitting still watching a movie as when swinging a tennis racquet with your right hand. Did you choose to be right-handed? No? Then prove it. (You can't; as one clinical researcher noted tersely, "Science can't 'prove' you don't choose to have appendicitis.") Just as obviously, an interiorly heterosexual person is not homosexual even in the midst of homosexual intercourse. Behavior (when it does not reflect the interior orientation) is irrelevant, and a homosexual is equally homosexual when having sex and when driving a car.

Another biologically significant similarity between the two is their ubiquitous and consistent presence across populations. "Of particular interest," researcher I. C. McManus writes of handedness, "is the absence of geographical differences, a finding compatible with handedness being a balanced polymorphism present in all cultures."

There is one interesting difference between handedness and sexual orientation, and that is that we actually know less about the biological origins of handedness than about those of sexual orientation. David Marash, science journalist of ABC TV, reported that a genetic locus linked to homosexuality "suggests that homosexuality may not be a choice," which was itself a suggestion that unless we locate a gene for homosexuality, unless we know where it is and how it works, then homosexuality is a choice (or, at a minimum, that we somehow couldn't know whether it is a choice or not). Unfortunately, this is the equivalent of saying that since we haven't found the gene that governs left-handedness—and we haven't—then left-handed people choose to be left-handed (or, at a minimum, we can't determine if they do so).

It is the equivalent of saying that if you don't understand the principles of how an internal combustion engine functions, you can't know if your car will run.

It is similar to saying that since physicists haven't found the cause and origin of gravity, Boeing and Airbus and NASA cannot build their flying machines. For gravity is a black box. It's a force whose characteristics we know well, whose trait profile we have measured so accurately that we can use it to guide and propel spacecraft between planets that are themselves traveling along trajectories we calculate by knowing the strength of gravity. We can describe it in exquisite detail: The force of gravity is equal to the gravitational constant (6.67 times 10 to the negative 11th Newtons) times the mass of the first and second objects divided by the distance between them. ($F = Gm1m2/r2$.) On the surface, we know everything about gravity.

And under the surface, we know nothing. We know nothing about gravity's origins or actual physical operation or "mechanism" (a word used by both physicists and biologists). Gravity may be created by curved space-time (which Einstein believed) or by Something Else (which quantum theory claims) operating through particles called gravitons. While physicists have found particles generating the strong, weak, and electromagnetic forces, the operation of gravity remains intractably mysterious. "We have demonstrated the existence of three of the four forces in particle accelerators," notes a physicist. "No one has ever physically demonstrated the existence of gravity. We've never seen a graviton. Yet I know of no one who would claim that because we haven't found the particle, we can't say definitively that gravity exists."

What would account for the baffling illogic of the media, the failure to grasp this elementary aspect of biological research? Perhaps modern manufacturing has turned our intuitive conceptions of genetics upside down. In an age of complex machines built from detailed plans that are human-designed, human-engineered, and under our control, perhaps we have forgotten that the most complex machine on the planet, the human body and brain, springs from a design plan of a complexity and nature we have barely begun to understand. Since we have become the creators of machines, our perspective is reversed; we understand them from the inside first and only then build them. We forget there are machines whose insides are at present far beyond our reach.

Or perhaps it is simply the refusal to recognize the evidence before our eyes since that recognition would ruthlessly change so much.

Genes are, as Harvard geneticist Richard Lewontin has pointed out, just bits of DNA, threads of lifeless acid molecules inside cells. They are design plans for the many parts of the human body. The only reason genes exist is to be "expressed" as thousands of traits, such as eye color, but, Lewontin notes impatiently, you don't look at genes to answer the question "What is the trait eye color?" Genes don't have eyes. Only people have eyes, and if you're interested in defining about eye color, look at people. Now if, Lewontin adds readily, you want to know where eye color comes from, you'd look at genes. But that's a different question. Genes will tell you how the trait was created, not what the trait is.

Contemporary research into what sexual orientation is began essentially in the 1940s when Alfred Kinsey launched his study of human sexuality. Kinsey's work asked only the first of the two biological questions: "What is the trait?" His methodology was appropriate to his purpose: He posed questions. Part of his research involved simply tracking down what most social scientists and psychoanalysts presumed would be a few isolated homosexual individuals. Since homosexuality was presumed caused by faulty parenting and psychological stress, or by bad moral example in unsavory quarters, it would only be found irregularly. Historian John D'Emilio wrote of the shock caused by what Kinsey actually discovered:

> Kinsey's findings on homosexuality departed so drastically from traditional notions that he felt compelled to comment on them. In his male study Kinsey acknowledged that he and his colleagues "were totally unprepared" for such incidence data and "were repeatedly assailed with doubts" about their validity. Checking and cross-checking their tabulations only widened the distance separating their results from the estimations of others. "Whether the histories were taken in one large city or another," Kinsey wrote, "whether they were taken in large cities, in small towns, or in rural areas, whether they came from one college or from another, a church school or a state university or some private institution, whether they came from one part of the country or from another, the incidence data on the homosexual have been more or less the same. . . . Persons with homosexual histories are to be found in every age group, in every social level, in every conceivable occupation, in cities and on farms, and in the most remote areas of the country."[1]

Kinsey was also the first to observe systematically homosexuality's immutability. Psychiatrist Richard Isay, with reference to attempts to change sexual orientation, has written of Kinsey's experience:

> Kinsey and his co-workers for many years attempted to find patients who had been converted from homosexuality to het-

erosexuality during therapy, and were surprised that they could not find one whose sexual orientation had been changed. When they interviewed persons who claimed they had been homosexuals but were now functioning heterosexually, they found that all these men were simply suppressing homosexual behavior . . . and that they used homosexual fantasies to maintain potency when they attempted intercourse. One man proclaimed that, although he had once been actively homosexual, he had now "cut out all of that and don't even think of men—except when I masturbate.[2]

In part it was the directness of Kinsey's approach to his subject that was shocking. D'Emilio notes that "Kinsey treated his sensitive subject in the matter-of-fact manner more typical of the bug collector than the writer on sex." (Actually, he *was* a bug collector; Kinsey was a highly respected zoologist who had made a reputation studying gall wasps.)

The quiet academic chose a quiet academic publisher for his sexuality surveys. The publisher ran a few marketing tests and estimated it would sell about 5,000 copies of the 804-page tome. The book was released on January 3, 1948. Within two weeks 185,000 copies were in print. The book shot to the top of the best-seller lists, where it encamped for months, alarming the country. Writes D'Emilio:

> . . . the picture Kinsey provided of the sexuality of ordinary white Americans must have been startling. Among men he found that masturbation was a nearly universal practice, that virtually all had established a regular sexual outlet by the age of fifteen, that half of the husbands in the survey engaged in extramarital intercourse, and that 95 percent of white American males had violated the law in some way at least once along the way to an orgasm.

The head of Union Theological Seminary, Henry Van Dusen, said at the time of Kinsey's work that it reflected "degradation in American morality approximating the worst decadence of the Roman era." Research into human sexuality seems to be perpetually alarming, as much at the end of the 20th Century as its middle.

Since Kinsey, extensive testing and observation of the trait homosexuality has taken place. The research of sexual orientation as a human trait is essentially complete, and we now have a detailed trait description. What this has left us with is the black box, and this is where we are today. We know the dimensions of the box, how tall it is and how wide; we know how much it weighs, what it is and isn't. We know how it is expressed. What we don't know is what formed it. The tightly closed lid hides the biological plan that created it, the hormones or genes or biochemicals that made it what it is. The biologists have now begun the next phase of their work.

Currently we have no medical test, no CAT scan, no radioactive counter, no X ray, no blood workup, no biopsy to identify either a left-hander or a homosexual. Both are indistinguishable from majority counterparts. Both can move among us secretly at will. They can hide their behavior, mask their true natures for their entire lives in societies that repress them. In many Arab cultures, there appear to be no left-handed people at all. But they are there, hiding. There is only one way to determine if a person is left-handed: He or she states it. The same is true of homosexuals. By the time the research is complete, we may be able to distinguish in some lab test the differences that make some of us homosexual.

The search for markers to identify what was previously unknowable and the definition of what has always been undefined means that when a child registers for elementary school, or a man or woman enters the army or gets an annual physical, and blood is drawn, someone in some laboratory may extract from it pure, whitish DNA and pour it into a bed of electrified gelatin, leaving that person's sexual orientation written clearly in the streaks of the autorad. It means that when a woman is pregnant, a few stray cells from her womb may tell her, or someone else, this aspect of her child. It means that gay people will no longer have the option of hiding.

The biologists have asked, "What is the box?" They are now asking "How does this box work? What genes or hormones or neurons make it tick?" Scientists have begun to open the lid of this black box and stick their hands down into the genes and dendrites and hormones that create human traits to try to solve this mystery, the origins of human sexual orientation.

TWO TRAIT PROFILES: HANDEDNESS AND SEXUAL ORIENTATION

	Trait:	
	Handedness	*Sexual Orientation*
Description	Stable bimodalism, behaviorally expressed	Stable bimodalism, behaviorally expressed
Distribution[1]	Majority and minority orientations	Majority and minority orientations
Population distribution	Majority orientation: 92%	Majority orientation: 95%
	Minority orientation: 8%	Minority orientation: 5%
Population distribution of orientations according to sex	Male: 9% Female: 7%	Male: 6% Female: 3%
Male/Female ratio for minority orientation	1.3/1 Minority orientation 30% higher in men than women	2/1 Minority orientation approximately 50% higher in men than women
Correlation data: Does minority orientation correlate with		
race	No	No
age	No	No
geography	No	No
culture[2]	No	No
pathology:[3]		
mental	No	No
physical	No	No
Age of appearance of trait	@ age 2	@ age 2
Is either orientation chosen?	No	No
Is either orientation pathological?	No	No

| | Trait: | |
	Handedness	*Sexual Orientation*
Can external expression be altered?	Yes	Yes
Can interior orientation be altered clinically?	No	No
Is trait familial/does trait run in families?	Yes	Yes
Pattern of familiality	"Maternal effect"— implies X-chromosome linkage	"Maternal effect"— implies X-chromosome linkage[4]
Parent-to-child segregation[5]	Little to none. Handedness of adopted—in contrast to biological—children shows no relationship to that of adoptive parents, indicating a genetic influence.	Little to none. Sexual orientation of adopted—in contrast to biological—children shows no relationship to that of adoptive parents, indicating a genetic influence.
Increased incidence of siblings of gay offspring having minority orientation?	Yes. Elevated rate of left-handedness in families with other left-handed children.	Yes. Elevated rate of homosexuality in families with other homosexual children.
Are monozygotic (identical) twins more likely to share minority orientation?	Yes.	Yes
MZ concordance for minority orientation[6] (Background rate:)	12% (8%, so MZ rate is 1.5 times higher)	50% (5%, so MZ rate is 10 times higher)

Sources: I. C. McManus, "The Inheritance of Left-Handedness," *Biological Asymmetry and Handedness,* Ciba Foundation Symposium 162. (Chichester) John Wiley & Sons: 1991, 251–267; J. Michael Bailey and Richard Pillard, "A Genetic Study of Male Sexual Orientation," *Archives of General Psychiatry* 48 (December 1991): 1089–1096; Dean Hamer et al., "A Linkage Between DNA Markers on the X Chromosome and Male Sexual Orientation," *Science* 261 (July 16, 1993): 321–327

[1]Both traits show a very small number of humans are ambioriented. Handedness shows almost none for both men and women—McManus: "Measures of handedness usually show a bimodal distribution with few subjects appearing truly ambidextrous." Sexual orientation, likewise, shows almost none for men but a still small though significant number for women.

[2]However, may highly influence *expression*.

[3]There is currently fierce debate over the existence of a correlation between left-handedness and certain pathologies, most notably schizophrenia. Some researchers assert that handedness, thought to reflect one aspect of brain lateralization, may be a result or a cause—in some manner a concomitant—of schizophrenia's etiology or pathophysiology. A study done by Charles Boklage ("Schizophrenia, brain asymmetry development, and twinning," *Biol. Psychiatry* 12, 19–35, 1977) powerfully developed the hypothesis, and Nancy Segal ("Origins and implications of handedness and relative birth weight for IQ in monozygotic pairs," *Neuropsychology* 27, 549–561, 1989) also supports some form of correlation. On the other hand, Luchins et al. (1980) and Lewis et al. (1989), in their respective replication attempts of Boklage's work, found little support, and Gottesman et al. ("Handedness in twins with schizophrenia: was Boklage correct?" *Schizophrenia Research* 9, 83–85, 1993) conclude that there does not appear to be an association between handedness and schizophrenia. (See Gottesman for a more complete bibliography.) The point, however, is the distinct difference between the trait profile of handedness and that of sexual orientation: while there is clinical debate in scientific and research circles over whether handedness correlates in some way with psychobiological abnormalities, no such debate exists regarding sexual orientation, and neither heterosexuality nor homosexuality are implicated in any mental or physical pathology.

[4]A subset of gay men show the maternal effect. It does not appear in women.

[5]"Segregation" is a genetic term of art meaning the way the trait shows up in individuals down through generations.

[6]Indicates that genetics play a significantly greater role in sexual orientation than in handedness.

PART II

WHAT WE DON'T KNOW: OPENING THE BOX

"There are always two sides to each story," goes the maxim, but the problem is that as far as biology goes, it is false.

Present a thesis, and biology usually can be used to build two iron-clad cases, one for it, one against, one yes, one no. Reporters and politicians always cleave the research of sexual orientation into black or white, up or down. Reporters, who instinctually offer every study and fact as supporting the political Right or Left, refer in serious tones to their two-sided journalistic construction as The Debate. But this debate is science not as scientific process but as political vehicle. Biology is presented as unambiguous because the two hostile political camps that fuel The Debate are unambiguous. It is either/or because either/or is the easiest way to report things. It uses headlines like "Homosexuality: Biology or Choice?" It proves one of the two sides, true or false, right or wrong.

But science is process. Science is ambiguity. Science is not two sides but millions.

Still . . . if you wanted to tell a story and make it all black or all white, if you wanted to present your viewers with a yes (or a no) because they wanted a yes (or a no), if you wanted to give political supporters a rallying cry for whichever side you favored, you could use biology to build a "definitive" case.

There are two stories you could tell. The first is titled, "Definitive Proof that Homosexuality is Biological." It goes like this.

Chapter Two

THE DEBATE: DEFINITIVE PROOF THAT HOMOSEXUALITY IS BIOLOGICAL

O N THE MORNING of August 30, 1991, in a time of furious political debate over the biology of IQ, violence, alcoholism, gender differences, homosexuality, and race, the world woke to front-pages headlines announcing the discovery of a physical difference between heterosexuals and homosexuals: A nucleus—a small cluster of cells—in the hypothalamus of the human brain was larger in straight men than it was in gay men.[1] The biological race for the origin of sexual orientation was on.

The study made a celebrity of its author, Simon LeVay, a young, gay, and, by all accounts brilliant neuroscientist at the Salk Institute, and it dropped the biological research of sexual orientation into the center of a growing political debate over gay rights like a small bomb. This was not, scientifically, a simple mystery that was being pierced. At a time when biology was daring to attempt to crack open increasingly complex mysteries in the human machine, here was sudden evidence that we were nearer than we realized to deciphering the most basic traits that make us what we are.

Dr. LeVay is an intense, wiry British neuroanatomist who made his scientific reputation in visual systems. (Two of his mentors, Nobel laureates Torsten Wiesel and David Hubel, were pioneers in the field.) LeVay had concentrated on a specific problem—the way the brain processes visual information—but over the years, he had grown interested in the biology of how he came to be gay. He decided to find out. Like other traits that were expressed behaviorally, sexual orientation was thought to be an infamously tough mystery. There were, however, already a number of clues. LeVay decided to go back and begin with a strange discovery that involved one of the complex cascade of changes that hormones create in rats' brains. In 1959, a UCLA researcher named Dr. Charles Barraclough found that if you injected a female rat shortly before or just after her birth with testosterone, a male hormone, it would for some reason sterilize her, and she would be unable to ovulate.[2] "Ovulation" in this case meant not only an ovary releasing an egg but the biochemical chain of hormonal switches that, when tripped, lead the ovary to launch the egg on its journey.

Unlike human females, who have monthly cycling, rat females cycle every four days when their glands start pumping estrogens (female hormones) into their bloodstreams. A small bit of hormone 1 feeds back and ratchets up production of hormone 2, which produces still more of hormone 1, like waves urging each other on with increasing violence. The estrogens pour into the blood, which flows to the brain where, when they reach a certain concentration, they kick on the hypothalamus (a small part of the brain responsible for regulating thirst and other functions), which in turn dispatches its own liquid signal to switch on the pituitary. The pituitary then sends a hormonal signal to the ovaries, which release the egg into the fallopian tubes. This domino effect leading to ovulation is called positive feedback, because higher estrogen levels "positively" trigger the hypothalamus, and on down the line, an amazing process of biochemical ovulation that begins life. Barraclough discovered that exposing female rats to testosterone around the time of birth destroyed the dominoes permanently.

Then Geoffrey Harris, a researcher at Oxford University, made another, stranger discovery: that it was possible to make *male* rats ovulate, at least in this biochemical sense.[3] Castrate newborn males to take away their source of precious male testosterone, and then in adulthood inject them with female estrogen. The estrogen floodwaters rise, switching on the hypothalamus, which kicks on the pituitary, and so on, and the result is "positive feedback"—ovulation, but in a male rat. Graft an ovary onto the male, and he will ovulate perfectly. Subsequent experiments revealed yet another remarkable discovery: a male-female asymmetry. When biologists deprived males of testosterone, they got male rats with femalelike positive feedback—males with female biochemistry. But deprive females of estrogen, and the females do not become like males. There is no comparable "masculinization." They merely develop quite contentedly as females with a positive female feedback.*

In the end, scientists realized that to become male, males had to have testosterone. Of course, they had to start with an XY male genetic blueprint, but the male Y was worthless without testosterone. Why? Testosterone, it turned out, actually *physically alters* the brain. Rats, it was discovered, are born with "gendered" brains—males with male brains, females with female brains—in exactly the same way they are born with gendered genitals, males with penises, females with vaginas.

But there was one catch: timing. The female brain needed no estrogen to become female, but for a male rat to develop a male brain, he has to be exposed to testosterone within five days after birth, the "critical period." After the fifth day the masculinizing window of opportunity slams shut forever, the critical period is over, and no matter what hormones he is exposed to, that XY genetically male rat will grow up with a female brain.

We now knew that the basic-model, default brain of the rat (and, it turns out, for humans too, during the first trimester) is female, and in order to make a *special* brain—that is, a male brain—and to develop a penis instead of a clitoris, a fetus must be exposed to testosterone.

This was the concept on which the neurobiological search for homosexuality would rest, the "sexual differentiation of the brain." In 1977, peering through his laboratory microscope at tiny gray-pink slices of rat brain, UCLA neurobiologist Roger Gorski discovered a nucleus in the rat hypothalamus that was five times larger in males than females.[4] The nucleus is huge—so large, in fact, that to any layperson simply looking at slices of male and female brains the difference is immediately obvious—and the brains can be sexed, picking the males and females, as easily as by looking at the rats' genitals. But then, from a biological point of view, brains and genitals are the same thing: features related to gender and sex that are physically molded by sex hormones.

Any part of the body consistently of one shape in women and another in men is termed sexually dimorphic, of two shapes according to sex. Hips and breasts in men and women are sexually dimorphic, as a glance at any of Titian's paintings will tell you, as are somewhat less obvious aspects of our bodies, such as fat-to-muscle ratio. Geni-

---

*Estrogen's role here is a matter of debate. Some biologists believe it is not necessary for normal female development, some believe it is.

tals are only the most obvious dimorphism. For Gorski's team, it was logical to name the new rat nucleus the sexually dimorphic nucleus (SDN).

Gorski has become the national authority on the SDN and its neurological and hormonal aspects. The study of the nucleus, which has an immense and dramatic impact on the behavior of rats, also led him, somewhat inadvertently, to become a filmmaker. On the seventh floor of the UCLA biology building, in a classroom, Gorski prepares to show his videotapes of the sexual behavior of rats. They have been previewed at neurological conferences around the world by Gorski, an enthusiastic, almost elfin man with a mustache. As a graduate student prepares the video system, Gorski discusses the sexual positions of his subjects.

While running experiments, certain scientists noticed almost incidentally that injections of hormones reversed the sexual positions rats normally assumed. It changed the way they ovulated yes, but that was an invisible, biochemical effect; this was out-and-out behavior, and much more dramatic: Give a female testosterone, and not only does she show less lordosis (the receptive female mating posture where she curves her back and lifts her genitals for mounting by the male), but she exhibits typical male sex behavior and begins mounting other females. The more testosterone they receive, the more female rats display typical male-rat sex behavior. In contrast, deprive males of testosterone by castrating them just before birth, and in adulthood their sex behavior is decidedly female; they not only mount less, they let themselves be mounted by other males. Through perinatal exposure to the brain-organizing effects of opposite-gender hormones, the animals have been made gay.

In the video room, the graduate student coaxes the machines to life and slips in the cassette. On the bank of monitors hanging down like boxy stalactites from the ceiling over the black lab tables, the tape begins to roll. Gorski enjoys screening his oeuvre. "There are six couples," he announces cheerfully, although in the first shot only one bored-looking white rat sniffs the set. "That's an unaltered female. They're going to put in another female that has been injected with testosterone." On cue, a hand reaches down into the frame and lets fall a second rat, black and white, which plops bemusedly next to her all-white companion.

The two females edge around one another vaguely, but within seconds the testosterone-injected female begins aggressively sniffing the unaltered female. In a flash she mounts her, her groin pumping away at the female poised in unmistakable lordosis beneath her. Up and down the room twenty video screens are jumping in unison with the image. The three or four bio students working at the tables don't bother to look up.

In the next segment, the film cuts to two males, one "perinatally castrated" (his testes were cut off at birth, taking away his source of testosterone) and one "intact," as Gorski jovially if unpoetically puts it. Again, the altered male arrives from above, and after some initial maneuvering, the castrated male responds to the intact male's advances by bending his back and raising his genitals in female lordosis, submitting as the intact male mounts him.

The tape continues with similar scenes. Its point is clear: Hormones determine sexual behavior. And this is an effect noted in other animals as well, such as songbirds. Among the zebra finches of Australia, for example, the male sings and the female doesn't. In the finches' brains, a certain nucleus controls singing. Birds with large nuclei sing, and the nuclei's size is determined by male hormones, so unless something goes wrong it is always larger in males. But give a newly hatched female testosterone, and she will sing like a male. Look at her brain after the testosterone has done its work, and her nucleus will have grown to the size of a male's.

We had known that hormones influenced our behavior and molded our bodies. With Gorski's work, we learned that they molded rats' brains. Neuroanatomists who worked on humans thus hypothesized that hormones were also molding human brains as well. Soon this was confirmed. In 1982 the British team of cell biologist Christine de Lacoste and physical anthropologist Ralph Holloway published a study in the journal *Science*.[5] They had found that like rats, parts of the brain in men and woman are shaped differently. De Lacoste and Holloway's sex dimorphism was in the part called the corpus callosum.

At almost exactly the length of a credit card, the corpus callosum is one of the largest single structures in the brain. It isn't a thinking or processing structure, like the frontal lobe, or one that helps react to a tennis ball or dodge a taxi, like the cerebellum, nor does it have a regulatory and processing function like the microprocessor-like hypothalamus, which eternally monitors the levels of the body's heating, cooling, and fuel. What it *is* is the brain's central wiring shaft, a big AT&T cable stuffed with information-carrying fibers called axons running through the center of the brain to connect its two halves and enable communication between them. De Lacoste and Holloway found that the shape of the corpus callosum's splenium (its bulbous posterior fifth) differed so dramatically between the sexes that, like Gorski's rat nucleus, complete novices could easily tell if a brain belonged to a man or a woman simply by looking at it.[6]

And then, in 1990, a Dutch researcher in Amsterdam named Dick Swaab announced something completely unexpected: He had investigated a nucleus in the human brain called the suprachiasmatic nucleus, or SCN, and found that it too was dimorphic—not with sex but with sexual orientation. The nucleus was twice as large in homosexual men as in heterosexual men.[7]

LeVay, his interest kindled, was led logically, almost inexorably, by these studies. He already had a host of clues from the "phenotype," the trait of sexual orientation as people expressed it externally, and these clues pointed toward biology. Now came still more clues, but this time from *inside* the body, at the level of neurons, leading to the same conclusion: that somewhere inside them, in their genes or hormone systems or—it now appeared—brains and hormone levels, gay and straight people were physically different from each other. But where to look for evidence of the biological origins of sexual orientation? LeVay wondered. Where would the substrate, the piece of brain molded by and/or molding sexual orientation, be?

Swaab's SCN seemed like an obvious candidate. But the SCN governs the body's daily rhythms, not its sexual drive. So LeVay began to look elsewhere. Laura Allen, a postdoctoral candidate in Roger Gorski's lab, provided the next clue. Allen had wondered if, like rats, we humans had an SDN, a sexually dimorphic nucleus. She gathered some human brains and opened them to the part of the hypothalamus where the SDN would be. She did not find an SDN. But she did find something else.

Allen located and identified four nuclei, each the size of the tip of a pin, in the anterior portion of the human hypothalamus. She and Gorski named them interstitial nuclei of the anterior hypothalamus 1, 2, 3, and 4—or INAH (pronounced "EYE-nah") 1 through 4 for short.[8] Allen had found that both INAH 2 and 3 were sexually dimorphic in humans, significantly larger in men than women. (The political reaction was swift. As evidence of brain differences between men and women, Allen's discovery was met with howls of protest from feminists, who saw it (correctly) as more ammunition for conservatives in the battle over gender; Allen was saying: Men and women are, in profound ways, different.)

"Hypothalamus" means under the thalamus, which is where this part of the brain is found. As far as we know, the thalamus, which sits in the center of the brain, is a relay station for almost all the information that comes into the brain from the outside world, receiving it, sorting it, and then routing it to appropriate areas for further processing. The smaller hypothalamus, tucked under the thalamus and cerebrum and next to the amygdala, is interesting in that it acts as an "internal" version of its larger, exterior-focused counterpart. It too relays signals, but its information-gathering and processing jurisdiction is limited to signals that come from inside the body—a sensory FBI to the thalamus's CIA. LeVay knew that the hypothalamus is a regulatory authority monitoring information from the nervous system, constantly evaluating and controlling (among other things) thirst, body temperature, and sex drive; it has the job of turning on and off the pituitary, the all-important gland whose hormonal products set off ovulation in rats and humans.

The hypothalamus itself is tiny, and INAH 1, which is by far the largest INAH, is smaller than the period at the end of this sentence. But perhaps, LeVay thought, given that the hypothalamus was involved in controlling sex, given Allen's discovery that INAHs 2 and 3 were larger in men than women, there was a dimorphism between gays and straights in one of these dots of brain tissue. This scientific reasoning led to a four-page paper entitled "A Difference in Hypothalamic Structure Between Heterosexual and Homosexual Men." Published in *Science* magazine in August 1991, these four pages exploded into world news.

LeVay began his paper with the logical hypothesis that "A likely biological substrate for sexual orientation is the brain region involved in the regulation of sexual behavior." Because the hypothalamus affected sex, Allen's "two [INAH] nuclei could be involved in the generation of male-typical sexual behavior." In other words, LeVay conjectured that a part of the brain regulating sex drive might differ according to different drives. Therefore, he wrote, "I tested the idea that one or both of these nuclei exhibit a size dimorphism, not [only] with sex, but with sexual orientation. Specifically, I hypothesized that INAH 2 or INAH 3 is large in individuals sexually oriented toward women (heterosexual men and homosexual women) and small in individuals sexually oriented toward men (heterosexual women and homosexual men)."

LeVay got brain tissue from forty-one subjects who died at seven hospitals in New York and California. Among his subjects were nineteen gay men, all of whom had died of AIDS; sixteen presumed straight men (the six of these who died of AIDS had a history of IV drug use); and six presumed straight women. No lesbian brain tissue was available because, LeVay explained, there was no disease he could use to identify lesbians. LeVay wrote: "Brain tissue from [men] known to be homosexual has only become available as a result of the AIDS epidemic. Nevertheless . . . it does not provide tissue from homosexual women because this group has not been affected by the epidemic to any great extent." As soon as lesbian brains could be found, the experiment would be expanded to include them.

In his study, LeVay used sections, slices of the brain itself. He carefully extracted the anterior hypothalami, which gave him forty-one tiny blobs of grayish-pink brain matter, and with a machine called a microtome that resembles a precision meat slicer he sliced those portions into razor-thin sections. LeVay meticulously described his methodology:

> The [sections of the] brains were fixed by immersion for 1 to 2 weeks in 10 or 20% buffered formalin and then sliced by hand at a thickness of about 1 cm. . . . Tissue blocks containing the anterior hypothalamus were dissected from these slices and stored for 1 to 8 weeks in 10% buffered formalin. These blocks were then given code numbers; all subsequent processing and morphometric analysis was done without knowledge of the subject group to which each block belonged. . . . The sections were mounted serially on slides, dried, defatted in xylene, stained with 1% thionin in acetate buffer. . . .

The staining renders each small, translucent section of cells visible as pale lavender, as if a single drop of India ink had been spilled from a fountain pen into a cup of water. Because the cells of a nucleus are denser than the brain cells surrounding them, they stain differently, and with a microscope they can be outlined to measure their area. LeVay outlined and measured each nucleus. Then, in dispassionate prose, he described what he had found:

> INAH-3 did exhibit dimorphism. . . . [T]he volume of this nucleus was more than twice as large in the heterosexual men . . . as in the homosexual men. . . . There was a similar difference between heterosexual men and the women. . . . The discovery that a nucleus differs in size between heterosexual and homosexual men illustrates that sexual orientation in humans

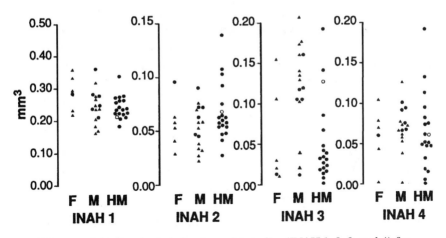

Volumes of the four hypothalamic nuclei studies (INAH 1, 2, 3, and 4) for the three subject groups: females (F), presumed heterosexual males (HM), and homosexual males. Individuals who died of complications of AIDS, ● individuals who died of causes other than AIDS, ▲; and an individual who was a bisexual male and died of AIDS, ○. For statistical purposes this bisexual individual was included with the homosexual men.

is amenable to study at the biological level, and this discovery opens the door to studies of neurotransmitters or receptors that might be involved in regulating [sexual orientation.]

Now that there was a difference in the brain, the question turned to the things that create the brain: genes. Did certain genes influence sexual orientation?

The answer was yes. In 1963 Kulbir Gill, a visiting Indian scientist at Yale, was looking in organisms for the genetic causes of female sterility and discovered, to his surprise, genetically homosexual fruit flies.

The fruit fly, *Drosophila melanogaster,* is one of the common workhorses of genetic research. Along with the gentle nematode worm *(Caenorhabditis elegans)* and the tiny colon bacteria *(Escherichia coli),* the fruit fly possesses just the right combination of job qualifications to fill a special niche in the geneticist's lab: a simple genome (*Drosophila* only has four chromosomes and an estimated 15,000 genes; humans have twenty-three chromosomes and around 70,000 to 150,000 genes), simple enough that we now know our way around it in great detail, including almost everything that its genes do; consistent and highly characteristic mating, eating, and living habits so that differences in behavior are easily detectible; and a short life span of two or

three months, which soon gives researchers many altered generations to observe. It is also easy and cheap to feed; if there were such
a thing as Purina *Drosophila* Chow, it would consist of wax and cheap
grain. *Drosophila* seems to have been tailor-made for science: The
chromosomes fruit flies carry in their salivary glands are giant, easily
seen with a simple light microscope, which made these creatures excellent subjects for genetic study before the age of the electron microscope. In exchange for all this, in the service of biology the fruit
fly is fed chemicals, frozen, starved, dissected, dropped on its head,
shaken up, and chopped into little pieces. Its main job, though, is to
be exposed to X-ray radiation and turned thus into, literally, thousands of mutants: Zap this gene, and the normally flat wings come out
crinkled, or curled, or missing; zap that gene, and you get a white-
colored body instead of the usual black. Expose the parents to X rays,
and the children, and their children, and their children, pass down
their mutated genes and the various strange effects that go with them.
This is the key to *Drosophila* genetics: The effect announces the function of the gene.

In the course of his research, Gill noticed one day that he had a
strain of sterile male flies on his hands, a fact he realized when some
mutant *Drosophila* that had been thrown together had generated no
offspring, a rarity for flies. The question was, *why* were these flies sterile? Observing them, he noticed that the mutant males were courting
other males, following them and vibrating their wings to make characteristic and unmistakable courtship "songs." Not one to let a little
homosexuality among his charges put him off, Gill tracked down the
mutant gene responsible—it sits on the right arm of the third chromosome at position 91—and casually published his finding in a short
note in *"Drosophila* Information Service," a clearinghouse of information on fruit flies.[9]

An informal publication, "DIS" can be slightly more casual than
its academic counterparts. It runs short pieces like "The Rate of *D.
melanogaster* Sperm Migration in Inter- and Intra-Strain Matings,"
which discusses fly sperm. Gill's contribution to "DIS," "A Mutation
Causing Abnormal Mating Behavior," noted, "This mutation, a third
chromosomal recessive, was X-ray induced and originally obtained as
a male-sterile mutation." Females with the mutated gene showed normal mating behavior, he said, but males that had it "actively court each
other," so while both sexes carried the gene, it was acting only in
males. "This homosexual behavior," he continued, "is observed even
in the presence of female flies." Gill suggested that this particular mu-

tation be given the name "fruity"—all mutations are named—and promptly returned to the matter of female sterility. (Mutant naming is a subject unto itself. One mutation knocks flies unconscious when the temperature drops below 76 degrees Fahrenheit; its name is Out Cold. A mutation that puts splotches of color on the abdomens of male flies but adds nothing to females is called Male Chauvinist Pigmentation [mcp]. One mutation that makes flies mate for only ten minutes instead of the usual twenty is called Coitus Interruptus; and another, having a somewhat opposite effect, is called Stuck.)

Jeffrey Hall, a geneticist at Brandeis University who has taken up Gill's work, discreetly redubbed the mutation, still somewhat tongue-in-cheek, *fruitless.* The bearded Hall has a professorial demeanor and a gruff but thorough manner. He also knows a staggering amount about flies and takes them quite seriously. He resents bitterly the fact that "the animal rights people will get on you about cats and mice, sure, but it's perfectly okay to smash and grind up any old invertebrate." Taped to the door of his office in the Brandeis biology department is a large poster from the movie *The Fly* along with a photocopy of a quote from Lenny Bruce: "I sort of felt sorry for the damn flies. They never hurt anybody. Even though they were supposed to carry disease, I never heard anybody say he caught anything from a fly. My cousin gave two guys the clap, and nobody ever whacked *her* with a newspaper." He delights (gruffly) in the various fantastic names given *Drosophila* mutants. Nematode worms never get wild names. "You can't dynamite an interesting name out of a nematologist," Hall says with a sniff. Hall likes things that are interesting. He began a recent paper entitled "Genetics of Sexual Behavior in *Drosophila*" seriously: "Courtship in fruit flies is influenced by a variety of genes. The range of genotypes studied in this regard is wide—perhaps too wide, in the sense that 'general behavioral decrements,' caused by several different kinds of pleiotropic mutations, lead to defective courtships." In small print on the previous page, discreetly printed below Hall's name, the paper's epigraph comes from Woody Allen: " 'My brain! That's my second favorite organ.' "

The *fruitless* gene mutation, Hall explains, produces two distinct behaviors. First, it makes males actively court other males. Second, *fruitless* males are not only receptive to but actually elicit courtship *from* males, something males without this mutation reject. At first, researchers were suspicious. Could something other than a gene be causing this phenomenon? Very young males, after all, do normally elicit courtship from other males, but this anomaly is probably explained by pheromones, chemicals that draw sexual attention. By the time normal males are three days old they no longer attract other males, so pheromones can't account for the fact that *fruitless* draw this male attention throughout their adult lives. "It remains," says Hall, "one of the most spectacular mutations ever studied in fruit flies. It's so totally different from what normal males do in this species."

This gay fly strain has persisted through hundreds of generations unchanged. Certain lethal gene mutations kill flies; *fruitless* is not one of these, nor does it cause illness. Rather it is, says Hall, a genetic mutation responsible for a nonpathological, consistent behavior. "Huntington's [disease] and *fruitless* are genetically analogous because both are caused by a single gene and have to do with nervous system functioning. And they're both mappable—you can track them down in the genome. The difference is that Huntington's is a pathology and *fruitless* is a true behavioral mutation." In fact, nature provides a good method of distinguishing sick mutations from nonpathological ones: Between fertilization of the egg and the time of birth, flies with sick mutations die at huge rates. "It is very common for a mutant to have a decreased survival in the womb," says Hall. But this is not the case with *fruitless*. As with any nonpathological mutation, there is no decrease in the prenatal survival rate of *fruitless* flies. "Sometimes a sick mutant doesn't mate because it barely moves. But *fruitless* moves all around and chases other flies. Sick means poor survival and sluggish performance of its activities. That's not the case here."

The *fruitless* gene also creates a sexual dimorphism in the organisms that carry it. In the abdomen of all male *Drosophila* is a muscle, called the Muscle of Lawrence, whose function is unknown, though it is not needed for mating. Female fruit flies don't have it, and neither do *fruitless* males. "Studies have demonstrated," Hall explains, "that it is the male nerves which are responsible for making the Muscle of Lawrence, so if *fruitless* doesn't have it, it means the gene might be affecting the nervous system, a neural etiology." And so, as Hall points out, the gene not only creates a behavior, it also, as may be the case of LeVay's nucleus in the human brain, determines anatomy. (One of the more curious historical aspects of *Drosophila* work is that there used to be female version of the *fruitless* mutation that produced lesbianlike behavior in females. The mutation was called, with no great originality, *lesbian;* discovered by Robert Cook in a lab in England, it was first described in a 1975 paper in *Nature* magazine. But the mutant was somehow lost in the late 1970s, and no one can find it.)

Knowing all this is one thing. No amount of verbal description of *fruitless,* however, can prepare one for the uncanny experience of actually seeing the results of this single gene mutation. At the National Institutes of Health (NIH), which conducts *Drosophila* work, a researcher demonstrates *fruitless* in action in the lab. She takes a small glass container of the flies (they are surprisingly tiny, perhaps one-quarter of an inch long at most), pops off the top, and sticks an ether-filled cotton ball into its mouth. Within a few seconds, the flies are lying stunned on the glass floor. With a plastic stick she separates out a few of the male flies—fruit flies are differentiated by the genitalia at the end of their abdomen, smooth and light colored for females, furry and dark for males—into a larger Pyrex jar. Seen through a microscope, their bodies vibrate violently in spite of the ether, huge red eyes bulging.

As the flies revive, she points and says, "Watch that one." Sure enough, that fly comes up behind another male fly, vibrating its wings. "This is courtship," she explains, as the dance continues for several minutes. "The mounting fly climbs up on top of the other male and hooks himself on with special hooks on his arms." Soon the two male flies, one on top of the other, are wandering around the jar.

They look bored. She smiles. "That's what flies look like when they mate." She explains the procedure. "In normal mating, the male taps his foreleg on the female's abdomen to let her know he's ready to go, and then she'll either perform rejection behavior or not. If she rejects him, he'll try tapping again, and if she gives him the green light, he'll scurry around in front and face her eye to eye and extend one or both of his wings, fluttering them at a rapid rate to 'sing' his courtship song. The song needs to be at a certain Hertz frequency, or else she'll reject him. It appears that females all have their own ideas of how the song should sound. One song won't work with all of them. And then he'll go around back and get ready to copulate. *Or* he'll head straight to the back. In that case, he sings his song from behind, usually, if she's not moving and is just standing there, for a few seconds."

The details of fly mating are somewhat exotic. The male fly first extends his proboscis ("All flies," the researcher explains, "eat by regurgitating enzymes onto food, which digest and liquefy it, and the proboscis is like a straw that slurps up this material and funnels it into the abdomen") with which he "licks" or kisses the genitals of the female ("He doesn't have a tongue so it's not really licking, but that's how it's referred to"). The female then opens her vaginal plate, which prevents the male from copulating when she's not interested, and he then very quickly curls his abdomen around and interlocks his genitals with hers, injecting her with sperm, although his penis itself is internal. The male will ride her for up to twenty minutes while she walks around the vial—which, if you consider that the life span of the fruit fly is thirty to forty days, is a significant portion of their lives.

In contrast to the heterosexual method, the researcher explains, "*Fruitless* males can't curl their abdomens to interlock their genitals as males do with females. We don't know why they can't. Maybe it's because they don't have this Muscle of Lawrence, but then the few normal males who happen to be missing it can still mate, so that's not it." She shrugs. "Also, when a *fruitless* male goes after other males, there are certain aspects of the courtship behavior that are exaggerated over what normal males do with females. He may tap the abdomen of the under-male, or he might head directly for the rear. I have observed that there is much more genital licking with male-on-male couples than with male-on-female couplings. Still, for flies with this gene, the mating process for these male-male couples is essentially identical to that of male-female couples."

Although we know exactly where *fruitless* is in the fly's chromosomes, we don't know much about it molecularly—that is, we don't know the sequence of the DNA molecule that makes up that gene—but Hall is pursuing this work now at Brandeis, trying to reproduce the DNA code for the enzyme that makes *fruitless*. Once he has figured that out, biologists can begin to ask whether a comparable chemical—a *fruitless* homologue—exists in human beings. There are DNA homologues for all sorts of fruit-fly genes in our own human genetic makeup, some startlingly (almost eerily) similar; despite the fact that they have only four chromosomes and we have twenty-three, as far as many of our genes go, we're amazingly similar. (For that matter, overall we are genetically quite similar to some other species; the human genome is only about 1 percent different from the genome of the chimpanzee.) We don't know if there's a human homologue for *fruitless,* but it's interesting to contemplate. We can already artificially manufacture the genetic product rhodopsin that allows us to distinguish colors. Perhaps soon we'll be able to manufacture the protein pumped out by the *fruitless* gene.

But if the *fruitless* fly is a dramatic example of homosexuality in animals, nature provides us with numerous animal species that, quite unaltered and uncoerced, engage in homosexual sex, from alligators to cows to geese to the mountain ram. Female sea gulls are known for forming long-term lesbian couples to raise children. Homosexuality in other species only underlines the naturalness of homosexuality in ours.

LeVay found a nucleus dimorphic according to sexual orientation, and rats and flies provided two animal models, hormonal and genetic respectively, for homosexuality, but the question of how much genes contribute to human homosexuality still was unanswered. Then, four months after the appearance of LeVay's work in 1991, a study of twins entitled "A Genetic Study of Male Sexual Orientation" was published in the December issue of the *Archives of General Psychiatry* by Michael Bailey, a psychologist at Northwestern University, and psychiatrist Richard Pillard of Boston University School of Medicine.[10]

Identical twins are natural clones, two human beings with the exact same genes, while fraternal twins are just like usual siblings; they share half their genes. Looking at twins should help researchers determine if a gene creates a trait, or *sometimes* creates a trait, or creates *part* of a trait. It's a simple principle: The more genes two people share, the more genetic traits they both will have in common. If a gene creates a given trait, a pattern should be visible: Identical twins should share that trait most frequently; fraternal twins and nontwin siblings next most; and adopted siblings, who share no genetic material, least frequently. Bailey and Pillard found exactly this pattern for homosexuality.

In their study, Bailey and Pillard compared fifty-six monozygotic twins (identical twins, from the same zygote or fertilized egg), fifty-four dizygotic (fraternal) twins, and fifty-seven nongenetically related adopted brothers. The results lined up in exactly the way expected for a genetic trait. Eleven percent of the adoptive brothers were both gay, which is fairly close to the average or "background" rate in the general population; 22 percent of the dizygotic twins were both gay, and 52 percent of the monozygotic twins were both homosexual. Another way of describing the finding is that identical twins, who share the same genes, are twice as likely to share a homosexual sexual orientation as fraternal twins, who share only half their genes, and *five* times as likely as adopted brothers, who were raised in the same family by the same parents—the same environment—but share no genes at all. So the difference must come from genes.

A year later Bailey and Pillard published a similar study of lesbian women.[11] The results were virtually identical. Forty-eight percent of the twin sisters of lesbians were also lesbians, and, as predicted, as genetic relatedness fell, so did the frequency of lesbianism; for fraternal twins the figure was 16 percent and for adopted sisters, only 6. The data from the studies suggested, as Natalie Angier reported in *The New York Times,* that "70 percent of homosexuality is attributable to genetics." And the 30 percent that is not genetic is, as the rat work demonstrates, hormonal, affecting the neural architecture, molding sexual orientation in each of us.

From all this evidence scientists have conclusively decided: homosexuality is biological.

～

That is one story you could tell using the biological evidence, one side of The Debate. The other story is called "Definitive Proof that Homosexuality Is Not Biological," and it goes like this.

Chapter Three

THE DEBATE: DEFINITIVE PROOF THAT HOMOSEXUALITY IS NOT BIOLOGICAL

ON THE MORNING of August 30, 1991, the world woke to the claim of a physical difference between heterosexuals and homosexuals: a nucleus—a small cluster of cells—in the hypothalamus of the human brain. This nucleus, it was asserted, was bigger on average in straight men than it was in gay men, although there was significant overlap in the data that the neuroanatomist, who was, it should be noted, himself homosexual, presented: some of the nuclei of the gay men were larger—not smaller—than those of the straight men. A frantic contest to discover the purported biological origin of sexual orientation was on. The study made a media celebrity of its author, Simon LeVay, a young scientist at the Salk Institute, who had measured these nuclei, using the traditional neuroanatomist's staining method. According to some researchers, the method was probably wrong.

Evan Balaban, a biologist at Harvard, raises the following question: What was it that Simon LeVay actually measured? "What are you really studying when you find a group of cells?" asks Balaban. "Simon used Nissl stain on these cells. How do we know when we stain something in the brain with Nissl that we can talk about those cells as a 'unit'?" LeVay's study begins to fall apart because of the way stains work on the brain of the Belgian Wasserslager canary.

Sitting at a table in a small coffeeshop behind the Harvard Museum of Zoology on a sunny, cold morning in December, Balaban points out an assumption in the title of LeVay's study. "Simon's title

is 'A *Structural* Difference in the Brains of Homosexual and Heterosexual Men,' " Balaban notes. "He doesn't make it explicit, but by saying that this nucleus he's measured is 'structural,' he is implying that
it's a distinct unit in the part of the brain's wiring that has an effect
on sexual orientation."

The brain is all about communication, with brain cells talking to
each other constantly through the day. (At night, these cells shut
down and other cells wake up and converse. These conversations are
dreams, and though we believe they occur because the night cells are
actively cleaning up after daily operations and reconciling the day's
accounts, their point remains rather mysterious.) Let's assume, suggests Balaban, that these conversations are controlled by a central
wiring system, a blueprint that determines which cells are hooked up
to which other cells. That system will govern which cells talk to which
other cells, which cells won't take an incoming call, how often the
caller gets a busy signal, whether the cell has call waiting or not, and
whether the cell is part of a local area network. The connections
and features make up the system. Control the way the system is set up,
and you control the brain.

"Let's then assume," Balaban goes further, "that if you get *one* con-
figuration of this system, you're a heterosexual, and if you get another,
you're a homosexual. We have reason to believe, from what we know
about the brain, that the system is fixed very early—somewhere from
the embryo stage to age four, when the brain usually stops adding
cells. The structures in your system—the nuclei and axons and neural
switching stations and signal boosters and fiber optic relay stations that
your body is building in your head—these determine how it's going
to work: how easy will it be to call from the rural outlying areas of the
cerebellum into the busy urban areas of the cerebrum, for example.
Clearly, we interpret structure as something that's going to affect
you. Simon interprets it this way, which is why he's measuring this
structure, the nucleus, this relay station or signal booster or switching station or whatever INAH 3 really is.

"So the question we have for Simon is this: Does the neural research technique he used actually show you this wiring? Or does
it show you something else?" Balaban pauses, raises an eyebrow,
and grins. "What if he's *not* looking at structure? What if he's looking not at structure but at something completely different that happens to be distinguishing the INAH 3 cells from other cells? His argument would fall apart, because then this nucleus wouldn't be a
relay station in the system affecting the conversations of the brain
cells, it would just be . . ." Balaban shrugs. "We don't know what. A

glob of cells maybe. But it wouldn't have anything to do with homosexuality."

And here we get to the problem. It has to do with the stains neurologists use. As mentioned, LeVay used Nissl stain to measure his nucleus, but the fact is, Balaban explains, we actually aren't sure exactly what Nissl colors, nor why it stains certain cells heavily and others not. It is thought that Nissl binds to certain molecules in the cell that read RNA to make proteins—these molecules are called ribosomes—so if cells are actively manufacturing proteins, they would contain a lot of ribosomes, and the Nissl ought to stain heavily. But some cells are very active and don't ever stain well with Nissl, while others aren't active and do stain well. And, Balaban cautions, "nucleus" is, as he puts it, "just a word." If we see a group of cells, what they might do, how they might function, *if* they do anything at all—we don't necessarily know.

When the first modern neuroscientists began using Nissl stain near the turn of the century, they were simply trying to create a basic roadmap of the brain, a general survey of this vast, uncharted territory. According to Balaban, they did not make the mistake of assuming that the things they saw—masses of cells here and shafts of axons there—were necessarily coherent entities with a role to play in a single mental function. They were simply engaged in making an atlas of the brain to give them a common frame of reference with recognizable landmarks. By the 1960s researchers had begun interpreting these landmarks as functional structures, assigning them meanings. "It's called reification," says Balaban, "giving what is essentially just a reference point, like a funny-looking rock that happens to be recognizable, a deeper meaning or purpose it doesn't have."

Balaban's point, which he deploys energetically across the tabletop, culminates with a young German graduate student named Manfred Gahr, who was studying bird brains at the University of Kaiserlautern. In the brain of the male bird exists one of the most striking sex differences known, a nucleus that grows and shrinks seasonally. In spring when the birds are breeding, this nucleus is massive, and in the fall when they aren't, it shrinks to a tiny speck. Gahr was the first person to get the idea of comparing the results scientists were getting from the Nissl stain, a technology created around the turn of the century (although still used today in labs), with results from two other more modern, more sophisticated stains. The three work very differently.

The first of these newer stains Gahr used is called a connection stain, and it takes advantage of an odd habit of brain cells: They talk to each other by squirting chemicals called neurotransmitters; essentially, brain cells talk by drinking from each other. If the telephone network works as an analogy for the brain, in this strange network the filaments carry electricity but operate, at their ends, with liquids.[1]

There are many neurotransmitters—acetylcholine, dopamine, glycine, glutamate and on and on—each a different signal. (The antidepressant drug Prozac works by stopping brain cells from reabsorbing the neurotransmitters they have sent to other cells.) The lab animal is anesthetized and the stain injected directly into its brain with a very fine, essentially microscopic needle. One of the more popular connection stains is horseradish peroxidase, an enzyme gotten from horseradish, to which is attached a neon pink or green carbocyanin dye. Active cells in the brain are busily pumping neurotransmitters to other cells; axons are for outflow, dendrites for inflow, and the dendrites of other cells drink up the neurotransmitters squirted out by the axons. They drink up the stain as well, turning themselves pink and green. Since connection stain only marks active cells, only neurons in active conversation will light up, giving the researcher a snapshot of connected neurons. They are marked by whom they talk to. (Balaban adds that one of the "neat special features" of carbocyanin dyes is that they can be used even on some dead brains, which is helpful in human research.)

The second stain Gahr tested is called antibody stain. It was created as a technological offshoot of immunochemistry, a field dealing with the body's immune system and its production of antibodies. (Antibodies are particles produced by the body that react to, and only to, one specific type of molecule or tissue. The body uses antibodies to fight disease-causing invaders.) Whereas connection staining marks parts of the brain talking to each other, antibody staining identifies groups of cells that perform the same function. Say, for example, researchers want to find the cells that have the estrogen receptor molecule on their surface. They take an estrogen receptor antibody and attach a dye to it—a fluorescent dye or radioactive marker. Then they kill the animal, slice up its brain, soak it in this chemical marinade, and place the slices on radioactive film. The antibodies will have homed in like tiny missiles on the estrogen receptor molecules, and the cells that are stained are the ones absorbing estrogen.

Gahr set out to compare these two new stains to Nissl. "He took Belgian Wasserslager canary males in the spring," Balaban explains, "sacrificed them, and stained their brains with all three stains. Then he did the same in the fall. What he found was that, in the spring, the three methods outline exactly the same group of cells, the same boundary. But in the fall, the area indicated by the antibody and connection stains was still big, yet the Nissl stain area was tiny." Balaban sits back, triumphant. "The only way to interpret the study is that there are cells that change their affinity for Nissl within individuals—seasonally, in this case. Simon called his study 'A *Structural* Difference . . . ,' but the moral of the Belgian Wasserslager is, of course, that if you think Nissl is showing you structure, you're wrong. Or at least you could be wrong. The structure can be the same, as we can see by these other stains. So what is the Nissl measuring? What was Simon measuring in his brains? The answer to that is, we don't know. Connection and antibody stains are better in that you *know* what they're telling you because you know what they do. As for what Nissl does, my hunch is that it just measures the metabolic activity in the cells, how active they are, not any structure they might or might not be a part of. And that's *very* different.

"I don't mean to say Simon's study is worthless. Nissl is *some* kind of difference; we just don't know what, exactly. You know," Balaban says in the end, shrugging his shoulders, "it could be that all Simon is measuring is some effect from an anti-AIDS drug or the result of stress."

Indeed, if one had read LeVay's study carefully, the possibility that AIDS might have changed the tissue was clearly if subtly brought up. And after that, there was a question of cause and effect: "[T]he results," admitted LeVay, "do not allow one to decide if the size of INAH 3 . . . is the cause or consequence of that individual's sexual orientation, or if the size of INAH 3 and sexual orientation covary under the influence of some third, unidentified variable."

~

But if Balaban's critique was of LeVay's methodology, Anne Fausto-Sterling, a developmental geneticist at Brown University in Rhode Island, and William Byne, a neurobiologist and psychiatrist at Mount Sinai Medical Center in Manhattan, take issue with the study at the even deeper level of its basic assumptions.

Fausto-Sterling has short, dark hair and a pleasant, serious manner. She teaches and lives at Brown with her lesbian partner, and has written extensively on what she sees as the speciousness of sexual orientation and gender research, particularly in her book *Myths of Gender*. She gives as one example a study done in 1908 by the morphologist E. A. Spitzka, who compared the brains of eminent scientists with those of white laborers and blacks, and concluded that great men had larger corpora callosa. (Naturally, Fausto-Sterling notes, the largest of all belonged to a morphologist.) Spitzka's research helped provide the basis for not only sex but race differences, among other things, the widespread confidence among educated people in the early 1900s in the "well-known" mental superiority of Caucasians. This confidence was despite the fact that as early as 1909 the anatomist Franklin Mall refuted Spitzka's findings, noting, for example, the extraordinary individual variation in size and shape of the corpus callosum (CC) *within* each sex. Spitzka had neglected to mention that the *range* of sizes between male CCs (not just the few largest exceptions to the rule) was almost completely identical to the range of CCs in females. The differences between men and women, if there were any at all, were both literally and figuratively marginal.

Fausto-Sterling argues that there is in fact no convincing evidence that the corpus callosum is different in men and women. To her the issue isn't merely a matter of whether this or that nucleus is sexually dimorphic, but rather that sexual dimorphism of the human brain, one of the key principles on which the entire premise of LeVay's research rests, may not even exist.

Owing largely to the cachet—some would say excessive domination—of *Science* (a publication with the clout in the science world of the *Atlantic,* the *New York Times,* and *Vanity Fair* put together), the de Lacoste and Holloway study quickly became famous and frequently cited. Overlooked at the time was the fact that its findings that women's callosa were larger than men's—were directly opposite to Spitzka's results. A more serious challenge to its validity, though, and also not much talked about was the fact that the study has had serious replication problems.

Replication in science is everything, the primary quality-control mechanism on science's intellectual assembly line. Researchers can do a study, find the answer to the question of life itself, but if no one can repeat the work and arrive at the same results, the effort is, for all practical purposes, worthless. And this replication must be accomplished by someone other than the original researchers, using the same methodology; the subsequent researchers can't slice their brains at 60-micron thickness if the original study used 32 microns, or soak their sections in 20 percent buffered formalin if the original study used 45 percent. Without replication, a "finding" such as LeVay's is nothing but an interesting possibility.

In the flurry of studies that followed, no one was able to replicate de Lacoste and Holloway's findings. No one found the callosum to be sexually dimorphic. Fausto-Sterling grimly lists the bottom lines of the subsequent research: from a 1985 study: "[N]o sex differences in . . . shape, width, or area"; from 1988: "three independent observers unable to distinguish male from female"; from 1989: "women had smaller callosal areas [but] larger percent of area in splenium, more slender CCs [but] more bulbous splenia"; and so on. Two studies from 1986 and 1987 concluded that sex differences in corpora callosa were not statistically significant in callosal area, width, or curvature.

Statistical significance measures any researcher's findings. Like replication, it is a standard tool of the industry. Usually represented by p = some decimal, significance is a somewhat arbitrary (in fact, a completely arbitrary) line scientists have drawn to decide if what they've found is "scientifically useful" or, in English, "real." The neuroanatomist measures the nucleus, but the statistician comes up with the numbers that tell her if her finding is meaningful. Significance— p value (the letter stands for "probability")—is the number that editors of the prestigious journals such as *Science* and *Nature* and *Cell* use to ask of a study: What are the odds that the authors just came up with this finding by chance? Or, as geneticist Dean Hamer explained it with admirable clarity, "Your 'statistical significance' figure just tells you how likely it is that your results are bullshit." The "chance" part is what Hamer means by bullshit, because if one flips a coin enough times, eventually, just by chance, one is going to get ten heads in a row. Ten heads in a row by chance is impressive, but it doesn't mean anything unless p says: It wasn't by chance, this really means something.

Corpus callosum studies find all sorts of differences between the sexes all the time, but unless the study's p value is less than .05 (meaning that if someone else tried to replicate the study, he or she wouldn't come up with a chance finding more than five times out of one hundred, twenty-to-one odds), no one will publish it. As it happens, the p values for the corpus callosum studies haven't been good: This supposed sex difference usually has been found to be *not* statistically significant. Add to this the fact that 90 percent of all we know about the brain has been learned in roughly the last ten years, a breathless rate of discovery that outstrips our rate of comprehension and only calls attention to the much vaster amount we still don't know about it— including the corpus callosum itself. And what is perhaps most striking is that the corpus callosum is the *most* researched sex difference in the human brain. (On top of this, Swaab, the Dutch researcher, claimed he had found the analog of the rat SDN in humans and christened it, rather unoriginally, the human sexually dimorphic nucleus [the human SDN]. The problem is, Swaab has not been replicated. Both Allen and LeVay, in separate studies, expressly found that Swaab's nucleus was *not* sexually dimorphic, and it now appears that the so-called human sexually dimorphic nucleus isn't sexually dimorphic.)

As for the hypothalamus, it turns out that it may not be such an optimal place to look for homosexuality after all. While Fausto-Sterling questions LeVay's study on its statistics, neuroanatomist Bill Byne attacks the subjects who were used. "Some—perhaps one-half—of LeVay's presumed-homosexual subjects had testicular atrophy, degeneration of the testes," Byne says, "which of course supply testosterone. The gerbil's brain has a sexually dimorphic nucleus whose size is controlled by testosterone. Considering that the way he identified his gay subjects was by the fact that they had AIDS, this could be an effect of three things."

Byne is a big man with dirty-blond hair and is sort of nervous, impassioned, aggressive, and apologetic all at the same time. When he gets impassioned, he tends to repeat the words at the beginning of his sentences, and he puts his three critiques out so quickly he almost stutters. "Point one: The AIDS virus has a direct impact on hypothalamic function, which could cause the INAH-3 nucleus to shrink. So what Simon may have measured is nothing more than an AIDS effect. Two, this shrinkage of neurons is seen in most wasting illnesses like AIDS—so, same problem. And third, some drugs used to treat opportunistic infections in AIDS patients—ketoconazole, for instance—lower testosterone, and if you fool around with any hormone, you can get changes in the brain. Steroids are used too, and they do the same thing.

"Simon claims he's found a nucleus that's smaller in some human brains. But if you find such a nucleus, it might be varying with testosterone in circulation and have absolutely nothing to do with sexual orientation at all. The scientists who do these studies put down 'WNL' for testosterone levels, 'Within Normal Limits.' But they rarely check the medical histories of the subjects." Byne shakes his head and, in his absentminded-seeming way, adds in exasperation, "Does WNL mean 'Within Normal Limits' or 'We Never Looked?' I mean, quite a few of these patients were getting *steroids.* "

What truly maddens Byne, however, is using same-sex sexual behavior in Roger Gorski's rats, a phenomenon created with biochemicals, as an analogy for human homosexuality. He rejects looking at human sexual orientation through "animal models." "They have a sexually dimorphic nucleus in rats, the rat SDN, which is five times bigger in the male. And they know that it's created by hormones. And with hormone injections Roger can make male and female rats mount other rats of their own sex. Well . . . *so what?*"

LeVay, protests Byne, is hypothesizing that because the SDN has an effect on sexuality in rats, it also has an effect on sexuality in humans. But Byne insists that if you use animals as a model, you must use them completely. It has been found that burning out a rat's SDN has no effect whatsoever on its sexual behavior. Monkeys appear to have an SDN (although this is still questionable), but if it is burned out, strangely enough, while the male monkey's sexual activity with females will decrease, he'll masturbate more. Ferrets appear to have a homologous structure to the rat SDN, but the hamster, which is more closely related to the rat than the ferret, does not. On the other hand, the hamster *does* have a sex dimorphism in its brain, but it isn't the same sex dimorphism as the one in the rat—or the ferret. Gorski found the SDN in rats, Byne notes, and in monkeys, but he couldn't find it in mice. Nor in gerbils. So if a nucleus exists in rats but isn't consistent across species even within rodents, how can it mean anything at all for humans?

As for the flies, the *fruitless* males don't just court other males, they court females as well. So at best they would serve as a model for bisexuality. (This applies to the rats as well: They'll have sex with males or females, which is unlike human sexual orientation.) Those eager to extrapolate sexual orientation from fruit flies to humans should consider that (for example) when a female fly isn't interested in a male, a repertoire of rejection responses comes into play, the most important of which is extrusion of her genitalia toward the courting male. In successful courtship, the male extends the proboscis and licks the genitalia of the female, at the same time flexing his abdomen in preparation for attempts at copulation. This is rather vastly different from human sexuality and, more to the point here, in another evolutionary phylum.

In fact, the scientist who knows *fruitless* best, Jeff Hall of Brandeis, is careful to separate fly from human, describing *fruitless* as "a mutation that leads to a mimic of homosexuality." *Fruitless* males never achieve intromission, as it is delicately put, with each other. They just court, an eternal and unconsummated dance. "It could be satisfying for them," he said. "It could be delicious." (Hall was referring to the pheromones.) "Or it could be frustrating, but this becomes ludicrous. How do you know when a fruit fly is frustrated?"

Hall adds, "*Fruitless* is an anomaly, but human homosexuality is not. You don't have to treat human chromosomes with X rays, an outside agent that is known to damage genes, to get human homosexuality. But that's how *fruitless* is created. *Fruitless* is a very rare occurrence, never found outside the artificial boundaries of the lab; human homosexuality is found everywhere in nature in *extremely* high percentages." Hall explains that biologists consider a trait present in even 1 percent of a population, one out of every hundred people, to be enormously significant, and the incidence of homosexuality appears far above 1 percent.

As with the flies, so with the rats. Gorski, who knows more about rat sexuality than anyone, makes it clear that if rats are a model for anything—and he thinks they are—it is not for sexual orientation. "The rat model," says Gorski, "is a model for transsexuality, the feeling on the part of a man or woman that they were born into the body of the wrong sex. With our hormonal direction, we've effectively rewired the male rat's sexual system to the point where most probably what he's thinking is 'Help! I'm a girl, get me out of this boy body!' "

Gorski's theory is convincing, but it points out the problem with interpreting humans through rats. Nancy Brackett of the University of Miami notes, "Female rats will go into lordosis if a male rat mounts them or if you touch their genitals with a cotton swab. Do they want to mate with the swab? I really doubt it. The biggest limitation is that we never know what a rat actually has on his or her mind. People, because they're compelled to by their bosses or priests or government or the dinner guests they sit next to, behave all the time in certain ways that have little or nothing to do with the way they really feel or think. Anyone who's mastered a social smile at a cocktail party knows this, gay people, who are oriented inside toward the same sex but because of social pressure outwardly behave otherwise and have sex with the opposite sex, are perhaps the best example of the phenomenon. But people can tell us, if we allow them, what's really going on. Rats can't tell you, 'I'm doing this but I really *want* to do that.' When a rat is mating, we assume it's because he wants to. When he doesn't mate, is it because he doesn't like the animal he's with? Is it because of his partner's gender, or is it his or her deodorant? Is he just not in the mood? Has a headache? Is he dissatisfied with the leading economic indicators or irritated by the budget proposal of the current administration? We don't know."

But it is Byne who brings up the most succinct and telling point. In Gorski's rat films, males let other males mount them and females mount other females, and viewers almost unavoidably think: "These animals have been made gay." Byne lifts an eyebrow and asks, "If these animals are homosexual, which one is the homosexual?" Is it the one on the bottom that's been altered? Or the one on top that hasn't been touched? "When a man, fully aroused, eagerly mounts another man who is also fully aroused and proceeds to have sex with him, only one of them is gay? And the female rats—the sexually receptive female who is allowing herself to be mounted by another female is considered heterosexual. A human woman who welcomes sexual activity from another woman is not considered heterosexual by anyone *I* know."

The data from the rat studies are a treasure trove of biochemical information on which hormones direct the rat's biological sexual system and how it functions. It gives us clues about the biochemical processes the rodent body goes through and takes us a step closer to unlocking their secrets. What does it mean for human sexual orientation? We don't know.

Because some animals engage in same-sex behavior, is that strong evidence that homosexuality is natural in humans? Byne's response (given with a sigh) is that animals don't cook. What does this say about the human practice of cooking? Nothing. Animals don't paint or sculpt or write poetry or pay taxes. Animals make war (ants) and construct buildings (beavers, birds, and, again, ants) and tend herds of domesticated animals (ants . . .), but they don't have alphabets or quantum physics or mail order catalogs, nor do they use telephones or faxes. Should we stop doing these things? No. There are animals that eat their young and their own feces and animals that bite off the heads of their mates during sexual intercourse, but this does not necessarily mean we should make such practices mandatory for ourselves.

Humans are one of the very few animals that have sex when not in heat and even when the female is menstruating. Nonhuman animals rarely do this. What does it mean? All it means is that we're not other species of animals. We're human animals.

And—finally—the twin study? Ironically, what the study actually demonstrates is that homosexuality is *not* purely genetic. Identical twins—clones—have the same genomes *exactly*. If sexual orientation were 100 percent genetic, then 100 percent of all identical twins would have the same sexual orientations. But they clearly don't. Only about 50 percent of them do, so the other 50 percent must be nongenetic.

Not only that, Bailey and Pillard did their study with brothers who had been raised together. All the boys—identical twins, fraternal twins, and adopted brothers—had been brought up in the same households. So how did they control for "environment"? Bill Byne points out that there is, in fact, no control for external factors. Were the identical twins treated more similarly than the fraternal or adopted brothers? Parents often dress twins alike and treat them alike, he notes. The problem is, this study can't tease these two factors apart. As Bailey himself has acknowledged, "I think that the most troubling methodological problem with the twins study is the 'ascertainment bias' problem: These were volunteers, and we don't know what went into their decision to volunteer. So your data can be skewed."

And thus, from all this evidence scientists have conclusively decided: Homosexuality is not biological.

~

This is The Debate. Yes or no, up or down, black or white. This is the way the biological research of sexual orientation is presented on ABC *World News Tonight,* in *Newsweek,* and on CNN. The studies are said to "prove" something one way or another about homosexuality. And these comments reflect a common misperception of science: that science "proves" things, that like some divine hand it separates things into True and False, Black and White, Right and Wrong.

"Proof" is a word that is rarely heard in laboratories and research institutes, nor do reputable scientists use it often. (Hearing it, they tend to become wary, as if someone had suddenly said something slightly dangerous.) A scientific study on any subject, ranging from those done by geneticists engaged in cancer research to physicists conducting experiments on quarks and leptons, doesn't "prove" anything. All a study does is, first, give you results, and, second, tell you statistically whether or not, if you repeated the experiment, how likely it would be that you'd arrive at those same results. Nothing more.

Science is a series of qualified assertions about findings on a subject, people saying to other people "Well, it *appears* that this frog nucleus is structural (although we've only used one staining method), and it's a nucleus that *appears* to be unique to this particular species of frog (although perhaps we simply haven't found it in other species of frog yet), and since it looks as if this particular frog lives an extraordinarily long time, we're hypothesizing that the reason the frog lives so long has *something* to do (although we don't know what) with the fact that it has this nucleus. But what did *you* find?" The Nissl stain and Belgian Wasserslager canary experiments are science. No one says, "Ah, proof!" They say, "Hm, it *appears* as if . . ."

Outside The Debate, outside the political battles and the Yes or No of journalism, the real scientific dialogues quietly continue among researchers in the field, grappling with complexities, conducted entirely in shades of gray.

This is the way the real scientific journey into this black box goes.

Chapter Four

HOW TO LOOK AT A BRAIN

*Neuroanatomy: the branch of anatomy dealing with
the nervous system, especially the brain.*

A T ABOUT eight forty-five on a cool, fresh morning, when New York's streets are still wet and clean from last night's spring rain, Bill Byne is in his office trying to untangle a slide projector. Outside, the residents of the Upper West Side haven't completely woken up yet, carrying their paper cups of coffee to work. Byne, who lives just across the street, is a controversial figure. He has two doctoral degrees, his first in neuroscience, where his doctoral research was on the development and function of structural sex differences in the guinea pig hypothalamus. He then entered medical school in 1989 and completed an internship in internal medicine, a residency in psychiatry, and a postdoctoral fellowship in neuropathology (he uses the office for his private psychiatric practice). The drawers of his desk are stuffed with diplomas and licenses he's never gotten around to having framed.

But Byne has made a national name for himself as a critic of the biological research of homosexuality, specifically of Simon LeVay's study. His critiques are weapons brandished by the Family Research Council, one of the largest conservative Christian organizations in the country and a group leading the political battle against civil protections for homosexuals. Because it calls into question not just LeVay's work but the very notion that sexual orientation has been proven to be biological, the Family Research Council's Washington, D.C., office regularly mails out a copy of a paper Byne published recently in *Archives of General Psychiatry* entitled, "Human Sexual Orientation: The Biologic Theories Reappraised."[1]

Byne carries himself almost awkwardly, like someone not used to his own large over-six-foot frame. While the habit sometimes makes him appear diffident, he is relentlessly dogged in pursuing problems with the research. He can seem a bit otherworldly. He is also exhaustively and sometimes stunningly well versed in both the animal and human literature, which is why, as much as his colleagues sometimes would like to, no one can ignore him.

One who would like to ignore him is Dr. Laura Allen, a neuroanatomist at UCLA and Byne's neuroanatomical opponent in this research. Allen is one of the preeminent neuroanatomists in the United States and the discoverer of INAHs 1 through 4, where Simon LeVay found his difference between gay men and straight. She also has found a human sex dimorphism in the corpus callosum (CC) of the human brain; her data find the CCs of men different from the CCs of women. Byne disputes the finding determinedly. He and Allen have tangled at various scientific conferences across the country, Byne's tentative voice launching critical points, Allen giving calm but insistent responses. It is she who, in these scientific debates, takes LeVay's side.

Byne and Allen have also been matched up in court. During the bitter, angry, nationally publicized battle over Colorado's Amendment 2, a constitutional amendment aimed at forbidding civil rights protections based on homosexual orientation, Byne was invited by Colorado for Family Values, the Christian fundamentalist group spearheading the antigay effort, to submit evidence in court against gay rights. The object of Byne's attack was Allen's and LeVay's work.

On first encounter Byne's viewpoint appears clear. The first time we met, he said in his intense, determined way, "You know, in the beginning, I bought the biological argument hook, line, and sinker. As an undergraduate, I could go from the molecular level to the social level all in one system, so it made sense. But as a graduate student, when I began to look carefully at the evidence, it didn't hold up. It was haphazard, and some of the basic tenets weren't supported by a shred of evidence. I became very skeptical. I think all behaviors can ultimately be explained in biological terms, but when it's sexual orientation there's hardly a shred of convincing evidence that supports current biological or hormonal theories. I mean, your sexual orientation must have some representation in your brain, but so does your choice of salad dressing."

Byne concluded with an earnest, worried look. "We know that there is a sexually dimorphic nucleus in the brain of the rat, and if you castrate the rat at birth and deprive him of male hormones, this nucleus remains the female size and the rat assumes the receptive, female sexual posture, swaying its back to allow itself to be mounted. We also know that the larger the nucleus is in male rats, the more mounting behavior these males show. But that's *all* we know! And that's about *rats*. We don't know if Simon's nucleus in humans controls sex drive. If you destroy what he claims is the analogous nucleus in rats, nothing happens to their behavior."

Byne's position seemed quite clear.

This morning, Byne is going to use the slide projector to explain how neuroanatomists look at a brain.

The brain is by far the most complex part of us, a labyrinth of cells and structures folded into levels and passages, many of which remain unknown, covered with more neurons than there are stars in our galaxy. And each cell and structure controls our movements, directs our emotions and desires. Untangling this phenomenal neural topography is, in essence, the science of neuroanatomy, the cartography of the mind. This is the territory LeVay explored in tracking down his nucleus.

Byne is going to explain LeVay's study with a series of slides. After fumbling with the uncooperative slide projector on his desk, he flicks it on. When the blinding white square is too low on the wall, he extracts a slim volume titled *Morbidity* from a bookshelf and slides it under the projector. The white square moves up a foot. "Okay," he begins, "we're going to start out kind of in reverse. First I'm going to show the final product. These are small areas of the human brain, what you use to measure the nuclei the way that Simon did. This is what a neuroanatomist sees when he or she measures a nucleus." He hesitates, then clarifies: "You wouldn't use the slides, of course. You use the segments themselves." He clicks the little plastic button. A bizarre scene is thrown against the wall.

The projection is completely disorienting. It looks like a photograph of outer space, a field of hazy purple stars and giant violet nebulae exploding in some galaxy light-years away. If Byne had said this was an astronomer's image, it would be credible. Slide 1 is, in fact, a tiny area of the human brain. It looks completely unfamiliar.

"The first problem inside the brain," Byne says, "it to orient yourself, to get used to the landmarks. It's as if you were out in the woods and navigating by the night sky: which astral bodies look familiar, where are you facing? The first thing to know is that these sections are cut front to back; you're facing the person."

Think of slices of the brain, what neurologists call sections, as slices of a fruitcake. Choose one end of the fruitcake as the face, or "front," take a very sharp knife, and start slicing the cake front to back in extremely thin slices, so thin they are translucent. (Brain matter sliced at a thickness of 52 microns—the thickness Simon LeVay used—are likewise translucent.) The first slice, the outside of the cake, is probably going to be dough, and the second the same. But by the fourth or fifth slice in, you'll probably see a few small dots in various places, indications of a cherry here and a raisin there and a walnut toward one side. As you keep slicing, you begin passing through these objects as a neuroanatomist passes through nuclei and other structures of the brain. The dots grow and then shrink in the sections until you reach the end of the cake, where your final slices will be, again, just dough.

Using a pencil to cast a pointer shadow, Byne points out the various landmarks in this neighborhood of the brain, the hypothalamus, which sits at the center and bottom of the brain. There is a large white oblong structure running north-south called the third ventricle. It looks exactly like the shore line of Lake Michigan, and instead of a field of stars, the scene now resembles a satellite photograph of the earth, with a vague dark blob where Chicago would be—perhaps slightly more inland than Chicago. "Below the ventricle is the optic chiasma," says Byne, continuing the orientation tour, "which is just above your mouth. Over here is the preoptic area, inside of which is the preoptic nucleus." He explains that a nucleus is defined as a dense cluster of cells, "*if* that same cluster is present in all brains." He notes that you can distinguish cells by varying size, and he traces their boundaries, which have been marked and highlighted with a stain. "And *here*," he announces with a sudden flourish, arriving at the tour's culmination, "is Simon's INAH 1."

The nucleus is a vague glob of stars, a thickness in a sea of stained brain cells. Its edges are not at all defined—it seems to melt into being at some point when its cells reach a certain density—but still, it stands out among the constellations of neurons and ventricles.

In Slide 2, the next section of the same area but one segment deeper into the brain, the nucleus metamorphoses slightly, a ghostly blob. Although the sections are 52 microns thick, the slides are actually taken at every *other* section, so we are moving at 104-micron-length steps into someone's head.

Slide 3. "See this?" asks Byne. "INAH 2 is just beginning. You always find it just below and to the right of INAH 1." A faint, dense mass has begun to emerge from the scatter of cells. The cells of INAH 2 are small and show up paler.

These nuclei, which are six inches across on Byne's wall, are in fact the size of the dot over this letter *i*.

Slide 4. Byne goes up to crouch by the image on the wall. INAH 1 is still clear, but INAH 2 is the merest smudge, its outline discernible only with his pencil. A few more steps into the brain, and with some amusement he traces the outline of INAH 3, which is even vaguer, as gossamer as the Magellanic Cloud. "INAH 1 is clearly there," says Byne. "It was actually identified in 1942 by H. Brokhaus and called the intermediate nucleus. But what about the others? Laura Allen was the first to describe 2, 3, and 4 in humans, although there's some doubt as to whether they exist." Byne adds, "I was doubtful myself at first whether INAH 2, 3, and 4 existed. I asked other anatomists, and they were skeptical. We thought they might be something else. The supraoptic and paraventricular nuclei, which are connected in early fetal development, move apart as the brain develops, like the continents of Africa and South America splitting apart hundreds of millions of years ago. Sometimes as they separate and migrate toward different areas of the brain, they leave behind them clumps of cells called magnocellular islands. I used to think that the INAHs were such islands, but they aren't."

He clicks the slide projector again. "INAH 3 was the one LeVay says is different in gay men," he says, gazing at the image on the wall. "INAH 1 is more than four times as large as other INAHs."

Byne clicks the slides forward, and as we proceed back into the brain the nuclei blossom and fade, 1 and 3 swelling in size and clarity, 2 disappearing, then, at the very end, 4 making a brief appearance before also vanishing.

The basic neural geography lesson is done. Byne takes out these slides and inserts a second batch, his ammunition against LeVay. He has basically three criticisms: first, that the methodology used can skew results; second, that applying animal neuroanatomy to humans doesn't work; and third, that the existence of sexual dimorphism in the human brain has not even been demonstrated—"which," Byne points out, "is, after all, one of the basic concepts on which LeVay's work rests."

Byne is fully prepared for his assault, with a small lecture on the three problems starting with methodology, or as Byne puts it, "how you measure this stuff." Byne thinks the Nissl stain LeVay used—a stain that loves acids—may have tripped him up.

He clicks in a slide, a close-up, magnified comparison of INAH 1 and 2, side by side, and points enthusiastically at their images. "See? See there? The cells are different." As he carefully delineates their boundaries with the shadow from his pencil, it becomes clear that, in fact, the cells in INAH 1 are larger than those in INAH 2, and they stain more darkly.

"Now," Byne asks, "why does Nissl stain brain cells?" The reason, it is believed, is that Nissl binds to acid in cells. The more metabolically active a cell is, the more acid it produces; the more acid in the cell, the more stain it attracts, and the deeper purple it stains. "But the problem," says Byne, "is that cells that are very active tend to get larger compared to cells that happen to be inactive at that moment. Simon is trying to measure these cells by dumping Nissl on them, but the stain could just be sticking to cells that happen to be very active at that moment—or to have been active at the moment the subject died—and are full of acid, whereas other cells, which just happened not to be so active at that moment and aren't acidic, don't get so stained. You look at one and you look at the other, and they appear different. But it could simply be that one has been active recently and the other hasn't; at a different time, maybe it would be the reverse. So how do you do your measuring? Your methodology, the way neuroanatomists go about doing this, is still extremely problematic."

Carefully selecting one of Gorski's slides, Byne slips it into the projector. "The sexually dimorphic nucleus (SDN) in the rat brain," he announces, and two rat brains, male and female, perch on the office wall, side by side. (A minor point: There are actually *two* rat SDNs and SCNs, Byne notes, one of each on either side of the ventricle. In humans, too, the hypothalamus sits in the middle of the brain and thus has two sides, left and right. So instead of one each of INAHs 1 through 4, we actually have mirror pairs of each, just as we have matched pairs of lungs or brain lobes or, in males, testicles, or, in females, ovaries.) He points out the nucleus in each brain, and in the giant image of neural cells splayed in light and shadow, the male's SDN is clearly much bigger than the female's.

But then Byne, almost gleefully, points out the rats' suprachiasmatic nucleus (SCN), the human version of which Swaab found to be larger in homosexual than heterosexual men. In the slide, it too clearly appears larger in the male rat brain. But a grinning Byne explains that it is not. "And that's the point," he says. "You can't trust what the slide seems to show you. I object to showing sex differences with these slides because the suprachiasmatic nucleus *appears* dimorphic; in fact, it's the same size in males and females. You have to be very careful with visual measurements."

Moving to his second criticism—that animal findings don't automatically transfer to humans—Byne offers the segmented brains of a male and a female guinea pig. This single slide shows two rows of ten sections of the brain running from front to back, like two rows of dominoes ten deep, stood on end. One row shows the AVPVN nucleus of a female guinea pig, the other, a male. Just as with the SDN of the female rat, this nucleus is of different sizes in female guinea pigs than it is in males (it's over two times larger in females). But while destroying the rat nucleus completely cuts off ovulation, destroying the guinea pig nucleus does absolutely nothing to ovulation in guinea pigs. "What can you conclude?" Byne asks. "If you can't extrapolate from a rat to a guinea pig, how can you extrapolate from a rat to a human being?

"Also the SDN doesn't even exist in mice. And if you can't go from rats to mice . . ."

Byne plasters another slide on the wall, a slice from the brain of a rhesus monkey in profile. It shows an SDN nucleus that is larger in the female than the male, unlike the case in rats. "LeVay said his nucleus is in an area of the brain controlling sexual activity in humans," Byne says, his voice rising with irritation. "Well, very precise work has been done in rats, cats, dogs, and monkeys to determine exactly which area of the brain is involved in male sex behavior in those species. In monkeys, it's this area outlined in green"—he flashes another slide on the wall—"which in humans corresponds, *I* believe, to the bed nucleus of the stria terminalis, a completely different part of the brain. LeVay has tried to say that INAH 3 has a sexual function in people, but that's based on work done in rats with the rat SDN when in fact we don't know *what* function the SDN has in rats. We *do* know that male sexual behavior in rats is linked with this green area, not the SDN. And frankly, Laura's standard talk makes it clear she is motivated by the rat evidence."

Byne is truly agitated by now. "This type of thinking pervades the medical field more than you can believe. When I arrived at the University of Wisconsin in Madison to work with Robert Goy at the Primate Center, he refused to allow me to take the U of W medical school course in endocrinology. He said I'd only learn about rats and screw up his research on primates. Six years later, when I entered medical school at the Albert Einstein College of Medicine, I was appalled to find that rat endocrinology was still being taught. I'm not faulting these institutions in specific but medical education in general. Even on the medical board exams, I would have to give the wrong answers—*rat* answers—to get credit. This was particularly galling for me, because one reason I went to medical school was that I was interested in human neuroendocrinology and was convinced that if you couldn't even extrapolate your basic findings from rat to mouse, you'd have to get medical training to learn about humans. And I found myself learning about *rats*."

Byne looked consternated. "You know, the media often calls on medical professionals to interpret sexual differentiation research for the public, but most of these professionals are not even aware that in medical school they were taught rat neuroendocrinology—not human. Not even monkey, nor guinea pig, nor even mouse. *Just* rat."

Byne's point about not extrapolating from one species to another is vividly illustrated by Scott Wersinger, a young graduate student at Boston University who works in the lab of BU endocrinologist Mike Baum. Scott studies ferrets under Baum, in particular the neurological mechanisms that trigger ovulation. "There are two types of ovulators: reflex and spontaneous," Wersinger explains. "Most people are familiar with the spontaneous sort because both humans and rats are spontaneous. So are dogs, sheep, almost all domesticated animals, and primates. But ferrets are quite different from us. Ferrets are reflex ovulators, like rabbits, cats, some species of voles, and camels. Where humans operate in twenty-eight-day cycles, ferrets come into heat once a year, which is called estrus, and they stay in estrus for four months. But they will never ovulate while in estrus unless a male mounts them and stimulates the vagina and cervix. This causes the endocrine system to kick on the process of ovulation, the male ejaculates, the sperm swim up—it takes them about fifteen to twenty hours to travel through the cervix, up the uterus and fallopian tubes to the egg—and the reproductive process continues on. I'm interested in how this purely physical sensory stimulation causes these hormonal changes in the brain."

Given Byne's lectures about all the differences between rats and humans, Wersinger's point that rats and humans are lumped together as spontaneous ovulators seems surprising. "But you see," he says, "they both are, and that's the thing: For any trait, you never know which group you're going to put any species in until you know. For instance, take two sexually related traits, ovulation and sexual receptivity. You'd put rats and ferrets and cats together for sexual receptivity on one side—species where the female isn't receptive unless she's fertile—and on the other, humans and primates and, in some cases, lions, animals that are sexually receptive regardless of fertility. For ovulation, it's totally different. In this case, rats are separated from cats and put with people, who are together with monkeys on one side, and cats, rabbits, and camels are together on the other side. Your dog is in one group and your cat is in another."

Brain development also often has nothing to do with the thousands of these differences, says Wersinger. Ferrets are more highly developed than rats in their cerebral cortex, almost like cats, but less developed than monkeys. And yet for some traits, rats may be in a biological group with ferrets, for others they might be grouped with monkeys. "Look at menstruation," says Wersinger. "That's another trait. Only what biologists call Old World—from Asia and Africa—primates and humans menstruate: chimps, gorillas, rhesus monkeys, apes—and us. All other species don't. In all other species, the uterine lining isn't shed. So here, we have us grouped with monkeys on one side and rats are with cats and everything else on the other." He laughs. "It's a mess. No, actually it's not. It's just saying that, look, in the long run, you have to take each species on its own." Even the way we carry our testicles differs: Elephants happen to be one of the very few species where the males have inguinal testes, which means they're inside the body cavity. Human males, of course, carry theirs on the outside. And where we have two sets of teeth, baby and adult, elephants have five.

The dictum of "taking each species on its own" is illustrated perhaps most dramatically by differences in the various species' immune systems. This becomes quite clear after a short amount of time in a college biology lab where graduate students castrate lab rats. The rooms are clean and well lit, with a lot of tile and stainless steel like the no-frills operating rooms they are, but unlike human operating rooms, sterility is utterly unnecessary. It is an interesting scene. The rats are unconscious, breathing at an extremely fast pace, and the students remove the testicles by making a small incision in the scrotum and then extracting them with tweezers. Rats' testicles are slightly elongated and startlingly huge, larger than their paws—if a man's testicles were of comparative size, they'd be as large as footballs—and the procedure can become messy. Yet within three hours after surgery, the rats are up and running around as usual, showing no aftereffects from the operation. Unlike surgery on humans, one of the students said cheerfully, holding a rat in one hand and a scalpel in his other, you don't have to sterilize the instruments when operating on rats.

Their immune systems are so highly developed, they can fight off infections that would quickly be fatal for us.

Another wide divergence is the senses. We have the well-known five; cats have six: touch, smell, sight, hearing, taste, and detection of chemicals from a nearly invisible system called the vomeronasal organ in their mouths. This organ picks up such substances in the air as sex pheromones, which, again, we humans may not produce, although this is a subject of great debate; one scientist thinks we do produce them, but, she believes, our culture teaches us to regard them as "offensive" smells, such as body odor. (Recent research has suggested that humans do, in fact, have vomeronasal organs, perhaps even in some sort of functioning form, which detects pheromones with receptors different from those that detect odors.)

Human senses are, in fact, inferior to those of most animals in many ways. Scientists Carl Sagan and Anne Druyan point out that bumblebees can detect the polarization of sunlight that is invisible to humans and that pit vipers pick up infrared radiation and temperature differences of 0.01 degrees Celsius at a distance of half a meter. Some African and Latin American freshwater fish generate a static electric sphere around themselves and sense intruders by slight perturbations in this field. (The African fish do this differently from the Latin American fish; apparently, the electrosensory systems evolved independently in these two continents.) Various scorpions have seismometers and hydrostatic devices to measure insect footsteps and water depth. Dolphins, whales, and bats use sonar guidance systems. Bacteria have in their bodies tiny crystals of magnetite, an iron mineral, used for positioning themselves in relation to the Earth's magnetic field. If we were to judge aspects of human beings by the presence or absence of such traits in other organisms, we would inappropriately make ourselves look both hyperevolved and hopelessly outmoded.

For all these reasons, looking to other species for a measure of "naturalness" or normalcy in our own is futile, at best. And yet people constantly do it. As UCLA researcher Richard Green put it, "There's a whole lot of things we don't find in animals. So what? 'Natural' is a word used in a meaningless way by most people. If to call something in us 'natural' we always had to find it in, for example, monkeys, we'd have one hell of a time explaining speech." And sex determination? The sex of *Homo sapiens* is determined by whether we have XX or XY sex chromosomes. The sex of the *Drosophila* fruit fly is determined by the *ratio* of sex chromosomes to autosomes. And yet some people have always made such comparisons, and some always will. It was said in this century that if God had intended for human beings to fly, he would have given them wings.

Byne looks at his watch. It's time to leave for his lab. As he steps outside onto West End Avenue and locks the door to his office, he mentions, since he is a thorough scientist, something odd, another piece of research evidence: "But then," he says in his deceptively casual, offhand way, struggling with the key in the door, "certain cells in the human hypothalamus have been identified as producing a hormone—called LHRH—that controls human ovulation. Rats use their hypothalami for the same function, and they also have LHRH there. So maybe human beings *are* like rats in this biochemical way."

He's been facing the door, and there is by this hour plenty of traffic on West End Avenue along with a group of movers working loudly at the curb, so perhaps it was the noise. What he's just said seems after all to contradict his position that you can't use animal models for humans, a position LeVay supports. Does Byne really mean to say that in some ways, rats *can* serve as biological models for humans?

Byne puts his key in his pocket. "Of course," he says mildly, as if it should be self-evident, and turns north toward the 90th Street subway station.

⟿

Byne's lab was at the Einstein College of Medicine of Yeshiva University in the Bronx, where he was a research associate. (He is now director of the neuroanatomy lab in the Department of Psychiatry at Mount Sinai Medical Center in Manhattan.) Sitting on the plastic seats on the number 2 train uptown, he begins to relate the saga of the corpus callosum. It is a story of one of the most antagonistic long-running scientific and political battles in biology, encompassing biological differences between men and women, feminism and conservatism, sex roles, education, professional opportunity, and, now, sexual orientation.

For years Byne has been waging war with UCLA neuroanatomist Laura Allen because the corpus callosum is the preeminent example of a brain difference between men and women. Listening to Byne above the rushing sound of the subway, it certainly sounds as if he is saying this sex difference doesn't exist, while Allen says it does. Notes Byne, "LeVay said that Christine de Lacoste's 1982 study, ['Sexual Dimorphism in the Human Corpus Callosum,' in *Science*, Vol 216, June 25, 1982], the first to find a sex difference in the corpus callosum, has been replicated by Laura in her 1991 study, 'Sex Differences in the Corpus Callosum of the Living Human Being.' [2] [April 1991 in *The Journal of Neuroscience*] And furthermore," he continues, "LeVay said that these sex differences were the same: The women had bigger CCs than the men. None of that," he says flatly, arriving finally at his third objection to LeVay's study, "is true."

He ticks off the points on his fingers. "Laura did *not* replicate de Lacoste's findings. De Lacoste said the splenium, the rear part of the callosum, was *larger* in women. Yet none of the twenty-four subsequent attempts at replicating de Lacoste find it larger, not even Laura. *She* found it more 'bulbous.' That's how she described it. That's different. Four other studies find the splenium more bulbous; of those four, two just eyeballed it, so only two were mathematically done morphometric studies. And of those two, one found it more bulbous in men, not women. De Lacoste stated that there was a 'dramatic' difference in the shape. Well, when Laura did a morphometric analysis, she found the difference in the bulbosity not particularly dramatic. When her lab tried to eyeball the difference—to tell the men from the women just by looking—they were successful about 50 percent of the time, which is exactly what you'd get by chance. In other words, they couldn't do it."

He pauses to take a breath as the train rockets past 149th Street and the Grand Concourse. "Eight studies found no difference in bulbosity. In rats where a sex difference has been established in the callosum, it's larger in males than females, the opposite of what they're suggesting in humans. Laura will tell you 'there is some disagreement' and 'the results have been mixed.' This is clearly misleading. She doesn't say, for example, 'One lab found it, twenty-three didn't.' I can cite for you a total of twenty-three studies of the corpus callosum, and de Lacoste is the only one that finds the cross-sectional area to be larger in women than in men. Two other studies find it larger in men than women. And the remaining studies find no statistically significant differences. Looking at these twenty-three studies, there's absolutely no reason to think it's larger in women, which is what Laura says."

To underline his point, he mentions Allen's finding of a sexual dimorphism in another structure, the anterior commissure. "Only one other lab has looked at the AC. They found it larger in men than women." Effortlessly, he reels off the citation from memory: Demeter et al., Brain Research Institute at Rochester, 1988, *Human Neurobiology*. "Laura references their study in her study as finding a sex difference, but she neglects to say that their results are the *opposite* of her own."

Byne doesn't merely think that Allen has yet to demonstrate that sexual dimorphism in the human brain really exists. He also raises the question of how you interpret one if it does. "Even if you do 'find' a difference between men's and women's brains"—"find" he says, his fingers making quote marks in the air—"so what? Let's say there is a dimorphism. What does that mean?"

It is a good question, one that plagues neuroanatomists; finding something in the brain only raises the question: What does it do? Apparently the size of the corpus callosum, the brain's main communications wiring shaft, affects the degree to which the two halves of the brain are able to specialize. But *how* does it do this?

Both brain halves, right and left, participate in every mental task, but each is preeminent in different fields; the right is generally more active in math and the left in speech, for example. The halves talk to each other through several bundles of fibers, the largest of which is the corpus callosum, and work on mice has led us to believe the corpus callosum plays some role in helping the brain specialize: The bigger the callosum in the mice's brains (and presumably in ours), the more the brain's halves split up chores. We have reached that conclusion from this observation: If a mouse is born without a callosum connecting the lobes of its brain, something that happens every so often, its two lobes will tend to work on the same problem at the same time. "How do I get the cheese out of the Tupperware?" and "Where is that damn cat?" But in the mouse born not just with a callosum but with one bigger than normal (mice are bred in labs for callosal size), a callosum bursting with axons connecting right and left, the halves will specialize even more than normal, divvying up various tasks as they come in through the senses from the outside world according to type. The right half of the mouse's brain might concentrate on figuring out how to get at the cheese, a job of mental calculation and reason, and the left will keep an eye out for Blossom, a sensory-monitoring job.

"So you see," says Byne, "if you extrapolate the animal research to humans, then if you have a large callosum that should mean that your brain's halves are more specialized." He pauses for a dramatic moment. "Well, this," Byne says triumphantly, "this is the *opposite* of Laura's model." He sits back in the plastic subway seat looking as if he has just hit a home run. "She adheres to the de Lacoste and Holloway interpretation, which said that in humans, a large callosum equals you've got more axons connecting your two halves, which equals *less* specialization of the two halves of the brain. The animal research suggests that a large callosum equals *more* specialization. Same anatomical observation," he notes with relish, "opposite conclusions."

The thing is, notes Byne, that all these equal signs are speculative. They're all interpretations of data. And there are always two ways to interpret any data. What is astonishing is how directly contradictory the interpretations can be.

Imagine that the corpus callosum is a transatlantic cable packed solid with fiber optic filaments. It connects two offices, yours in New York with a sister office in London where your colleagues work, and your two offices are working on a big, highly complex project: a new translation of the Bible. The cable allows you to stay in close touch with your colleagues in the London office, and the bigger the cable, the more closely your New York team can work with the London team. But it can allow the offices to work in two very different ways. Way 1: New York and London decide to *coordinate* their efforts, tackling Genesis together, then Exodus, and so on. Their big connection means they can be more symmetrical in getting the job done.

Or Way 2: instead of coordinating and working together, the teams could specialize and work separately. New York will do Genesis, London will take Exodus, and so on, in which case the bigger cable allows New York and London increasingly to be independent, to *dis*coordinate, so they aren't duplicating efforts. Neuroanatomists face exactly this problem of interpretation, and while a number of studies in women suggest that a larger callosum means the left and right halves of the brain are coordinating, the animal work shows the opposite: a larger callosum means they are specializing. What is the larger callosum actually doing, helping the halves coordinate or specialize? We don't know.

In fact, we don't even know how many fibers run through the corpus callosum, and this raises another question. The size of the callosum could depend on the size of the fibers inside it. Is the transatlantic cable bigger because it's packed with, say, 20 percent more fiber optic filaments? Or is it bigger because it has exactly the same number of filaments, but they're just 20 percent fatter around? Allen is measuring the area of the callosum, Byne points out, and assuming that just because one cable is bigger than another, there are more filaments inside it. But since no one has ever actually counted the number of axons in the corpus callosum, the larger area may be caused by more axons but it may also be because those axons are insulated with more myelin, the lipid (fat molecule) that covers axons. Although two different CCs have exactly the same number of axons inside them, the axons in one might happen to have a thicker fat coating, so that callosum has a larger diameter. We don't know. (Byne mentions that men's brains are, overall, larger than women's, even when corrected for body weight, and some of the more feminist-oriented neuroanatomists suggest that men have larger brains because their axons are more insulated with myelin; in other words, they're fat heads.)

And maybe, suggests Byne, a callosum is 20 percent bigger because it has more than one *kind* of axon inside it, and maybe the different kinds do different things. One kind might coordinate; another might not. "The animal literature suggests," he says, "that many of the axons that connect the brain's hemispheres are actually inhibitory: When one hemisphere is active, they *inhibit* the other from functioning." He finishes with a rush, holds his breath as he reaches the end of his mental checklist of points, then lets out a deep breath. "We just don't know," says Byne.

∼

Laura Allen sees things somewhat differently. "If," she notes in her sweet, soft-spoken, yet very firm way, "Bill had read my paper carefully, he would know that, in fact, I mention that the area of the corpus callosum may not reflect differences in the *number* of fibers but"—she enunciates it clearly—"in their *size.* "

"Basically our corpus callosum study *does* confirm the de Lacoste study," Allen says amiably, contradicting Byne. "You see, this is the first study on the CC that actually used the same measurements that de Lacoste did."

Allen sits on a stool at a black-topped laboratory bench, high up in UCLA's Department of Anatomy and Cell Biology in the Laboratory of Neuroendocrinology, where she does her research. (It is the same building that houses the rat movies; Roger Gorski's office is a few hundred yards from hers.) The lab's southern wall, completely glass, lets in the Los Angeles sun, and the room and its plastic tubes and wires and metal shelves and glass beakers radiate a coating of light.

Allen is completely at home in her lab. With her shoulder-length blond hair and relaxed demeanor, she looks very southern Californian. She lives with her husband and young children in nearby Seal Beach. Her voice, which is deceptively soft and brightly enthusiastic and is her most striking feature, hides the strength of scientific purpose that emerges when she has a point to make. Allen has a Ph.D. in neuroanatomy and is quite used to controversy. As a young researcher in 1984, she discovered the INAHs and, what was more, claimed that one of them—INAH 3—differed in the brains of men and women. It was a wildly unpopular claim; only a few areas of research—notably sexual orientation, race, and violence—rival the political explosiveness of the research on gender differences, which feminists decry as sexist. It can trip up or even demolish the careers of scientists who publish too soon, too dramatically, or even at all.

Roger Gorski, in whose lab Allen was doing her postdoctoral work, all but refused to let her publish the INAH study because of the political repercussions. "He really counseled me against going with it," she says. It was finally published in 1989, and two years later Simon LeVay turned to it for what it might say about a gay-straight difference, testing "the idea that one or both of these nuclei exhibit a size dimorphism, not [only] with sex, but with sexual orientation," and the rest is now history. As if that weren't enough, in 1991 this easygoing neuroanatomist studied the corpus callosum, the brain feature held to be different in men and women; this area is now, as Simon LeVay put it, "the longest running soap opera in biology" and the bloodiest neuroanatomical battlefield, adding another controversial area of biological research to Allen's professional portfolio.

Allen has had to be intellectually well armed, which in biology means being rigorously versed in the literature. She is. And she dispenses with Byne's claim that she "adheres" to "larger callosum equals more specialization" by noting "This is factually incorrect, as is clearly stated in my paper: 'It is unknown whether sex differences in regions of the CC correspond to sex differences in myelination, in the number of fibers, or to a different arrangement of axons coursing through the CC.' " She then effortlessly reels off the cite, *Journal of Neuroscience* 11 (4) p. 940, and ends briskly with "I adhere to this paper and not to what Bill thinks I 'adhere' to for the sake of criticism."

As a note in the margin of a letter, Byne once scrawled out that Allen had only gotten 50 percent correct when trying to identify corpus callosi as male or female. Allen got hold of the letter and scribbled in the other margin: "Absolutely untrue. Paper clearly states 66% of adults' CCs and 67% of children's were correctly IDed. That's a 3 out of 10,000 chance we're wrong. Bill claims 50%. Why?"

Byne has said cynically, "You know you've gotten to a bad place in science when you don't need to read the papers to find out their conclusions, you just need to know who wrote them."

"He would make fewer mistakes," Allen responds with a brief smile, "if he read the papers he uses for his arguments."

But Byne does have good ammunition from the literature—for instance his list of twenty-three studies that have failed to replicate Christine de Lacoste's corpus callosum study. Allen, reminded of them, smiles—she has heard this from him many times before—and explains her position gently but evenly. The problem with replicating the de Lacoste study, she says, reflects the myriad factors faced by biologists who work on the brain. The problem is also deceptively simple: How do you measure something the same way each time? "Many studies have contradicted her," says Allen, referring to Christine deLacoste, "but they measured the CC differently than she did so of course they got different results." She sighs, familiar with the problem. "The CC is very curved, and the entire CC may not be sexually dimorphic." She takes out a diagram of the callosum and places it on the black lab table. Sitting in the center of the head, the CC resembles a very bent, very irregularly shaped Frisbee that has been cut in half.

"To show it's sexually dimorphic," Allen explains, "you have to divide the CC up into different regions, and everyone's been dividing

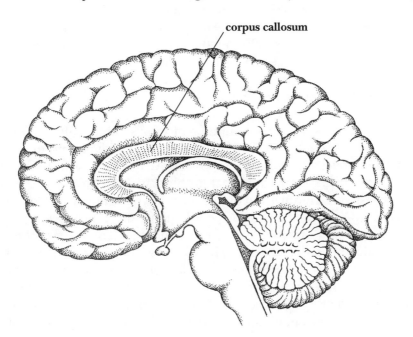

corpus callosum

it up differently, which is part of the controversy. Some scientists take into account the curvature of the CC, determining this line"—she draws a line with a finger—"and then dividing it into five different sections. Others use the straight-line method and just measure the anterior and posterior sections and divide that straight line into fifths. But that gives you anatomically different components."

Although Allen's measurements and results replicated those of de Lacoste, her methodology—the way she did the study—was different. "She used brain sections of the CC from postmortem tissue, as Simon did with the hypothalamus," said Allen, "and we used MRI scans of living human beings."

MRI—magnetic resonance imaging—is a scanning technique that delivers an image of a person's insides. While X rays give an image of bone and hard matter, MRI allows you to do something that X-ray technology can't: look at soft tissue in living people. This means mostly the brain. Typical MRI machines, which cost from $7 million up, look at signals from hydrogen atoms. Different tissues in the body, from fat to neural matter to blood, have different densities due to their varying amounts of water, and the protons in the hydrogen atoms of the water have slightly different motion dynamics depending on the tissue. The hydrogen atoms in a tumor, for example, look different to an MRI from those in healthy tissue, which is why doctors use MRIs to search for brain tumors. Neuroanatomists like Allen use them to study the structure of the healthy brain; the machine receives signals from the corpus callosum, the thalamus, and the medulla, and puts together an image of the living brain. The machine's parameters can be adjusted to focus on fast-moving atoms or slow ones. One MRI spectroscopist commented, "It's really an art as well as a science."

The machine was also something of a godsend for Allen. "Simon used postmortem brains," she explains, "and so he had no medical histories, no lesbian brains, and possibly AIDS effects, which is probably the primary methodological problem with the work at the moment. Using MRI, you avoid disease effects, you get medical histories because you can interview living people about their sexual desires, and you can get gay women—you just ask them if they'll do it. An MRI scan isn't bad. You don't feel a thing. It's like an X ray, only less radiation."

Allen places two MRI scans on the lab table, black-and-white "photos" of the inside of the brain. One splenium, it is immediately clear, is bulbous and the other one not. "This," she says, pointing at the large one, "is the callosum of a woman and this"—the smaller one—"of a man." She cautions that "these two were put together for illustrative comparison. Most of them are not this distinctively different. When I look at a series of unidentified callosa outlines, I'm able to determine about two-thirds of them correctly."

Having soared through the advantages of the callosum research with MRI, Allen then plunges into the disadvantages. They are legion.

Consider the plethora of CC replication attempts, Allen says. One found a difference between men and women but suggested it might be caused by cell death; another found that differences were actually greater *within* each sex than *between* the sexes, which is completely unhelpful; and a third obscurely found a greater callosal area in mixed-handed than in right-handed people. Part of the confusion with the research, she explains, is how to go about studying this fleshy tube of axons. As a researcher, every time you make a methodological choice, you move further down a path that is then more difficult to change. Deciding between corpus callosum slices cut from dead brains or live MRI images is just a first step. If you go with the slices, then you have to decide how thick to slice them. Allen says that Swaab was looking at his human SDN at a thickness of 6 microns when he found significance; when she attempted the same study, she used 60-micron-thick slices, and she didn't find any significance. This thickness issue is a very basic yet important difference that can change results: With thin-sliced tissue, a larger number of cells is needed to pick out the nucleus boundaries.

From that point on, a thousand decisions must be made. You can choose male and female brains of different ages or age-match your subjects. "And if you say 'Oh, well, we'll age-match them,' " explains Allen, "the problem is we have evidence that the corpus callosum develops and grows at different rates in males and females, which means if you want to age-match your subjects *developmentally* rather than simply chronologically, you have the problem of calculating what stage of development in males corresponds to the same stage in females." You then need to create an entire methodology just to carry out this study to allow you to do your *real* study. And then do you match for race? And if so, do you match for economic class and the way the person lived?

Then there are the histories of the people—complicated, living human beings. Don't forget to consider left- or right-handedness (which may affect the callosal structure), environmental factors such as stress, diet, or nutrition (which are known to change brain structure), subjects' health and history of medical care, cause of death (if you're using sections), drug user? drinker? vegetarian? smoker? and on and on. Each decision carries advantages and disadvantages and can change the results of the studies.

And then, for scientists facing the challenge of replicating a study, there's the problem of the English language. Allen relates that de Lacoste's three investigators were able to identify their brains with 100 percent accuracy, which seems like a startling figure until she explains that "that's probably not realistic. It could mean they used subjects of different ages, different races. Age is an important factor. So ours is actually the first study with age-matched subjects." But then it's impossible to know whether de Lacoste's subjects were age-matched "because," Allen says with a sigh, almost wincing, "it was a science paper, and science papers often get abbreviated." The three prestigious publications of the science world—*Nature, Science,* and *Cell,* where everyone is desperately trying to publish—require extremely short manuscripts, and their editorial departments wield extraordinary power. So out go all the details, the charts with "extraneous" data that happen to be clarifying, the specifics of what was done and how. "De Lacoste and Holloway looked at the posterior fifth of the callosum using the curved-line method," notes Allen, "but they didn't explain it in their paper, which has caused a lot of problems in the literature. If it was their writing or *Science*'s editing, I don't know. We had to go in and ask them for clarifications personally." She finishes glumly, "Papers don't usually give a lot of detail."

Given the nature of science, Allen acknowledges that replication is highly problematic, but she then turns the criticism around and aims at all scientists. "When de Lacoste did her study, she got $p = .08$ for the *area* of the splenium, the posterior fifth of the callosum, which doesn't achieve significance, but people read it and thought it was [significant]. And so when they weren't able to replicate it and get a significant result, they were very critical. I know the writing wasn't crystal clear, but . . ." She hesitates, uncomfortable with being overly direct about her colleagues. "I think," she tries delicately, "that a lot of people didn't read the paper carefully."

("If Laura would read the paper carefully," says Byne in New York, "she would see that de Lacoste interpreted $p = .08$ as significant. She has reported a sex difference in shape, not in size." "I have read the paper carefully," says Allen in Los Angeles. "Several times.")

If the corpus callosum has made Allen a magnet for controversy, the INAHs, as a sex *and* a sexual orientational dimorphism, have twice the political explosiveness of the CC. Some scientists argue that the INAHs—at least 2, 3, and 4—don't exist. Byne, for example, notes, "I went through every published neural atlas, and the INAHs weren't in there." Allen grants that 2 through 4 are hard for some researchers to identify, but suggests, "It means their histology [preparation of brain tissue] probably wasn't very good. It's very unlikely that these nuclei don't exist and that we're just seeing an artifact." Still, she admits that the nuclei are mostly mysterious. "Nuclei are clusters of brain cells. A lot of the brain is made of neuropil, just axons and dendrites, extensions of brain cells which are like tentacles, really, or I guess antennae since they send out and suck up information. A nucleus is a cluster of the bodies of the neurons without all the antennae, the powerhouses of the cell."

Neuroanatomy is a young science, and Allen is quite willing to admit that the hypothalamus, home to her four nuclei, is to a great extent uncharted territory. "We don't do surgery on the human hypothalamus, because we know so little about it, and partly because of that it hasn't been mapped out yet and remains largely unknown," she says. "Neurosurgeons operate on the cerebral cortex, the 'thinking' part of the brain involved with voluntary motor functions and the region of the human brain that really separates us from other animals. They also operate on the thalamus, a relay station to the cerebral cortex from the rest of the body. But not on the hypothalamus. Frankly, we know *much* more about the rat hypothalamus than about the human."

When LeVay's study finding a smaller INAH 3 in gay men came out, many of the news reports suggested that INAH 3's size "caused" a person to be homosexual. In fact, says Allen, that's quite possibly not what the study means at all. If the dimorphism truly exists, the nucleus might not be creating a person's sexual orientation but rather resulting from it. The name of Allen's lab explains it: Laboratory of Neuroendocrinology. While "neuro" refers to the structure of the brain, "endocrinology" is the study of hormones, and one of Allen's working hypotheses is that hormones might create people's sexual orientation, and sexual orientation would in turn influence people's brain structure. But, she says, the reverse could be true.

"Remember," she warns, "nobody knows what these nuclei actually do. In fact, nobody really know what *most* of the brain's parts do, and certainly not how the brain really works. Many studies show that you can remove parts of the brain and the person will not lose any particular function, so either the brain doesn't use all its parts or it has an amazing capacity almost instantaneously to rewire itself. Which is possible. But we speculate that the INAHs have a function. They're located in a part of brain that regulates hunger, thirst, sexual functions, temperature, and the release of certain hormones. Basically I believe that sex hormones during life, perhaps before birth, determine whether the brain is masculine or feminine for a number of characteristics, including sexual orientation and the size of INAH-3. Low levels of testosterone, for whatever reason, entering certain brain cells during early life would result in a small INAH-3 and a sexual orientation toward men, as in the case of heterosexual women and homosexual men, and higher levels would create the reverse in straight men and gay women.

"Of course," Allen finishes, sitting back with a smile, "we don't know."

~

Surprisingly, despite the war they wage over nuclei and axons, both Allen and Byne end up not at any firm disagreement but at "We don't know." But the war is certainly real and has taken on another dimension, one unfortunately not rare in science. On various op-ed pages, Harvard's Evan Balaban and feminist Anne Fausto-Sterling of Brown rip into LeVay's and Allen's research. By now Byne has become a well-known personality with a gadfly reputation, one he frankly does not much enjoy but that he sustains despite himself through passionate letters to the editor of the *New York Times* decrying much of the research on sexual orientation as reported in the *Times'* Tuesday Science section. Byne's usual demeanor on the subject, like Fausto-Sterling's, is intense frustration, partly because, despite his academic understanding of the subject and his familiarity with the literature, he is truly maladroit at communicating to others—both colleagues and the press—what he really means. This in turn makes his professional relationships suffer. *"Oh,"* said LeVay to me once, "you're talking to *Bill,*" and made a show of rolling his eyes. Byne's scientific efforts suffer as well. Byne wants to attempt a replication study of LeVay's INAH study—actually, to try to find Laura Allen's sexual dimorphism *and* LeVay's sexual orientational dimorphism. He asked LeVay for the brain sections LeVay used in his study, a request LeVay could easily enough honor: The sections, which are still at the Salk Institute, will, once they've been "fixed" with formalin and mounted on glass in a sterile, airtight environment, last indefinitely. All LeVay would have to do would be to pack the dozens of small pieces of glass with preserved brain matter on them and mail them to New York. Standard procedure.

LeVay (according to Byne) agreed, and then later changed his mind "because," says Byne, visibly pained and frustrated, "he said I wouldn't be objective." LeVay (according to LeVay) was only interested in having someone look at them who was not "biased." For his part, LeVay has little tolerance for Byne's charges, like his charge that the press, including the science press, is biased toward biological explanations of human traits. "Bill tried to replicate the CC stuff, and his study found no difference," says LeVay crisply. "And then he got antsy because *his* study wasn't published in *Science* and so he felt they were prejudiced toward nonbiological explanations. Well, *Science* rejected Laura's paper on the sexual dimorphism of the anterior commissure."

Perhaps because of Byne's persistence, as Laura Allen tactfully calls it, others share this opinion of Byne, although it probably comes down not to objectivity but to personality. Reporter Elliot Marshal once wrote in *Science* magazine that as a critic of sexual orientation research, Byne was "outside the field." Byne took it personally. "The article in *Science* said that I had an ax to grind," he says, clearly anguished, "that I have a chip on my shoulder, and that my measurements wouldn't be more objective than Simon's. In fact, *my* measurements would be blind because Simon has the codes for the sexual orientation for his brain sections, and I'd have to give him my measurements for him to decode them and get the results. And furthermore, if I didn't replicate his measurements, and if others too didn't replicate them, how can that be biased? Having seen a few of LeVay's sections, I am more skeptical than ever."

And so Byne has found his own brains and is proceeding with his own study in his lab, which is where he is now heading on the Number 2 train.

Fausto-Sterling believes that because of the way politics can skew and misuse LeVay's work—"This stuff just lends itself to the promotion of sexism and racism and homophobia and always has," she once said—biologists doing sexual orientation or gender research would have to consult feminist theorists, gay theorists, sociologists, and historians to get any useful data.

LeVay's response to this is measured somewhat. "Anne's criticisms are, when you get down to it, political, not scientific; Anne is adamantly opposed to the notion that biological mechanisms underlie sexual orientation and indeed many aspects of human personality like gender. As a liberal, she believes that social forces mold people in all sorts of respects. I am absolutely sympathetic with her in that I do think in the long run you have to have people with different points of view do their own investigations and hopefully reach a consensus. And it's reasonable enough for her to say that I'm coming from a certain point of view, that I wanted to find structural differences in the brain, that I had that hypothesis and if I hadn't found them, I would have been at a loss. But let's be frank. Almost all of us scientists are biased when we do experiments. Can you honestly say we'd all be equally happy with a positive or negative result when the negative is pretty much meaningless and uninteresting and the positive is informative and may lead to a great scientific discovery and wealth and fame and Nobel prizes and all that nice stuff? Of course not. You're biased—*emotionally*—toward the positive result. But that's natural, and in the end if you're a good scientist and you come up with a negative result, you shrug your shoulders, toss out your hypothesis, wipe your hands off, and start over. Because that's science. What I find difficult is Anne and Bill's view that you've just got to automatically reject these notions of inherent differences. Maybe I'm wrong about them, but that's the way it comes across. And that's *their* bias."

He ends with a grin. "Among scientists, I'd say we're pretty much in opposing camps." Since their last disastrous attempt, LeVay and Fausto-Sterling will no longer do radio shows together.

Some are less charitable than LeVay. One scientist, who requested anonymity, complains, "The way to fight sexism is not to pick apart the research. All you do is create an information mess that doesn't say anything. Laura Allen is *the* current expert in sexual dimorphism in the human brain and one of the experts on the brain today. She's the one who's done the most exhaustive and expensive studies and has had access to the best technical equipment. Who are these people to come in here and, often without doing their own work, deny that there's any worth to any part of this body of work? Anne makes a big deal about 'science' in her critiques, but her objection is on feminist grounds, it's purely ideological. I'm left with the impression that a lot of feminists have decided that this can't be true because they don't want it to be true. Instead of saying 'Wow, this is interesting, let's look at this critically and see where the weak points are and how they could firm up their analysis,' they make a career as Anne and Evan Balaban have of criticizing other people's work. This research could be happening in a collaborative atmosphere instead of a contentious one. The thing that has always struck me about the loudest voices criticizing LeVay and Allen is that they always accuse them of bias, and yet the critics are, I think, more biased than anyone."

⌒

Still a bus ride away once you've gotten off the subway at East 180th Street Station in the Bronx, Einstein is a gigantic box of metal and glass. While perhaps once a daring design achievement, today it is a textbook specimen of 1960s architecture. Byne's lab on the fourth floor is a traditional one, virtually identical to Laura Allen's except that outside is not a sunny, manicured Los Angeles but a slightly shabby, comfortably quiescent Bronx landscape. On the shelves are books with titles like *Degeneration and Regeneration of the Nervous System* and *The Brain of the Opossum (Didelphis Marsupialis)*. It is in this room that Byne is doing his study of thirty brains—ten homosexual men who died of AIDS, ten heterosexual men, five of whom had AIDS, and ten presumed heterosexual women, five of whom also had AIDS—in a conscious attempt to replicate LeVay's results. Byne has arranged to demonstrate his procedure, the first step of which, after the brains are selected, is to slice them with a microtome.

Down the hall, Dr. Linda Mattiace, an attractive, pleasant blond woman with a startlingly thick Brooklyn accent, cut and mounted the sections that Byne is using in his study. She sliced them with a microtome at 52 microns, the thickness LeVay used. Instead of the high-tech delicatessen meat slicer one expects, the microtome looks more like a small, old lathe from high school shop class. It has one very thick blade in the shape of an ax head, which doesn't look particularly sharp. Byne and Mattiace laugh; Byne says he knows a technician who was working with the microtome blade, dropped it, and tried to catch it. The blade instantly sliced her thumb off on its way to the floor. "Of course," he adds nonchalantly, "it was in a hospital, so they just reattached it." The thumb was fine.

The next stop is to look at some stained but unmounted sections stored in plastic vials bound together like a twelve-pack, each filled with clear liquid in which floats a thin, whitish rectangular film of brain slice. Again, there's a surprise: They are tiny, each slice of neurons maybe half an inch wide. They are slices from the hypothalamus.

Byne sits down at the bench and pulls a stool up to the counter in preparation for mounting the sections on the glass slides. He picks up the plastic twelve-pack where the sections are floating, and with a tiny camel-hair watercolor paintbrush, "like you get in any art supply store," he says, he delicately lifts a section out of one of the vials. It instantly clings to the brush like a particularly flimsy contact lens. He lays (almost pours) the brain segment onto a glass slide, which already has been dipped in clear gelatin, gets it centered and, after a little effort with the brush, lying flat, then puts the slide in a small rack to dry.

Skipping ahead, he picks up a slide, this one with a brain section already mounted and dry. An amazing amount of anatomy in the section is visible with the naked eye. Byne holds it up to the light, revealing lines and dots and the beginnings and ends of what are clearly different parts of the brain. There are tiny holes where the blood vessels were.

After the sections are mounted in gelatin on the glass slides, they are left in xylene, which is basically paint thinner, for three days. The xylene leaches all the fat out, which helps the sections take the stain better. Fat, an oil, rejects stain. After three days the xylene is washed out of the section with successive baths of alcohol, first pure, then a 70 percent solution, then 50 percent, and so on, until they equilibrate. ("Equilibrate" means that the tissue stabilizes.) "Then," Byne narrates, "they're stained." Neuroanatomist Norman Carlin commented: "What's so bemusing about this business of staining tissue sections is that the technology is well over a century old (the microtome and the successive baths of alcohol and xylene and the camel's-hair brush all look much as they would have in a histology lab circa 1895), and they still haven't worked out what it is they're staining, or what the 'perfect stain' might be.")

Although Byne usually uses a stain called cresyl violet, for this study he's using thionine, "because Simon used it." Thionine and cresyl violet both bind to acids like DNA and RNA, resulting in a two-color Jackson Pollock splatter made of human brain tissue, the brain cells violet and nerve fibers white. The process is similar to that of developing a photograph in its various chemical baths. The sections soak for fifteen minutes in the stain. "When they come out," says Byne, "you wash the excess stain off with water and then put them into another series of alcohols going back up 50 percent, 70 percent, to 100 percent pure alcohol, then back in the xylene, which gets out the alcohol and equilibrates the section with the glue that will be used to mount it. If you don't have all the water out the section will turn green instead of the violet it's supposed to be." Byne looks a little concerned about these sections. "Linda reminded me that formalin can be converted to alcohol, which would be a problem because it might suck 70 percent of the water out of the brain segments." Something else to worry about.

(Finding the perfect stain is becoming the key to the new neuroanatomy. Stains have now been developed that reveal *only* and *exactly* certain cells for which the neuroanatomist is looking. A Toronto lab has created a stain that only attaches to receptor molecules for androgen and estrogen, which happen to cover the SDN, and it may thus be the most accurate way of measuring the SDN so far; the "stain" itself is actually a colorant attached to an antibody for a specific molecule or tissue. Another such stain, the Alz-50 antibody stain, colors only one very particular group of cells in the hypothalamus. Finding the right antibody is the trick. "With Alz 50," Byne says, "they made hundreds of antibodies, and after going through them one by one for forty-nine tries they found one that coded *only* for the Alzheimer brain. And they consider themselves lucky it was only forty-nine false starts.")

"The brains" are then coded in two ways, each with a medical record and an autopsy number. (At this point neuroanatomists refer automatically to the sections as "brains.") Byne had a research assistant pick out candidate brains for the study, and the assistant gave him only autopsy numbers so that he had no way of knowing their sex or their sexual orientation. Byne then chose the brains that were most usable, restricting the samples to subjects between eighteen and forty-five years old with no documented testicular atrophy, no steroid or ketoconozole use within three months of death, and no neuropathologies, all factors of which he is critical in LeVay's study.

Byne then took them back to Mattiace, who had only the medical record number, not the autopsy numbers, so she couldn't go back to find out who was who. She gave the brains new numbers "so I am even another step removed," notes Byne, treading quickly along the polished linoleum hallway. He is on his way to the floor above where he does the measurements. *"LeVay,"* Byne stresses the name, "did all the work himself."

Turning a corner near the stairwell, Byne explains that he is measuring the sections himself in collaboration with Cliff Saper, an anatomist at Harvard, and is restricting his survey to INAHs 1, 2, and 3. "I don't think 4 is actually a nucleus," he confides. "LeVay says 4 is only present in about one-half of his brains, anyway."

The actual measuring is an exercise in geometry. "You look at each section and trace the nuclei, calculate an area, multiply that times the thickness of the section—52 microns—which gives you the volume of the nucleus. After you measure a nucleus's volume on each slice, you add those all together to get the total volume of the nucleus. It might take twenty sections to get through each nucleus. I will not be doing the analysis of the data myself. I just do the measurements." He repeats pointedly, "I'm removing any possibility of inadequate blinding. LeVay controlled for none of this."

And then, suddenly, and (typical of him) almost absently, Byne, midstride and without missing a beat, casually drops a comment that seems to reverse completely his position on LeVay's study: "You know, the only problem I really have with LeVay is that he didn't have sufficient medical histories. I think that at a minimum that needs to be clarified."

It sounded as if he said the "only" real problem with the study. And it's a "problem" relatively simple to correct, at that.

Asked about this, Byne acts surprised. And yet before, it seemed he believed not only that LeVay's findings were suspect but that the entire idea of researching sexual orientation and all the fundamental concepts based on it, such as sexual dimorphism in the brain, were profoundly ludicrous. Now he is saying the "only" problem is a simple, correctable part in the way it was set up?

"Of course." Byne blinks. "LeVay will tell you that he looked at things like medical histories, that these factors applied equally to his homosexual and heterosexual subjects, but he didn't do it *systematically*. The size of the nuclei in the hypothalamus varies in some species with the amount of testosterone in the bloodstream. Perhaps it does in humans too. He just needs to control for that." He looks as if the answer were obvious.

⸝

Upstairs in the measuring room, Byne starts turning on a series of machines. This is a state-of-the-art lab, filled not with the usual sinks and test tubes and shelves jammed to the limit with jars of chemicals but with industrial power cables and electronic devices and computers. As he flips switches, Byne explains how precomputer neurologists used to measure the area of a nucleus. "They would mount the section and then place it on the microscope, which we still do. Through a tube attached to it, the scope would project a greatly magnified image of what it saw—the nucleus—onto the tabletop. The researchers would then outline the image of the nucleus by hand, essentially tracing its shadow. They'd trace this outline on vellum, a paper with a highly homogenous thickness across its surface, cut out the outline, and weigh the pieces of vellum. Comparing the weight of the papers gave you your data. Today," he says, flicking on an oversized screen for an Apple computer and sitting down at a two-toned microscope projecting a laser green glow, "it's a little different."

Byne opens the hinged lid of a box. It is filled to capacity with lines of crystalline glass slides so clean they glint like rectangular chips of diamond in a display case. Small, almost invisible sections of human brain float frozen on their surfaces in clear gelatin. From near the middle he carefully selects the next slide to be measured, extracting it delicately and squinting at it, then slides it under the microscope. "The microscope is hooked up to the Apple," he says, squinting into the lens and pointing to the monitor at same time. "The Apple uses NIH imaging software to make the calculations." He flips a last switch, hits a few keys, and there on the screen INAH 1 unfolds and buds electronically, an immense amoeba in shades of lavender and black.

This image of the nucleus is infinitely sharper than in the slides on the wall of his office. It looks on the screen as if it were alive. The cells are more articulate, the ventricle is sharper, and the colors are vibrant, as if designed by a Disney animation team. This bud of dense cells is small since it is the outside slice of the fruitcake, and the nucleus is just starting to peek out from the dough of neurons around it. Byne says there are labs bringing sections up here and generating images, arguing about which is the prettiest nucleus.

Byne reaches for the mouse, moves the cursor to the edge of INAH 1, and clicks. The cursor begins to generate a thin black line, which he runs around the perimeter of the nucleus, closes it, and clicks the mouse again. The line, a necklace around the cells, pulses snakelike, and the computer instantly calculates the area, which for this nucleus is .12 square millimeters. The data appear in a box on the screen that also tells the number of pixels inside the image.

One of the benefits of this new system is that, instead of packing up hundreds of slides to mail across country, Byne can simply save the images from each section on disk and ship them to Laura Allen, along with his results, for her examination. But while more effective than primitive cutouts on vellum, this technology still depends a great deal on subjective judgment. It is, in fact, a technology already on its own way to oblivion. MRI technology is proceeding apace, and the new multimillion-dollar machines of just ten or fifteen years from now will have such fine resolution that even INAH 4 may be measured through a simple electronic scan of your brain: a few minutes and you're on your way. No more dead brains, no microtomes or formalin-drenched brain cells, no more boxes of glass slides.

There are 54 slides in the series that Linda Mattiace has prepared from this brain. By no. 26, INAH 2 has vanished and 3 and 4 have not appeared yet. Only INAH 1 is present all the way through. Byne's numbers for area, as calculated by the Apple, go up and down, although within a fairly steady range. He gets one notable leap, from .14 to .27, and then back down to .16, but that's an anomaly. The machine demands constant concentration. It is tedious, tiring work.

When he's done with this brain and has turned off the computer, Byne agrees to a look at the brain storage facilities. Down one flight and across the hall is the office of Dorita Thompson, a pleasant woman in her mid-fifties. She manages the brain bank, and cheerfully agrees to guide the tour. "Just two places to see, really," she says with a clipped Spanish accent.

One is close to her office, and she leads the way into what looks exactly like a restaurant refrigerator. In this bright metal room kept just above freezing are, floor to ceiling, rows of pinkish-gray human brains stacked on metal shelves floating in Plexiglas tubs of formalin, their fixing substance. These are the "intact" and structurally undamaged brains, the better-quality specimens usually not frozen and used for the delicate work of neuroanatomical research. Some neuroanatomists spend hours in this chilly, isolated cube.

Farther down the hall is a room containing five very large upright freezers. With a bit of effort, Thompson opens one of them. Inside is a miniature Antarctica, a wall of whiteness of frost and ice laid on so thick it looks like an overdone movie set. Freezing white vapor languidly uncurls itself and spools tentacles onto the floor. The frozen brains inside are kept at −70 degrees Fahrenheit, Thompson says.

Wearing latex gloves, she rummages through the various shelves, chatting pleasantly while occasionally removing a cardboard box full of brains stored in plastic Zip-Loc freezer bags. Byne explains that these frozen brains are put to a very different use from those in formalin. They are often damaged, some only half a brain or a frontal lobe, but their material is what's needed, not their structure, for testing antibodies. When Thompson finds two boxes acceptable, she takes them out and breaks them apart by whacking them on a big metal drum labeled "Warning! Radioactive material!" At −70 degrees, they seem hard as granite. After selecting some brains she likes from the boxes, she removes them from their individual bags and smacks these cheerfully a few more times to separate them. Shards of frozen brain spray out, and the top of the drum is covered with a fine silt of grayish pinkish snow.

Byne, also in gloves, eagerly picks up a brain and begins pointing out its features. The details are remarkably clear, though he is hampered somewhat by the fact that the brain is so cold; although he brushes away the frost from the corpus callosum, clearly visible and tucked into the center of the brain, the rock-hard, blood-red surface immediately frosts up again. Byne wears the gloves not because of the cold but because scientists who work with brains can be exposed to viruses or deadly diseases such as kuru, which lives in human brains. New Guinea cannibal headhunters get it from eating the brains of their victims. Although we know kuru is deadly, we don't know exactly what it is, but it may be a prion, Byne explains, a mysterious microscopic creature about which we know almost nothing. Some people think prions are infectious proteins; others, that they are viruses but— bizarrely—without the genetic material of a normal virus; and still others, that they are a type of parasite. "They're still working on them," says Byne. Meanwhile, the pink snow is melting, and the top of the barrel is flecked with human brain and blood.

Kuru is, Byne says, as he peers closely and with great interest at the brain in his hands, also resistant to formalin and can survive it quite well. In fact, he says cheerfully, kuru may be able to survive −70 degrees.

On the subway hurtling back to Manhattan, this critic of Simon LeVay's describes what he hopes to find with his own study. From his passionate arguments against LeVay's conclusions, one would expect him to be confident of confirming there is *no* sexual orientational dimorphism and that LeVay is wrong.

What Byne actually says, congenially, is: "I *hope* I find at least a sex difference." Why? "Just because it would be nice to get some consistency."

Again, this is confusing; does he not believe such biological differences between men and women, gays and straights, don't exist? And then, above the loud rush of the subway noise, Byne says something, casually, that reveals a great deal about him: "What I find out about human neuroanatomy will have no bearing on the way I live my life and shouldn't have any bearing on how anyone else lives theirs."

At the time, I was sure this was the most naive thing I had ever heard.

⁓

In the spring of 1993, the State of Colorado introduced Dr. William Byne's work in its brief to the state's Supreme Court in Denver in the politically charged legal case over Amendment 2 to the Colorado Constitution. The amendment, which forbade legal protections for homosexuals, had been supported by the conservative organization

Colorado for Family Values and had recently been passed by a majority of Colorado voters.

Because of the strictly scientific way that Byne expresses himself and because of the media's misrepresentation of the research of sexual orientation, the State of Colorado and conservative Christian groups thought that with Byne, they were getting an expert who was asserting that homosexuality was a "lifestyle," not a sexual orientation; that homosexuality was a chosen, alternative behavior and/or a mental pathology unworthy of civil rights protections; and that homosexuality was not immutable but, rather, easily changeable. (Journalists who report on Byne, sexual orientation research, and Amendment 2 thought they were getting the same thing.)

What they actually got was this.

Back in New York, Byne didn't realize he was being cited as evidence for choice, lifestyle, and behavior. He was informed of this in a tense phone call, by Dr. Dean Hamer, a molecular geneticist at the National Institutes of Health. Hamer, who also researches sexual orientation, had been subpoenaed as an expert witness by the opposing side, those attempting to halt Amendment 2. On receiving the briefs of the case, Hamer was stunned to see Byne cited since he knew that Byne did not, in fact, think that homosexuality was a choice and, furthermore, that Byne was completely opposed to the idea of legal discrimination against gays and lesbians. "They're going to use you," Hamer said when he phoned Byne in New York. Hamer had already been upset with Byne for making statements that the press, simply because the statements were critical of the research, mistakenly reported as "antigay." From his seat on a panel at a meeting of the Academy of Sex Research at Asilomar near Carmel, Hamer had commented acerbically (and with clear reference to Byne, who was sitting in the audience) that given the political aspects of the research, scientists had a responsibility to make themselves understood.

Byne was stung. What happened next surprised almost everyone, particularly the antigay conservatives. When the State of Colorado officially asked Byne to be a pro-Amendment 2 witness, he not only declined vehemently, he wrote a letter, dated July 2, 1993, and addressed to Hamer, to be used in the Amendment 2 case by gays and their supporters.

"Dear Dean," it began. "This letter is in response to your statement at the Academy meetings that I must take responsibility for the misuse of my work by the anti-gay political and religious right."

Byne proceeded to reverse everything that Colorado for Family

Values and the Family Research Council had mistakenly believed he was saying. "Some have apparently misinterpreted my recent review with Bruce Parsons as disproving the biological hypothesis for the [origin] of homosexuality," he wrote. "Then, because they have been duped into believing the false dichotomy presented in news media headlines such as 'Homosexuality: Biology or Choice?' they erroneously interpret my review as supporting the view that homosexuality is a chosen lifestyle."

First of all, Byne stated with controlled fury, he said what he meant, and meant exactly what he said: He did not reject even the simplest, most reductionist hypothesis that homosexuality was out-and-out biological. Quoting from his own writing (which the Family Research Council had been enthusiastically mailing out to Christian conservatives at its own expense), Byne noted that he had merely concluded that "there is no evidence at present to substantiate a biologic theory." He wasn't claiming it *wasn't* biological; all he was saying was that we don't *know* yet that it is. And the Christians and the press, had they been on the ball, would have noticed that he was equally critical of all existing environmental theories, concluding that "there is no compelling evidence to support any singular psychosocial explanation."

Second, he would never "imply that one consciously decides one's sexual orientation. Had I been able to foresee how the paper would be misinterpreted and misused, I would have amplified that last statement." He mentioned a *New York Times*/CBS News Poll showing that people who believe homosexuality "is a choice" are less tolerant than those who believe it "cannot be changed." "No one who has spoken in depth with a number of homosexuals," Byne continued tersely, "could conclude that they . . . choose their orientation. Moreover, a survey of the literature on attempts to change sexual orientation leads inevitably to the same conclusion. Very few of even the most highly motivated to change sexual orientation have been able to do so despite investing tremendous emotional effort, many years, and many thousands of dollars in psychotherapy. In fact, the legitimate literature on attempts to redirect sexual orientation suggests that not only is sexual orientation not chosen, it is, by and large, immutable."

In response to Hamer's comments at Asilomar, Byne also stated—a little defensively (this comment he aimed at the press)—that the idea that only biological traits are immutable was completely incorrect, chiding "those . . . responsible for the dissemination of this sloppy logic to the public through the news media. If I am to be held responsible for the misuse of my paper (which has occurred through misinterpretation), then you must also hold those . . . responsible who have participated in the popularization of the view that if homosexuality is not biologically determined, it is chosen. They set the stage for my criticisms of the biological research to be misinterpreted."

Why was "Biology or Choice?" the central idea of the *Washington Post*'s reporting on the research, an incorrect headline? Scientists have long known, noted Byne, that many traits purely social in origin are immutable, like a person's native accent, which can never be removed once it has been implanted in the brain. You don't choose it, your biology didn't create it, but most people retain accents throughout their lives despite their best efforts. Environment—the accents of West Texas or Shanghai or Odessa—have changed the brain, and these changes remain there forever, unchosen and unchangeable.

"In closing," wrote Byne, "I would suggest that the most responsible message that we can relay to the public is that we do not know what causes sexual orientation—but whatever the causes are, sexual orientation is not a matter of choice. To present a simple dichotomy— that sexual orientation is either biologically determined or a matter of choice—may recruit political activists to each side of the debate, but it will not advance science.

"Regards, William Byne, M.D., Ph.D."

And this was the desperately obvious yet unrealized truth: None of the participants in the scientific battle—LeVay, Allen, Byne, Fausto-Sterling, Balaban—disagrees on any of the "political" questions. Had, for example, any members of the press bothered to ask Anne Fausto-Sterling, they would have known, as she has made clear, that she too is appalled by the media's misinterpretation and misuse of her critique. "Don't make this more than it is," she warned of her "antigay" criticisms of LeVay's "progay" work. "All I'm saying is that at *this* point in time in *this* culture and *this* political climate, researching things like gender and sexuality, which Simon is doing, is going to be so difficult to do objectively that it approaches the impossible. Look at history. Whenever these things have been done, they've been biased by sexism, racism, homophobia, classism, and every other prejudice. So of course the results didn't stand up. And because these biases are as pervasive among scientists as anyone else, I'm saying that it would be tough to do right." She added carefully, "If the study were done right, then you do the science and you find what you find, of course. I'm a scientist. I'm not opposed to science."

LeVay and Fausto-Sterling and the others are not arguing about what sexual orientation is—which was the anachronistic crux of the Amendment 2 trial—but about how to research it. Byne and Allen dispute the methodology and measurement of sexual orientation. Not its existence. Nor its nature. In fact, both go to lengths to make very clear that although they don't agree on the question of what *causes* sexual orientation, they agree on the question of what it *is:* a sexual, erotic, romantic orientation toward either the same or the opposite (or sometimes the same and the opposite) sex, unchosen and immutable. As neuroanatomists, they are extremely sensitive to whether disease or disorder is associated with the traits they study—Byne, who does schizophrenia and other mental illness research, in particular. Both make it clear that whatever the biological cause turns out to be, neither orientation of the trait sexual orientation—heterosexuality or homosexuality—is a disorder or a "neuropathy," a mental disease. Byne is meticulous about pointing out that the history of research into homosexuality, which was for thirty-one years held to be a mental disorder, is an embarrassing and disturbing compendium of bad research and biased medicine. But then psychiatrist Robert Spitzer of Columbia Medical School points out that homosexuality is hardly the only trait that has suffered from distorted use of the label "disorder." The history of science is riddled with such cases, notes Spitzer, in part because of how hard it is to define "disorder."

"We really don't have a satisfactory definition of what is a 'disorder' or 'disease,'" Spitzer said in his office on 168th Street in Manhattan. "And we probably haven't had one in the history of medicine. Is baldness a disorder? If you're a dermatologist, it is. Does it impair your ability to do a job? No. Does it cause you distress? Not physical distress, but it can cause you lots of distress if you're attracted to some person who loves hair. Is a disease something that causes distress or impaired ability? Is nearsightedness a disease? Uncorrected, it impairs your ability to do a job, corrected it doesn't, but it doesn't in and of itself cause distress. Is depression a disease? Is anger? (It impairs your ability, causes stress, and can be healed.) Is left-handedness? Is a disease something that causes a natural, evolutionarily evolved function to stop functioning? Then how about having an appendix, which has no evolutionary function we can figure out? Basically, a disorder has been whatever people want it to be because of their prejudices, which leads to all sorts of ridiculousness."

The term "disorder" has been applied to virtually every group, condition, or behavior imaginable. A relatively recent version of the *Diagnostic and Statistical Manual* of the American Psychiatric Association created two disorders that affected only women, "premenstrual stress disorder" and "self-defeating personality disorder." Women were quick to point out that the first pathologized menstruation and the second pathologized women who remained in abusive relationships with men, removing responsibility from the abuser. Before the Civil War, "drapetomania," a mental disorder diagnosed by Samuel Cartwright, a Southern doctor, was a disorder that only affected blacks who were slaves. Its symptom was running away from the plantation.

Says Spitzer, "If you want a definition of disease or disorder that's going to be even somewhat useful, start with the precept that your definition can't be a tautology: 'Left-handedness is a disorder because left-handedness is a disease' or 'because I say so.' Or 'because God says so,' which is fine in theology but not in science. Next, there has to be some objective mark, maybe that the thing impairs a person's ability to perform a job. By virtually any definition, homosexuality doesn't fit. Physical or mental impairment, suffering (other than from outside prejudice), deterioration of function—none applies."

Why, then, the bitter three-decade battle over the psychiatric classification of homosexuality as a disorder? The little-known story of the central role homosexuality has played in psychiatry's rise in the twentieth century is as fascinating as it is quixotic and perverse. The man who in 1940 began the story was named Harry Stack Sullivan, and the woman who effectively ended it in 1956, Evelyn Hooker.

Sullivan described himself as a "slight, bespectacled mild-looking bachelor with thinning hair and mustache." As a psychiatrist, Sullivan was much anguished by the popular view of psychiatry among medical professionals, who dismissed it as a half science. Sullivan fervently believed in psychiatry's value and sought a way to elevate it to the medical mainstream. World War II provided the way.

In May of 1940, when President Roosevelt, in response to escalating warfare in Europe, asked Congress to expand the American military, Sullivan initiated psychiatry's first serious bid for respectability. In late October, he and two psychiatrist colleagues met in Washington with the Selective Service director and members of the War Department to conclude the details for a psychiatric screening plan Sullivan had created. If used by military medical boards across the country, Sullivan and his colleagues assured the War Department, psychiatry would be uniquely capable of weeding undesirables and potential mental casualties from the thousands of fresh inductees just beginning to pour into the armed services. The screening plan, as Alan Berube has written in *Coming Out Under Fire,* "called for the appointment of over thirty thousand local board examiners who would conduct psychiatric interviews ... six hundred Medical Advisory Boards with one psychiatrist on each to review problem cases, [and] a second psychiatric interview at Army induction stations. . . ."

Things got out of control almost immediately. Although Sullivan himself believed sexual orientation played only a minor role in mental disorder and had not initially included homosexuality, as the plan was revised, the screening process became increasingly dedicated to the identification of homosexuals. And psychiatry's dominant theory of the time—that homosexuality was a mental disorder—was increasingly implemented in practice. Homosexual people were inherently unstable, said psychiatrists, and thus security risks and bad for military morale and discipline. It was asserted that gay soldiers would crack under fire, endangering their comrades and the Allied cause in general. In November of 1941, only a year after the process had begun, Sullivan resigned from the advisory board. The military hardened its screening procedures against homosexuality.

By the war's end, psychiatry had made itself indispensable to the American armed forces and had introduced its basic principles to the mainstream medical profession and American society in general. Of course, all this necessitated something: that the medical view of homosexuality as a "pathology" be continued. The only problem was that psychiatry not only consistently failed to demonstrate that homosexuality was a pathology, it actively demonstrated the opposite.

In the early 1950s, Evelyn Hooker, a young psychologist teaching at UCLA, got to know one of her students, Sam From, who was homosexual. Hooker, a tall, commonsense midwesterner, was intelligent, curious, and quite unintimidated by the unknown. She was also familiar with the nature of entrenched ideas; she had wanted to pursue graduate studies at Yale, but her professor had told her he felt he couldn't recommend a woman—and she soon introduced Sam to her husband, Dr. Edwin Hooker. Sam in turn introduced the Hookers to a large group of his gay friends, whom they entertained at their Los Angeles home. It was an era in which newspaper headlines such as "Perverts Are as Dangerous as Reds" were common, but Hooker was struck by how happy, healthy, and entirely normal this group of homosexuals was, an observation entirely at odds with her psychological training. Her casual observation was put to the test when Sam said to her, "Now, Evelyn, it is your scientific duty to study people like us." She demurred—"You're my friends," she objected—until a fellow scientist remarked to her, "He's right, you know. We know nothing about them."

So Hooker applied to the National Institutes of Mental Health for a grant, got it (to her amazement), and began putting the study together. "I was prepared," Hooker said later, "if I was so convinced, to tell these men they were not as well adjusted as they seemed on the surface." At the same time, she added, "I didn't put too much credence in what the books said because I knew no one had really studied [homosexuality]." In fact, what was remarkable about Hooker's research was that it was the first to examine homosexuals outside a medical setting or in prison, where they had by definition already been declared mentally ill or pathological.

Hooker began searching for subjects. There were some amusing glitches in setting up the study. She went to a meeting of the Mattachine society, an early gay organization, to look for men to participate in her study. Because she wanted homosexuals who had never engaged in heterosexual sex, she asked if any of them were or had been married. Almost all raised their hands. Oh dear, said Hooker, I can't use any of you, and there was some confusion until she realized that they had taken "married" to mean a committed relationship with another man. Those she worked with were in constant danger of being identified as homosexuals. "It was a very perilous study," said Hooker. "If a man walked down the hall to my office it would immediately be known '*Oh*, that's the Hooker project,' and that would be that for him." In fact, finding heterosexual men for the study turned out to be even more difficult. Hooker eventually resorted to propositioning male visitors to her home on Salt Air Avenue, and then accosting the mailman and the paper boy, which led her husband to remark dryly, "No man is safe on Salt Air." Finally, she assembled thirty homosexual men and thirty heterosexual men, and administered to them identical psychological tests, including the Rorschach ink-blot test, producing sixty psychological profiles. She removed all identifying marks, including sexual orientation, and, to eliminate her own biases—this is standard practice—gave the profiles to three other psychologists to interpret. The first was the nationally renowned Bruno Klopfer, who was well known for claiming he could easily identify a person's sexual orientation by his or her Rorschach.

In fact, Klopfer could not, nor could any of Hooker's colleagues. In side-by-side comparisons, the tests of the heterosexuals and homosexuals were indistinguishable, demonstrating an equal distribution of pathology and mental health irrespective of sexual orientation. Dr. Edwin Schneidman, reviewing Hooker's results from a test in which the subject creates pictures with cut-out figures, discovered one subject's orientation only when he came across a scene in a bedroom. The subject had placed two men in the bed. Schneidman remembers, "I said to Evelyn, 'Gee, I wish I could say that I had been getting indications of it up to that point, but I wasn't.'" The third psychologist was so incredulous that he insisted on reevaluating the entire test a second time; he arrived at the same result.

In 1956 Hooker presented her paper, "The Adjustment of the Male Overt Homosexual" ("overt," in Hooker's day, meant out of the closet, openly gay instead of hidden)[3] at a meeting of the American Psychological Association in the ballroom of the Sherman Hotel in Chicago. It generated furious debate, which would continue for almost two decades, but it was the beginning of the end for the belief that the minority sexual orientation was a disease. Hooker observed tartly and would continue to observe throughout a long career that for psychiatry to be even minimally scientific, pathology must be defined in a way that was objective and empirically observable through the scientific method. In 1973 the American Psychiatric Association removed homosexuality as a disorder from its official *Diagnostic and Statistical Manual.*

Says Byne with a sigh, "If anyone had bothered to be the least bit objective, this whole dismal, embarrassing chapter in American medical and scientific history could have been avoided. Above all, Freud and psychoanalysis have wounded themselves immeasurably through the preposterous theories about homosexuality. In the 1960s psychoanalyst Edmund Bergler held that gays were 'incapable of true creativity.' That this dialogue was going on in serious discussion—I mean, did they not consider Michelangelo or A. E. Houseman?" Byne, the doctor, shakes his head and adds, "Another belief was that homosexuality was a 'very rare' aberration, yet many of these people were doctors who, at a minimum, should have known the contents of their own Hippocratic oath, which, when the Greeks created it, was aimed entirely at men. It ends with a statement warning against sexual involvement with your patient, 'no matter whether your patient be a woman, Free Man, or slave.'"

The final irony is that Henry Stack Sullivan, a widely respected doctor instrumental in creating the illness model of homosexuality for the good of his profession, was himself gay and lived a comfortable and relatively open life in Bethesda with his devoted male partner.

～

As Byne and I walk up a freezing sidewalk on the West Side in the winter sun, he says, "I can't stop saying what's accurate." He is troubled and angry at the way his critique is misinterpreted, frustrated at his

own difficulty in explaining himself. "All I'm saying is that the biological evidence is preliminary, that we don't understand it. Which is
true. But that's *all*. How could they *possibly* think I'm 'antigay'?"

The fact is, I tell him rather heartlessly, he is terrible at giving
sound bites. He doesn't keep politics in mind when he answers questions but rather sticks strictly to science, and unfortunately the Family Research Council and *Newsweek* magazine and the *Washington Post*
hear his comments politically, as "the antigay view," because they
don't really care about axons and formalin and the tough methodological problems at the heart of real science. They care about politics. And they cast his strictly scientific comments according to their
political needs.

And still he publicly says things like: "What I find should have no
bearing on the way anyone lives their life."

"I simply meant by that," Byne explains, defensively, "that whether
there is a sexual dimorphism or not in the human brain won't mean
that people will suddenly get out of bed differently, think differently,
buy their groceries or talk with their friends or eat their dinners or
fall in love differently. It doesn't mean *anything* for the way anyone
lives their life. It just means we know one more thing about our biology."

In front of the Advent Lutheran Church at 93rd and Broadway,
Dr. Byne tries, once again, to make clear what is to him obvious, the
scientist struggling to explain his world to the political world. "People have to take the studies for what they mean. It's absurd to say that
a biological study is 'pro-gay' or 'antigay.' What does it matter if a trait
is genetic or environmental? It's the trait that counts. If it hurts people and they suffer, it's bad, and if not, then it's just human nature.
I'm interested in the biology because I'm a biologist. If people are
going to discuss the research, they need to discuss it accurately," he
says. "All the rest is just—politics."

Byne stands shivering slightly in the cold winter sunlight looking
baffled and mildly angry. "I'm talking about *science*," he says, brow fur-
rowed. "Biology. Research. What does *politics* have to do with that?"

~

Recently, at a conference at Cornell Medical School in Manhattan
where Allen was speaking, I saw Byne through the crowd in the lobby
where he was waiting in line to register. He greeted me happily and
then asked, "Have you seen Laura?" He scanned the crowd.

I said I had seen her inside setting up her overhead projections.

"I need to find her," he said. "I have some rhesus monkey brains up at Einstein, and I'd really like her to take a look at them."

He found her in the auditorium, where they chatted about rhesus brains as people wandered in and took their seats. During the question-and-answer period following Allen's presentation they argued fiercely about the anterior commissure, Byne launching critiques from up in his seat on the steep cliff of the amphitheater, Allen down at the podium armed with her data. The audience watched the facts shooting back and forth, rapt as at a tennis match.

Afterward, the two broke out date books and confirmed that both were free that afternoon to go to Einstein. Allen asked him how his replication attempt of LeVay's INAH study was going. Byne has not found a dimorphism for sexual orientation, but he *has* found a sexual dimorphism in INAH-1 in the human brain, confirming Allen. He hopes to publish it soon. They are both quite pleased, although he still firmly believes she is wrong about the corpus callosum. "Well," she says with a calm smile, "we'll see," and they leave together to go look at rhesus monkey brains.

Chapter Five

BIOLOGICAL ARCHAEOLOGY

*Endocrinology: the science dealing with the
endocrine glands, such as the thyroid, adrenal, and
pituitary, and the hormones they secrete
directly into the blood.*

ONE OF THE inconveniences facing the biologists who look at
sexual orientation in humans is that they cannot kill their research subjects.

Endocrinologists are biologists whose specialty is digging delicately inside us, gently brushing away the layers of cells and tissue that,
like strata of silt and dirt, cloak a multitude of neural and physical
changes that occurred long ago in our fetal brains and bodies. Endocrinology tells the history of the chemistry that molded us inside
the womb. Endocrinologists attempt to reconstruct this forgotten Pa-
leocene age, examining the hormonal fossils and biochemical painted
cave walls created in a now-ancient embryologic time that flowered
and died before our current consciousness began.

That the digging areas and excavation sites in which endocrinologists camp happen to be living human beings is a problem. Endocrinology is a field of frustration. Much of the scientific debate over
sexual orientation has centered on when, exactly, this trait appears
in us. Endocrinologists, battling over whether hormones have a role
in shaping hetero- and homosexuality, can point to a trail of tanta-
lizing clues that have led them to a theory about how hormones lay
down heterosexuality and homosexuality in their liquid tracks. But
they have no way of testing these hypotheses directly, no way to plunge
their shovels and picks into our brains and excavate the evidence they
believe is there.

The first clues came once again from rats and birds, which we
could kill.

Bird biologist Evan Balaban poses a question that sounds like the Sphinx's riddle: What is the difference between an "innate" trait and an "immutable" trait? While both words are thrown around carelessly as synonyms in The Political/Media Debate about sexual orientation, a semantic distinction between them actually illustrates an important facet of biology. Something can be immutable and not be innate at all. Balaban's answer comes from birdsong.

Birds sing to attract mates and to mark territory, and their characteristic songs identify them to other members of their own species. Most birds "bring their songs with them" in their genes, singing them correctly and automatically without coaching from their parents. The songs are both innate *and* immutable. "The only way to erase them," says Balaban, "would be to remove their brains." But a baby robin must hear the song of its own species to learn it, so the song is not innate. Once it *does* learn it, the song does not change; it is immutable.

There is, however, a catch. The bird must hear its own song within a certain, precisely determined number of days or it will never learn it. And if it hears the song of another species first, it will learn that song instead and can never unlearn it. This, in turn, means it will never be able to attract a mate or produce offspring. This period of time, this window of opportunity, is called the critical period. And there is also the case of the white-crowned sparrow, which, if it hears the song of another species during the critical period, does *not* learn that song; instead, it never sings. These sparrows have a genetic template for learning their own species' song and can learn only that song, but they can't learn it if they are not "taught" it by exposure.

These developmental periods open and close on cast-iron schedules, biochemical opportunities that come and go and never, ever come again. The windows are controlled by hormones, and rat brains held the clearest example: Withhold testosterone from a male rat for the first five days of life while the clock runs down and his narrow window of opportunity slips away, and his brain will no longer masculinize. Pump him full of testosterone after the fifth day, and it won't matter. He will never develop a large SDN, and his sexual behavior will remain throughout life irreversibly female.

This is the critical-period phenomenon, the idea that genes set clocks in each of us, biochemical timing machines that open windows to masculinization, sexualization, growth, and mental development. These genes start the timer and then stand there, holding the window open, counting backward to zero. From the word "go" till time is up, the hormones urgently run their biological obstacle course while the clock ticks down on this fantastically complex drill by which we all develop: Cells have to grow at the right time and on the right plan; glands have to manufacture, invoice, and deliver into the bloodstream the right hormone products in the right amounts, which have to arrive on time in the right places and work their changes in the body, get the organs up and running so they can string the wiring, weld the connections, and fuse the circuits together. Through it all hormones are delivering the messages, hormones are giving orders, hormones are directing the show. All is deadline.

If the job isn't done by the time the genetic clock slams the door shut, if the rat does not receive the splash of testosterone in time, there is no appeal. The brain will never change, the nuclei will never grow, the organs will never function properly—at least in rats.[1] Gorski's rat nucleus (whose critical period, we now know, opens just before birth and slams shut five days after) established the principle that prenatal exposure to hormones could create sexual behavior in rodents. The question was: Did critical periods, windows that opened and closed inside of us, exist in *humans?*

There were increasingly strong signals that they did. Studies of very young children were indicating that biology organized the brain, but these signals were not coming from research into sexual orientation. They were coming, rather, from scientific inquiry into human speech.

In his fascinating book *Genie,* journalist Russ Rymer explores "the language question"—how is it that we human beings are able to speak? For millennia, the problem has been approached by looking into language acquisition in children. The answers used to be simple. People simply thought languages were composed of random words and syntaxes that societies had chanced to come up with over time. After all, languages were so different and so obviously unrelated. There was vocabulary, and there was grammar into which the vocabulary fit, all of it different, all of it purely the random result of environmental and cultural influences. How you spoke depended purely on your social environment. And in one sense, this is still true; the language you speak—Urdu or Cebuano—depends on where you live, what you study. But it doesn't answer the question of *why* you speak. Forget French versus Hindi versus Russian—why does the human ability to speak *any* human language exist?

Then in 1957 Noam Chomsky published a book called *Syntactic Structures,* and suddenly, violently, bizarrely, human language was *not* produced by culture. It was, Chomsky conjectured, organic, inside us, in our genes and the cells of our brains. It was biology. "Language is a tool," Chomsky, now a professor at the Massachusetts Institute of Technology, told Rymer. "The tool is endlessly useful—in the sense that we commonly create and understand sentences we have never heard before. How do we do it?"

How indeed . . .

Chomsky proposed that we can generate sentences we've never heard before—we do it every day—because while we learn vocabulary and accents and all the stylistic touches of language, the structure and skeleton of language itself is in our biology. We carry this infinitely malleable tool with us: Language is genes, an organic, living neurological plan for communication.

Chomsky launched a war with the old guard who held that the ability to speak language was purely a learned, environmentally determined behavior. Linguists who aligned themselves with him and biology called themselves rationalists; those who said environment decides, empiricists. In *Genie*, Rymer sets up the language debate this way: "The study of language acquisition in children," he writes, "turns on a single simple idea—one that I heard most succinctly expressed in the keynote speech at the 1989 Stanford Child Language Research Forum."

Lila Gleitman, a professor of psychology and linguistics at the University of Pennsylvania, was the speaker. "In her late fifties, with close-cropped dark-gray hair and wearing an orange-patterned frock and sneakers, she managed to give the impression, as she leaned on the lectern, of a truant leaning against a gymnasium wall smoking a cigarette. On the movie screen behind her appeared a slide of the front page of a supermarket tabloid, with a headline reading MOM GIVES BIRTH TO 2-YEAR-OLD BABY, beneath which was the subhead, CHILD WALKS, TALKS IN THREE DAYS." After the audience laughed, Gleitman finished arranging her papers, looked up, and said that she was, as they probably knew, Lila Gleitman. "And basically what I want to talk about is this." She walked over and hit the screen sharply with a pointer. "What took three days?"

What Took Three Days has been Gleitman's obsession for the last several decades, an obsession that pushed her from her place in the heart of the culture-centered empiricist camp over to becoming an ardent Chomskian. "People say, 'That Lila, she's just this crazy rationalist,' " Gleitman told Rymer. " 'She thinks everything's innate.' But I started out as a hard-core empiricist, honest! I designed my studies to prove the empiricist position, and I couldn't ignore it when they showed me to be wrong." Rymer explains:

One of the experiments she designed was directly inspired by empiricism's patron saint. "Locke said, 'Look at blind people—there should be some things they can't learn,' " she told me. "So we did the experiment. We thought, We'll see how experience guides language learning. But what happened was that the blind children learned things they shouldn't have been able to. They knew the answers to things beyond their ability to experience. That was very upsetting. Well, we were happy at this victory of the human spirit but unhappy at having wasted our time with blind children. I figured the experiment had failed—simple as that! I went to my husband, Henry"—Henry Gleitman was then the chairman of Penn's Psychology Department—"and he said, 'So how *did* the kid learn the answer?' I said, 'Oh, that's not important,' and I went to Cambridge to talk to Chomsky. He was very interested. He said, 'So how *did* the kid learn the answer?' This was a little epiphany to me. I said, 'Oh, boy, I'm in trouble. Chomsky the mad rationalist and Henry Gleitman the mad empiricist agree on this.' So we went back, and the only explanation we could find was that the child was being guided by syntactic rules within the question—rules he already understood. The syntax tells the answer."

Ultimately what Gleitman was trying to figure out was, How is it that children respond quite early to rules of grammar and syntax so complex that even many adults can't explain them? Where do small children get the ability, from the finite number of sentences they've heard, to create an infinite number of new sentences?

Very young children understand intuitively that some verbs are different from others. There are verbs whose action is direct, like "push," and others where it is ballistic, like "throw," and children as young as three and a half somehow know the rules that govern the different ways they are used. They also know that irregular plurals such as "teeth" can make compound words but regular plurals that end in "s" can't, so you can have your baby sister's teeth marks on your arm, but not the cat's claws marks.

Could this be learned? By three-and-a-half-year-olds, when plural compounds appear once in a million words? Gleitman did tests on prematurely born children and turned up fascinating results. Korean culture counts a child as nine months old at the time of its birth, and it turned out that biology counts similarly; although the premature children had spent as much time in the world and been exposed to the same amount of language (the environmental measurement) as had their full-term counterparts, they were physiologically (a biological measurement) younger, and the age from conception is, Gleitman found, a better indicator of language performance than age since birth. (This is a not entirely subtle point in the debate over abortion and the beginning of human life.) At conception, a timer had been set, and it led Gleitman to a conclusion: "Children aren't learning language from experience. They learn *words* from experience. They bring the sentence with them." This was Gleitman's answer to her three days' question: Human language is a genetic possibility. Its stimulus is not hormones but words. We learn vocabulary. We learn idiom and accent and the myriad touches of interior decoration of the room of language. But the ability to speak human language—the room itself—is in our genes. Or, as one scientist commented, "Of course the ability to speak is genetic. That's why dogs aren't doing so well at it."

And the children are racing against time, we found, hurrying to activate and develop the capacity for language before their biological windows for language close. We discovered this from "wild children"; one, named Victor, the "Wild Boy of Aveyron," was discovered in eighteenth-century France running with the wolves that had raised him. Another was Genie. Here was a human being who had lived her entire existence virtually without human contact. Genie's father had kept her tied up in a straitjacket in a back bedroom of their suburban Los Angeles house, naked and harnessed to an infant's potty seat. She was unable to move, had nothing to hear, and she was kept this way for fourteen years, at which point Genie had heard almost no language and, in fact, had received practically no auditory stimulation. When she made noise, her father beat her with a paddle he kept in her room. She learned to be silent.

When at age fourteen Genie's almost blind mother escaped with her (Genie's physical condition was so pitiful the social worker who first saw her diagnosed her as a six- or seven-year-old autistic), her language window, like Victor's, had closed. Genie had no language, no grammar, no syntax, and almost no vocabulary. She could speak two words, "Stopit" and "Nomore," and understood fewer than twenty. Her ability to speak would never change; she would go on to learn more words but never mastered grammar and syntax, was never able to string the words together to make language.

Genie never learned to speak. Nor did Victor. Their brains, like

the brains of the rats in Gorski's labs, were never exposed to orga-
nizing stimuli: hormones for the rats, and, for the children, words.
And we discovered that the words did not have to be spoken. Because
deaf children often are misdiagnosed as mentally retarded, they frequently wait for years as their critical period ticks by before receiving
the stimuli, the words, that they can understand. In some rare cases,
the period expires and with it the ability ever to master sign language.
One deaf couple like this, both misdiagnosed, were never exposed to
sign, and both had poor grammar throughout their lives. But their
deaf son was from birth bathed in their sign, when his window was
wide open, and although his only input had been a flawed model, he
signed proper ASL grammar and syntax. It seemed impossible: He
could sign better than anyone he had ever known, and every day
he created and understood sentences that he had never seen before.
He had done something he shouldn't have been able to do, just as
Chomsky said he could.

But this discovery left the scientists with another question. If the
window for the language critical period opened at conception, when,
exactly, would it close?

If rats could speak, this experiment would be easy. You would raise
groups of rats from the day of birth in different language environments, some where they were exposed immediately and fully to a language (Finnish or Cantonese or Italian, it wouldn't matter), some a
little later, some later than that, and some in complete isolation. After
they reached adulthood, you would test your subjects: Does the rat
exposed from the beginning speak well? How about the rat exposed
late? If those rats who are not exposed to language until after twenty
days never learn language, but those exposed by nineteen days do,
you have determined the critical period for language.

The sole problem is that the experiment needs to be tested on
animals that actually can speak. And there is only one of those. Of
course, it is impossible to conduct the rat experiment on children.
Unlike the birds, we cannot slit open their brains in laboratories.
"Suppose that a child hears no language at all," Chomsky said to
Rymer. "There are two possibilities: he can have no language, or he
can invent a new one. If you were to put prelinguistic children on an
island, the chances are good that their language facility would soon
produce a language. . . . And that when they did so, it would resemble the languages we know. You can't do the experiment, because you
can't subject a child to that experience."

"Of course," he added, "there are natural experiments."

Chomsky was referring to Victor and Genie. Dr. Heino Meyer-Bahlburg has made virtually the identical observation about another, rather singular group of human beings, except that it was not about language acquisition but the development of sexual orientation.

An endocrinologist at Columbia University, tall and thin with a crisp Teutonic accent and efficient manner, Meyer-Bahlburg is a biochemical archaeologist. From his office on 168th Street high above the Hudson River's upper Manhattan edge, which overlooks the seemingly unending construction on the West Side Parkway, he searches not for sherds of pots from ancient civilizations but for traces of the androgens and estrogens that coursed through us when our bodies were made of fetal material, worked their chemical transformations, and then were gone. Meyer-Bahlburg's quarry is elusive. If, in fact, hormones have made our outlooks, our personalities, our sexual orientations what they are, they have left their fingerprints only in the faintest chemical grooves worn into the cells they touched and the neural tissue they molded and changed. According to the theory, these changes directed how the fetus would develop into the person it has become.

The first theories of how hormones affected sexual orientation were simplistic: Scientists speculated that men were gay because they didn't have enough male hormones.

Actually, the theory wasn't completely unlikely. In mammals, including humans, the main male hormones—called androgens, from the Greek *andros,* man—are testosterone and, present in smaller amounts but even more potent, dihydro-testosterone, which a biologist once described as "a sort of turbo-testosterone." This supertestosterone masculinizes the external genitalia, developing them into a penis and scrotum. Since male sex hormones make a body male and, in animals, control certain male sex behavior, scientists conjectured that adult gay men would have lower levels of testosterone, or perhaps higher levels of estrogen (a female hormone), than adult straight men, and the reverse for lesbians and straight women. This was called the adult hormonal theory of sexual orientation, and it was a forgettable disaster.

In the late 1960s, German scientist Gunther Dorner claimed that he had gotten just these results in his lab,[2] and a number of scientists spent much time and energy during the 1970s trying to replicate his work. In 1984 Meyer-Bahlburg decided to look back at twenty-seven of their studies and see what conclusions could be drawn. Did the theory hold up? No.[3] The results were somewhat schizophrenic: Twenty of the studies found no difference between the testosterone or estrogen levels of gay and straight men at all. Three studies did show that homosexuals had significantly lower levels of testosterone, but two of these were methodologically unsound while the third was derailed by psychotropic drug use among its subjects; such drugs depress testosterone levels. Two studies actually reported *higher* levels of testosterone in gay men than in straight men, and one very unhelpfully found them to be higher in bisexuals than in either straights or gays. The theory was clearly a nonstarter. One of the most interesting of these studies, done in 1976 by R. C. Friedman and associates, involved a set of identical (monozygotic) twins.[4] Theoretically, identical genomes should produce identical hormone levels in both twins and, if the theory were right, the same sexual orientations. But the twins, while they had the same hormones, had different sexual orientations: One was straight, one gay. By the late 1970s scientists had concluded that gays and straights have indistinguishable hormone levels and dropped the adult hormone theory altogether.

Of course, as Byne notes sardonically, anyone with a knowledge of high school biology who had given the matter more than a few seconds of serious consideration would have realized—or *should* have realized—when the theory was first proposed that it was absurd. The reason is obvious: Men with higher than normal levels of estrogen would be growing breasts, and women with abnormally high testosterone levels would have beards and other male characteristics. Researchers knew, had they bothered to remember, says Byne, that there are no such differences between gay and straight people, the fact that made the institution of the closet possible in the first place. And any medical doctor should have recalled that the sex hormone injections that had been used for years to treat adult cancer patients had never had any effect on their sexual orientation. But, Byne concludes, Dorner was more attached to his theory than to common sense.

Back to the drawing board. The next theory became the essence of the endocrinological search for sexual orientation, presenting Meyer-Bahlburg with the difficult work of excavating our primordial hormonal past. What if, scientists wondered, hormones created sexual orientation in utero? What if they were produced by the fetus, worked their magic, and disappeared before birth? When you were a fetus, many of your glands were already up and running, busily churning out various hormonal stimuli. These tiny hormone factories bathed you in a chemical soup that molded the person who emerged at birth. Perhaps, endocrinologists suggested, gay people became gay by being exposed prenatally to hormones during some critical period of development.

"We ask, 'What are heterosexuality and homosexuality and when and how do people get it?' " Meyer-Bahlburg says, "What we've observed of sexual orientation suggests to us that hormones are involved. But all the hormonal manipulation we do on rats in the laboratory, the injections, the castrations, we could never, of course, do on people."

It poses a significant problem.

Being an endocrinologist working on the problem of human sexual orientation is, in a sense, like being a specialist in internal human anatomy who can never actually examine the interior of a body. In this situation, the anatomist could only guess from external observation of how we move and breathe and eat and have sex what organs might be inside us—something that pumps blood, maybe here, and organs that digest food or absorb oxygen, perhaps there— but he or she could never slit open a human body to confirm the guesses.

And yet there *have* been living bodies slit open: Chomsky's natural experiments, the abandoned Victors and battered Genies and the dispossessed deaf. "There are ways around the problem," says Meyer-Bahlburg, echoing Chomsky, "if imperfect ones." These ways are people with endocrinological disorders. "Nature provides you with natural experiments, its mistakes," he explains. "When things go wrong developmentally, the result gives us a window into a biological process that happened years ago inside the womb."

One of these mistakes has the lengthy name congenital adrenal hyperplasia, or CAH. CAH is an endocrine disorder, an accident of nature present in both boys and girls, but in girls it can be used to study the effects of an abnormally large amount of male hormones on females before they are born. Normally, the adrenal gland works away dutifully pumping out cortisol, an important hormone, while the brain stands by, waiting for the feedback signals that will let it know when enough cortisol has been pumped into the bloodstream and the adrenal can take a break. (Since the feedback makes the brain order production shut *off*, it is called *negative* feedback.) That is the system. Usually it works quite well.

But not if the fetus has CAH. In some fetuses, a defective gene fails to produce an ingredient needed to make cortisol. And since this means that the adrenal can make barely any cortisol, obviously there's never enough, and the brain never gets the "negative feedback" telling it to shut down the adrenal. The brain, standing by for a signal that doesn't arrive, becomes confused. As far as the brain can tell, the adrenal is slacking off on the job, and so, via the pituitary, it orders the adrenal to step up production. The adrenal, doing its job down by the kidneys, obediently churns out what is, as far as it knows, perfectly good cortisol. What it's actually producing, however, are the cruder, unrefined raw materials—called precursors—of cortisol. By chance, these precursors are converted not into cortisol but into androgens. In effect, the adrenal spends weeks dumping male hormones into the female fetus's bloodstream.

It is a strange sensation, seeing the result. Roger Gorski, in an office chock full of paper and books and a slide-viewing machine, is talking about fetal development, peppering his conversation with quips about rats. In the middle of a sentence, he dives enthusiastically into a desk drawer, grabs a few slides, and slaps them on the viewer. He turns it on. "What sex is it?" he asks with a sideways look. There between the baby's legs is a penis, clear as day. "It's a boy," I answer confidently. Gorski's eyebrows go up with amusement. "Where are the testicles?" he asks.

What it was was a CAH baby. In the case of this child, the delivering doctor mistakenly decided it was a boy with undescended testicles, a relatively common minor condition, but in fact this baby with the penis was a genetic girl.

Plastic surgeons can, and usually do, make a CAH girl's exterior genitals look female, undoing what the hormones have done to them in the womb. Her internal reproductive equipment (ovaries, fallopian tubes, uterus) are in place, unaffected by the androgens, and she'll be raised as a girl. But the prenatal hormone theory postulates there is an area where the hormones have already done their work, one that plastic surgery can neither touch nor undo: the brain.

What could have made LeVay's INAH-3 nucleus so much larger in straight men than gay ones? The male rat gets his giant nucleus—and his masculine sexual behavior—from prenatal hormone exposure to androgens that change his brain. Perhaps androgens would have the same effect in these women.

Dr. John Money of Johns Hopkins University and other scientists looked at the sexual orientation of CAH females.[5] The current estimate of women in the general population who identify themselves as lesbian is around 2 to 3 percent; Money, in a 1984 study, found that among CAH women, the percentage was stunningly different: 37 per-

cent identified as lesbian or bisexual. There are other findings of significantly increased bisexual fantasy in CAH women. So far so good.
The prenatal theory seemed to be leading somewhere.

And there was a second "window" to look through—a hormonal
effect called LH feedback that seemed to be a sort of "gay blood test."
Feedback is the biochemical switch used all over by the body to switch
things on and off, and LH (luteinizing hormone), a chemical made
by the pituitary gland, tells the ovaries to release their eggs into the
fallopian tubes to await sperm. Women aren't the only ones who produce rising and falling levels of estrogen; men do too, although in
men these hormones have a very male function: controlling how
much sperm and androgen their testes make. In men, rising estrogen
generates "negative feedback"—it turns *off* the pituitary—where in
women it gives "positive feedback," turning *on* ovulation. The sym-
metries and asymmetries were intriguing.

We could make male rats go through the biochemical motions of
ovulation by injecting them with female hormones; we could make
them display same-sex sexual behavior, we knew they have a huge nucleus in their brain created by exposure to male hormones in the
womb—and thus endocrinologists wondered if, when gay men were
in the womb, they, like straight women, failed to have their brains
male-organized by male hormones during some critical period. If
this were so, LH would be the key to excavating prenatal hormonal
effects: Gay men would show a *positive* feedback like straight women
and not a negative feedback like most men.

This was the theory. It was tested, and as Meyer-Bahlburg writes in
his review of the hormonal literature, the positive feedback this theory predicts "has in fact been demonstrated by Dorner's group . . .
using a single intravenous injection of estrogens. As a group, gonadally
intact homosexual men showed a *positive* feedback . . . whereas intact
heterosexual men did not." This was extremely interesting. True, the
gay men's reaction to the estrogen took twenty-four hours to develop
and was weaker than it was in straight women, but there it was.

Or was it?

As Meyer-Bahlburg explains, there are two major problems.

First is that, after all the observation and theorizing and testing,
LH feedback in human beings probably reveals nothing at all about
what happened to us biochemically in the womb. Why? Because of
Byne's law: Human evolution—and the peculiarly human blend of
prenatal hormones—lays down no rules for us for after we're born,
while rats, as another species, are another hormonal story. Rats use

prenatal hormones to hard-wire the cycling of their ovulation, but rhesus monkeys rely on a less rigid design to control their ovulation, a more evolutionary sophisticated interplay of adult sex hormones in circulation in the bloodstream. Or at least this is what some studies on rhesus monkeys have shown. Feedback just doesn't mean as much for monkeys. And, Meyer-Bahlburg asks, if this is so in monkeys, would it not be even more so in humans?

Second, even if LH feedback does tell us something about our prenatal hormones, it's doubtful that human LH feedback is related to sexual orientation at all. Again, "different species, different biological rules." In rats, LH feedback is a program that controls both ovulation and sex behavior; for an animal that only has sex when it's fertile, as do rats, there must be a direct linkage mechanism between the two. Kick one off, and the other is automatically kicked off as well. So the malelike LH effects scientists were obtaining when they injected female rats with hormones were simply caused by the fact that the estrogen was interfering with the rats' control systems. In primates, which constitute a more biochemically complex model, basic reproductive functions such as sperm production on the one hand and mating behavior on the other are not directly linked. So LH feedback is far from the home pregnancy test kit of sexual orientation that some scientists have taken it for.

On top of all the technical problems, a warning must go with all hormonal work. The animals used for these studies are surgically castrated, ovarectomized, lesioned with chemicals and electricity, have nuclei in their brains burned out, and are injected with chemicals to the point that the final observed effects can be so artificial as to be scientifically useless. For obvious reasons, scientists term these effects merely "pharmacological," more crassly known as "swallowing a drugstore." And there's the question of interpreting reactions. Inject dopamine into the brain of a male ferret, put him with a female, and he will mount in a frenzy, but give him the same injection and put a bowl of food in front of him, and he will eat in a frenzy. The biochemical is not stimulus specific. The author of one study reviewed by Meyer-Bahlburg notes (not, apparently, humorously) that a woman's positive estrogen LH feedback can be observed in both "intact and castrated" men if one injects estradiol, a kind of estrogen, into their bloodstreams and artificially maintains their estrogen levels at those of peak menstruating women for four or five days. One wonders ultimately if there is anything that could *not* be observed, given enough chemicals and surgical manipulation, considering the potency of even tiny amounts of human hormones. The French use pigs to find truffles because the little black fungus produces a chemical identical to 5-alpha-androstenol, a type of testosterone, and the pigs can smell this sex hormone buried several feet deep in the ground. These examples do not mean the endocrinological work is useless. Far from it. But pharmacological effects are one of the inevitable dangers in a field that sometimes can only guess at the "normal" levels of certain hormones it's trying to compare—not only between males and females or even between different species but between unknown and mysterious neurodevelopmental phases.

This is, in the end, the endocrinology problem: Meyer-Bahlburg cites LH studies of two genetic XY men, both of whom had a condition called extreme androgen insensitivity, which meant that their bodies were completely incapable of responding to the masculinizing hormones their glands produced. He reached over his desk, retrieved a well-used book, and riffled through it for a moment till he found a page with a photograph. "This is what a man, with male XY sex chromosomes, looks like when he has androgen insensitivity," says Meyer-Bahlburg.

Both men in these studies lived as women and married men. But

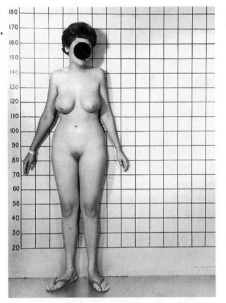

Copyright John Money, Johns Hopkins University

when their LH responses were tested, they were masculine, not feminine. So the question arises: What is the sexual orientation of these men? It is simple enough to ask.

The answer is that it would be heterosexual if you measure it in relation to their anatomy (female), homosexual from the point of view of their genes (XY male), heterosexual according to the gender they feel they are (female), and homosexual according to their LH feedback (male).

~

With the theories currently boxed into a labyrinth of uninterpretable endocrinological results, scientists must retreat to a more elemental question. We know that hormones mediate everything, carrying signals, directing biological development. Before getting into which of them might create sexual orientation, the groundwork has to be laid by figuring out how these complex and powerful molecules interact with the cellular machine to drive living tissue.

One of the most intriguing investigations into the impact of sex hormones on cells is being conducted behind a metal fence high in the dry, sunny hills above the University of California, Berkeley. A team of three Berkeley researchers—Steve Glickman, a psychologist and the project's director; Paul Licht, an endocrinologist; and Laurence Frank, a zoologist—is running the project, studying a species of animal in which the female is larger than the male, more violent and aggressive, and has a penis.

With his somewhat wild, thinning black hair ringing his head like fire and his laid-back academic demeanor, Frank resembles a young, very cool, very California Einstein. The license plates of his silver Porsche 944 read "Crocuta," the Latin name for hyena. As he rockets it up the asphalt road that snakes into the hills among Berkeley's administrative and academic buildings and toward the top of the ridge, Frank explains that what we are on our way to see began in Kenya, where he was living alone in a tent under a fig tree.

Frank was in Africa for his Ph.D. in zoology, doing a field study in the wild and retreating to Nairobi every now and then for supplies. The study was the first attempt to look carefully at the social behavior of hyenas, which, it was known, were social animals in a matriarchical society. Nobody, however, had ever investigated how their society functioned or how the hyenas were organized. Frank set out to gather that information. And of course he had heard the stories. "This business about their masculinized genitalia has been known for a long time," he said, "so I had heard about it and I wasn't particularly interested in it."

He encamped in Africa and began the study. "It was known that the females were at the core of everything," he says. Female hyenas were hyperaggressive, and they dominated the males. Frank observed that hyena society was matriarchal, organized into families of related females. And in the process of these observations, he discovered something else. When he began tagging the animals, he found, to his surprise, that although he usually could tell whether an animal was male or female, frequently he couldn't. "And when you start handling animals, when you've caught an animal, you've marked it, you've seen it running around for three months, and you *still* can't figure out whether it's male or female, it's mind boggling. You get very interested."

Back in Berkeley, Frank began talking to endocrinologists and physiologists, having in mind some rather simple field experiments that "turned out to be completely impractical." Undaunted, he and his colleagues started dreaming of using hyenas as experimental animals to answer some of our basic questions about sexual differentiation. Steve Glickman wrote the grant to the National Institutes of Mental Health, which funded the project, and the group set off to investigate "the question," says Frank, "that is probably as old as biology: how do sex hormones affect animals?"

The road up the ridge passes the Berkeley football stadium, a huge green oval sitting snugly in the hills. Frank remarks that seismologists are expecting the next big quake in the Bay Area to be on a fault line that runs directly under it. As the Porsche ascends, nearly flying, its suspension registers every jolt with exquisite clarity. The Berkeley hills are golden dry and covered with sunlight and eucalyptus.

It was known from experimental evidence that androgens, both pre- and postnatal, can strongly influence aggressiveness in mammals. Here was an animal where the female was not only highly aggressive compared to the male but had a complete set of erectile male sexual organs. "You can't help asking," says Frank, "if there is a relationship between these masculinized genitalia, which were made by hormones—since we don't know of any other way to make male genitals except with hormones—and this aggressive behavior, which must come from the brain. Are the hormones sculpting both the animal's body and its mind?" Here was an amazing opportunity, a natural experiment in the wild. "Virtually everything we know at this point about how hormones mold sexual differentiation *and* aggres-

siveness and sexual behavior comes from decades of artificial manipulation of animals in the lab. Well"—Frank raises his eyebrows, an
understated movement of great emphasis—"this isn't the lab."

Near the top of the ridge, Frank turns right onto a gravel road and
taps a button on the sun visor, and a tall metal fence topped with
barbed wire slides slowly open. Just inside are two prefab buildings
with extremely heavy doors; beyond them and somewhat obscured by
the trees is a large complex of what oddly enough appears from this
distance to be tennis courts, several courts deep, sitting up here in the
middle of nowhere on the steep hills. "C'mon," says Frank, and starts
walking toward the tennis courts, slightly down the hill.

The hyena project is aiming to confirm one of endocrinology's
central theses, one directly implied in the research of sexual orientation, not to mention the research of gender and of violence, with
all their politicized ramifications. If, notes Frank, as currently appears likely to the team, these female hyenas are being masculinized
through androgens, that proves that nature makes unusually malelike
females using the same systems it uses to make normal males. "But
then you get to the extremely interesting question of where the hell
do you get the necessary levels of androgens in a *female* to produce
this effect? That's the mystery. I mean, all sorts of tissues, particularly
the adrenal gland, make androgens in females, but mostly in very low
amounts. Only a testis can churn out the gobs of androgens you need
to get a solidly masculinizing effect on a fetus. And these females definitely don't have testes. So," he asks the hills, "where in the world do
the androgens come from?"

It is utterly silent, except for a fitful breeze brushing the eucalyptus
leaves, which rustle around us. As we get nearer, the tennis courts reveal themselves to be a tall, metal chain-link fence enclosing more
chain-link fences in a perimeter that stretches down the hill and out
of view. Some are covered in the green nylon mesh windbreaks used
on tennis courts; under the hot sun in the cobalt blue sky, the complex seems like a madman's eerily empty country club. Frank undoes
the large lock on its heavy chain and swings the gate open. Inside are
hot, dusty areas surrounded by more fence. And inside those are, finally, the hyenas.

They are shockingly different from the expected. We have all
seen photographs of hyenas roaming the African landscape, but the
photographs are no preparation for the animals themselves, which
watch us in absolute silence. Up close they are surprisingly and alarm-
ingly large, menacing, much larger than the biggest German shep-

herds, and—another surprise—deformed-looking. Their backs are inclines, sloping downward so that their haunches are well below the height of their shoulders. But their most disconcerting feature is their silence. They watch Frank intently, monitoring him with total concentration, trotting at a clipped pace back and forth along the fence on powerful legs and making a scuffing noise in the dust with their paws, but they make no vocalizations at all. Frank, on the other hand, is completely relaxed and enters the outer enclosure greeting them loudly. "Hi, you guys! Hey, you guys! Hi there! Where is everybody?" The animals watch him with feral eyes and no visible response.

In the area between the outside perimeter and the inner inclosures, he points out two hyenas in a shaded area, Talek and Tumo. "These two came from the Talek River in Africa, and the little den hole I dug them out of was on the hill right across from my tent. That's how all these guys were originally captured."

Where were Mom and Dad?

"Dad has nothing to do with anything, and Mom was elsewhere."

And if she'd come back while you were there?

"Oh, they did frequently, but they don't mess with people."

But hyenas have such a bad, violent reputation.

Frank responds primly: "People have bad taste in animals." He passes on to another pen. "They're happy to see people. They usually want to be scratched. It's easiest to domesticate social animals, and hyenas are intensely social. Tuffy here is one I partially raised in my tent. Dos Equis over there was raised by his mom and had less human contact, and he's less friendly."

He walks on, talking to them through the fence. "I have less to do day to day with the animals than I used to, and some of them I don't know, like the younger ones. Some have artificial marks so we can identify them," he explains, and points out a young hyena with a dyed stripe down its back. "The oldest ones in the wild live till about sixteen and seventeen, but they'll probably go double that in captivity." Small high-tech–looking metal rods stick up every few yards along the path with the red glass eyes that constitute the infrared alarm system similar to that in Jurassic Park. What would happen if the system breaks down and the animals get out? Frank rolls his eyes. "The alarm system," he says, "is mostly to keep people from getting *in.* "

The next area holds three large animals. "These guys here are ten years old." Frank talks lovingly and a bit sarcastically, through the fence to one of them, a truly huge beast whom he identifies as Mouse. He scratches Mouse's head through the slightly opened gate. When

I stick my fingers through the fence and make an encouraging petting motion, Frank glances over, then away. "I wouldn't do that if I were you," he says.

The hyenas move sleekly, cautiously about inside their area, galloping one way and then turning suddenly, like chimeras, part great cat, part horse, part wolf or maybe bear, still completely silent. Their unsettling agility throws into relief their mulelike power and immense weight. They in fact are, Frank explains, their own family: Hyaenidae. "They look more like dogs but they're more closely related phylogenetically to cats." (Biology's basic phylogenetic categories are kingdom, phylum, class, order, family, genera, species, subspecies.) That's Mouse," he says, pointing, "and that's Owl, and Eeyore. They're being a little bit squirrely because you're here—a stranger—but this is a very good group for seeing the genitalia. They get excited when I go in, and one of the standard hyena social behaviors is a greeting ceremony where usually everybody gets an erection." He smiles. "They'll be initially interested in me and then they'll probably get just generally wound up and decide they have to greet each other."

Frank undoes the gate and lets himself in. "Hello, Mouse! Hello, Owl and Eeyore!" He shuts the gate behind him and sits on the hot, dry ground, embracing the hyenas as if they were huge dogs, which is the way they behave with him, bounding about and smelling him. "Yeees, I didn't take a shower especially for you." And now, only now, they begin making their first sounds, like the soft mewling of kittens crying, very faintly at first, then, volume rising as they become visibly more excited to be with Frank, like the strange staccato chattering of gibbons, high pitched, and then, finally and loudest, the full, low bellowing of whales cut with the deep bass, grating gutturals of lions. With mouths open, they swarm over Frank, talking to him in this eerie, penetrating, jagged symphony.

"Okay," Frank says from inside the pen, "Owl here is starting to greet." Hanging below Owl's underside, his penis is now clearly visible, dancing beneath him as he dodges around Frank like a boxer. "I don't know if you can see anything," says Frank, but the long and very dark-colored organ is growing as Owl moves his large, powerful head back and forth. His testes are visible as well. I say that I certainly can see him.

"Yeah, well. *Him.*" Frank nods sardonically at the animal, a hulk with a big black penis between its legs. "That's a girl. And she's not even fully erect, just partially."

Owl is a girl? How about Eeyore over there with his long erection, swinging from side to side like a vine?

"Eeyore's a girl too. *That's* a boy." One eyebrow raised in cool amusement, Frank points casually at Mouse, who is a little smaller than the other two—which, Frank adds, is usual.

Mouse also has an erection, but it looks virtually identical to the females'. "Eeyore here is the dominant female," observes Frank, "so she's less likely to show an erection. Getting an erection tends to be a component of subordinance, an appeasement display." He laughs. "People are always surprised that it isn't a sign of macho dominance." Frank is sitting on the ground in his jeans and T-shirt, holding and energetically rubbing Owl in a very rough way as she vibrates vocally in the bone-crunching bass range. When Mouse makes it clear he wants attention, Frank moves to him. "Mousy-Mouse, you are so fat, you are such a grease ball. You're really an embarrassment, Mouse." The three hyenas, having swarmed enthusiastically over Frank, suddenly decide to take off for the fence separating them from three other hyenas. "Now they're going to start an argument with the neighbors," Frank says cheerfully. The six hyenas, thoroughly wound up, quarrel across the fence.

Watching the melee, Frank adds that Eeyore is pregnant. This is extremely confusing since Eeyore has testes, or at least it certainly *looks* like she has testes, but no, Frank says, "what you're seeing *is* a scrotum, but her labia have fused to create the scrotum, which is what a scrotum in fact is: fused labia. She has no testes. But she has no vulva. She has no ordinary female genitalia at all."

Female hyenas mate through their penises. The female's penis is really her clitoris—biologically the female penises are, as are the scrotum and labia, formed from the same tissue during the process of sexual differentiation in the womb. She retracts her penis the way one would push a sleeve up the arm and makes a functional vagina into which the male can insert his penis. The urogenital canal runs the length of this penile clitoris, and the female mates through it and gives birth through it.

"People always ask us," says Frank, " 'Why the hell did they become like this in the first place?' We think the reason might be a response to competition in feeding. Few animals have the competitive feeding situations that social carnivores do where they're in almost violent competition for a small chunk of food. In lions, as an example of another species of social carnivores, a tremendous number of young starve to death because they're the last to eat. So we think nature's solution to this problem in hyenas might have been to make sure that Mama is aggressive enough that she can ensure that her kids will eat. And you make an animal aggressive with hormones. The question arises, of course, why lions haven't done it then, and we certainly don't know. But, you know, nature has a lot of different solutions to various problems, so it's not unduly strange that one species wound up with this one."

In terms of sex roles, male hyenas, who are not feminized, show no maternal behavior, which is generally typical of mammals. Both sexes are interested in the opposite sex when it comes time to mate, although Frank pointed out that there is the classic male/female disparity: "The male hyenas are almost always interested in sex and the females almost never are. And since they're dominant, the males just have to stay pretty hard up most of the time."

Later Frank runs a videotape of a female giving birth. Her penis is swollen with the dead infant stuck inside it; it's her first pregnancy, and the baby is forced through the narrow, suffocating birth canal of its mother's penis and must rip open the end before reaching the light and air of the outside. The first birth frequently ends up killing the infant. After that one, deliveries are easier. "The baby has to travel all the way from the womb, make a 180-degree turn in the pelvis, and make it out of the penile clitoris. The trip from the uterus to the outside is about 60 cm long and the umbilical cord is only 15 cm long, so it has to make the trip quickly because most of the way it's cut off from its supply of oxygen. It's not easy for the mother either; almost 20 percent of the births here at Berkeley have required veterinary intervention to save the life of the mother, and we think that a significant percentage of them must die in the wild.

"The only time," Frank adds, "that I ever heard a female hyena cry was during a birth. She whimpered a little from the pain."

~

The three-man hyena team is having lunch at the Yenching Chinese restaurant on Shattuck Avenue in Berkeley, and while walking there on a path through the Berkeley campus, they argue zoological trivia. The trivia, however, which is about penis size, derives from an essential biological question—one bordering on the metaphysical—that must be answered by every biologist studying any trait: Why would evolution produce this? The hyena's genitals spark the discussion, and Dr. Paul Licht notes, "Some animals have different size genitals for different biological reasons. Monogamous animals tend to have smaller testicles than herd animals because, we believe, herd males have to mate with a lot more females and so need a lot more sperm. Chimpanzees have very large testes. But gorillas have small testes and tiny little penises." He pauses. "Laurence," Licht, the endocrinologist of the team, turns around to quiz Frank, the zoologist, over his shoulder, "do you remember why?"

"Oh, sure," says Frank, "there's no sperm competition. It's a monomale group, one male has total sexual access to all the females and no other males have any chance of mating with her, so there's no need to swamp somebody else's sperm with your own."

"Well," Licht objects, "but that's *testicle* size, not penis size."

"I thought that's what you were asking about," says Frank.

"No, no. Penis size." Licht and Frank debate the point, Licht trying to recall a seminar where the speaker argued that penis size in the gorilla was a function of some aspect of the mating system. He notes that in the case of humans, the evolutionary question would be "Why has evolution given human males, compared to other primates, exceedingly large genitals?" Another evolutionary question for humans, one that must eventually be plausibly answered, is "Why would evolution make some people gay?"

Inside the Yenching, they discover three of their colleagues eating at one of the tables. One is an anatomist who is helping the hyena team. He dissects the animal models they've worked on, and they are excited to run into him. "We've been trying to talk to you for a week about the latest results," Steve Glickman, the psychologist, reproaches him.

"This guy," Frank says loudly. "You can't ever get him on the phone."

The anatomist grins and details the results so far of his work. "I only found one ovary," he tells them, slightly worried.

"No, that's good," says Frank. "There should only be one. We removed the other for Paul's studies." They seem partly relieved, partly just fascinated, and listen intently to the rest of his preliminary report.

Seated, Licht accepts his Chinese menu and says, "Ah, MSG, the chemical that destroys your brain and makes you sterile." He explains that while you can use MSG in food to increase flavor, you can also use it in the laboratory to sterilize animals for experiments. "You inject monosodium glutamate into the young animal, and it selectively destroys certain neurons in the brain that, eventually, control the reproductive hormones. The animal becomes sterile. It's very effective."

"Of course," adds Licht, pouring green tea, "that's obviously at levels much higher than you'd expect to get from a Chinese meal. And it's being injected."

The three of them—psychologist, endocrinologist, and zoologist—are out to answer (as Licht describes it) the "Why, Where, and How" of the aggressive, masculinized female hyenas. "Why, Where, and How" is the colloquial way of expressing the range of explanations for a phenomenon, running from ultimate causation—the "widest" or broadest reason for the thing scientists are seeing—to proximate cause, the narrowest, most close-up explanation of why the thing is what it is.

Ultimate causation, the broadest, macro question facing them, is the evolutionary question: Why have factors in the environment pushed the hyenas in this particular evolutionary direction? Why did this strange system evolve in reaction to evolutionary pressures? ("With birds, for instance," notes Frank, "the ultimate explanation for this ability to fly that they've evolved is that it allows them enormous mobility to find food, escape predators, and seek better climates.") The proximate causation for the hyena phenomenon will answer the question: How, exactly, does it work on a cellular, hormonal, and even molecular level? How are these hormones generated, how do they interact with cells? ("It's the mechanistic stuff," explains Frank. "With birds, you'd be determining exactly how the respiratory system and bone structure of a feathered, aviary creature work, muscle contraction and oxygen metabolism. The micro stuff.")

In between, there are the questions of where these two ends of the spectrum fit together. This is standard biological procedure; those researchers tackling sexual orientation will eventually have to answer these same questions, from ultimate all the way to proximate cause.

Endocrinologists have long observed the effect of steroids on organisms. Steroids are the sex hormones refined from raw cholesterol (in chemical etymology, the "ster-" of "steroids" is the same "-ster-" of "cholesterol"). Aggression flows inside all of us, distilled into chemicals, a reservoir of pure liquid violence measured out in droplets of driving rage, or retreating, defensive panic, or random destruction. "The connection between steroids and aggression," Carl Sagan and Ann Druyan have written in their fascinating *Shadows of Forgotten Ancestors*, "applies with surprising regularity across the animal kingdom. Remove the principal source of sex hormones, and aggression declines, not just among the mammals and birds, but in lizards and even fish." In species from pit vipers to primates to certain African fish, losing in ritual battles for leadership means a marked decrease in testosterone, a sort of biochemical loss of self-esteem. In contrast, give males testosterone and aggression jumps. The more macho the androgens make males, the more they neglect their family responsibilities; raise the testosterone levels of male birds, and they decline to feed their hatchlings.

Give an animal estrogen, and aggression falls; administer progesterone, and the female inclination to care for young increases. (The words mean, respectively, "estrus-maker" and "gestation-promoter.") From rats to dogs, virgin females ignore or avoid newborn pups, but prolonged treatment with progesterone and estradiol that bring their hormone levels up to those typical of late pregnancy results in virgin maternal behavior. Rats that have high levels of estrogen are less anxious, less fearful, and less likely to engage in conflict. And hormones also determine differences in violence, aggression, and maternal behavior between the sexes. This is not, Sagan and Druyan point out, the universal case. Male and female wolves, short-tailed shrews, ring-tailed lemurs, and gibbons are almost equally aggressive, but these are the exceptions. We humans are the rule. Males are more aggressive than females, and blood plasma testosterone is about ten times greater in men than in women. Compare levels of male sex hormones circulating in our bodies, and you find these figures:

	Men	*Women*
TESTOSTERONE	**22**	2

Compare estradiol, an estrogen, and maternalizing progesterone, and the difference is monumental:

	Men	*Women*
ESTRADIOL	.15	**1.7**
PROGESTERONE	.7	**32**

[Source: Bruce McEwen, Rockefeller University, "Circulating Levels of Sex Hormones in Humans." Measurements in nanograms of hormone per ml of plasma per person.]

The riddle that sparked the project was that the hyena is the only known female mammal that has this extraordinary masculinized genitalia. "Was it androgens there that were doing this," Glickman asks, "or—and this would have been a big deal in biology—were we going to discover a completely *new* hormonal mechanism for making male genitals that we didn't know about?"

At first they wondered if they were simply seeing massive CAH, the hormonal disorder that pumps male hormones into the bloodstreams of fetal girls, masculinizing their genitalia. Perhaps genes or diet or something else had given these animals chronic congenital adrenal hyperplasia. Looking into this, they found nothing abnormal in this way about the hyenas, and further study has shown that normal androgens are what is masculinizing the females' genitals. Yet these androgens are present in highly abnormal amounts and were producing an extremely rare effect. Where were they coming from?

The answer, says Licht, appears to be from an unlikely source. "The study did not start out as a study of the placenta," he says, surprise evident in his voice. "We had no idea the placenta was involved." From what they are seeing now, it appears that the tissue of the female hyena's placenta is churning out huge amounts of male hormones during pregnancy. If, in fact, the placenta is masculinizing these animals, the team will have discovered an organ that, bizarrely, produces vast amounts of androgens in the hyena while producing next to none in virtually every other of earth's species that has been studied.

But if they had found the source of the hormones, the puzzle became how, exactly, they were masculinizing the genitalia. Was it hormone levels? For females to have more male hormones than males do is unusual in any species. When the team ran early hormonal blood tests on young hyenas, the males had, as would be expected, more testosterone than the females. (The females had high levels when they were pregnant, which was quite surprising.) So the effect wasn't being created by testosterone.

Then they looked at androstenedione, a "male" hormone little understood by scientists. They discovered something very odd: Throughout life female hyenas have *higher* levels of androstenedione than males do. "Typically males would have more of this stuff," says Licht, "not females. So we thought it might be responsible for the higher level of aggression and dominance we see in females."

Besides the reversed levels, the scientists were puzzled by the nature of the hormone itself. "I mean, *androstenedione?*" said Licht. "This was very confusing because you'd figure some big, powerful, heavyduty steroid like testosterone would be creating what we were seeing here, and it wasn't. As for androstenedione—well, it's a so-called weak androgen." The hyena team was aware of some androstenedione studies done on male lab rats castrated at birth, further piquing their interest. "Even the way this stuff is made is odd," observes Licht. "You get a lot of male sex hormones from the adrenal gland, and they're called, logically enough, adrenal androgens. But not androstenedione. *It* happens to be made in the ovaries, because it's the raw material for estrogen. Every mammalian female is probably producing a lot of androstenedione in her ovaries. But we thought it was all being converted into estrogen. And we had absolutely no idea what to make of all this."

Then, by chance, a friend of Licht's attended a large endocrinology meeting at the National Institutes of Health. A group of scientists there was presenting a large data set on behavior in human adolescents. The study included interviews with the teenagers and their families and teachers, and they had also taken regular blood samples. "They were trying to find a correlation between any of the huge set of steroid measures they were taking and some behavioral variable," says Licht, "but they were coming up with nothing, no correlation between testosterone or estrogen or DhT or progesterone and this whole raft of behaviors they were looking at. Except for one hormone: There were significant correlations between levels of androstenedione and all sorts of teenage behaviors in girls like aggression or getting into fights. In the study they called it 'nastiness,' essentially what people complain about in their teenage girls and the ways in which they're hard to get along with.

"The woman presenting the paper said, 'We frankly have no idea what this means because as far as we know, androstenedione doesn't *do* anything.' So my friend walked up to her and said, 'Listen, I know some people who think that androstenedione might have a role in producing highly aggressive females in hyenas.' "

But if androstenedione is playing some role in *human* personality, teasing out what, exactly, that role is will be excruciatingly difficult because the organism in question is human. We have castrated male domestic animals for millennia, and there is, of course, no question that reducing testosterone via castration reduces aggression. But we also have discovered that in highly social animals, hormonal cause-and-effect is vastly more complex. Castrate a juvenile monkey who has never mated, and he will never show sexual interest in females. But castrate an adult monkey who's done some mating, and he will continue to mate, perhaps at a lower level, but he will show erections and sexual interest for months or years afterward. The more one looks at animals in increasingly complex social systems, the less one can say given behaviors are simple products of hormones. "It is clear androgens strongly influence behavior in animals," says Frank, "and I would be astounded if there were not also some influence in humans, but it will be an influence profoundly complicated by social effects."

One of the early studies of how androgens influence behavior was done on monkeys by Goy, Resko, and Geral, who showed clearly that in primates, male juveniles typically show higher levels of rough-and-tumble play than females.[7] Human behavior among the young reflects this, with boys more likely to beat each other with swords and girls to play with dolls. Goy then decided to give androgens to pregnant monkeys, exposing their young to these masculinizing hormones while in utero, and looked to see what happened.

What happened was that there was almost no effect on males, who were already being exposed to androgens from their own testes as they develop, but, strikingly, the females were born with significantly masculinized genitalia, and when Goy placed them in social groups, they played almost as aggressively as males.

This hormonally influenced male-typical, female-typical behavior evident in very young animals beckons toward a hormonal mechanism for sexual orientation. Many researchers have studied the question of when typical male-female behavior and evidence of heterosexual-homosexual orientation show up in humans, and the answer is: very young. Richard Green, a UCLA psychiatrist, studied effeminate and masculine boys over a period of fifteen years.[8] He interviewed the boys' parents and found that from a very young age, a certain percentage of children clearly behaved atypically for their gender. The boys who behaved markedly like girls did so from such a young age that there was no possibility of their having learned the behavior, no role models, no environmental inputs.

> MOTHER (speaking of her seven-year-old son): The first thing I noticed, one day at a friend's house—she has four little girls, he likes to go back in their room and play with their dolls. He went back there and grabbed a doll. . . . He took Kleenex and . . . folded them and poked holes in them and made a very attractive dress for those dolls . . .
>
> GREEN: How old was he?
>
> MOTHER: He was four, four and a half. . . . He would tie aprons around himself . . . to make a nice long skirt. . . . It got so that we were taking aprons and just sticking them places. . . . Even today I go around and find aprons wadded clear back on the closet shelves where we had tried to hide them, just in a fit of panic.

Or from a different interviewee:

MOTHER: He was playing with dolls, playing dress-up, playing school in nursery school at the age of three. By the time he was four it didn't diminish.

Green wrote, "The extent of play with two toys differed significantly. . . . 'Feminine' boys played about four times as much with the doll [but only] a third as much with the truck. . . . The 'feminine' boys' selections were like those of girls of the same age." Such differences are not particularly subtle; in a story on homosexuality, the *MacNeil/Lehrer NewsHour* presented two brothers—fraternal twins, no less—in a Christmas photograph posing with the gifts they'd requested: David, the heterosexual brother, with his war toys, a rifle, a tank, an army tent, and Eric, the gay brother, with his baby doll, kitchen set, and cardboard house.

Indeed, Green's most striking finding was the correlation between this gender-atypical behavior and sexual orientation. Interviewing his subjects as adults in their late teens and early twenties, he found that 75 percent of the men who had been effeminate boys had homosexual sexual orientations. Other researchers have seen similar correlation with girls who behaved like boys.[9] As with children learning the behavior we call speech, these gender-atypical children were not "taught" their behavior. They behaved like no one they'd ever seen and persevered despite harsh, constant social pressure to change. Like the genetic capacity for human language, the inescapable conclusion was that they were born with both their gender behavior and their sexual orientation.

Sociologist Fred Whitam compared the childhood experiences of 375 homosexual men in Guatemala, Brazil, the Philippines, Thailand, Peru, and the United States, and found in them "culturally invariable" characteristics that remained stable across geography, class, and time zone: playing with toys of the opposite sex, being regarded as a sissy, and preferring the company of girls and older women.[10] "There can be little doubt," Whitam wrote, "that this [early-childhood] behavior is linked to adult sexual orientation."

∼

In his paper "Culturally Invariable Properties of Male Homosexuality,"[11] Whitam delineates six phenotypic aspects of homosexuality remarkable across cultures:

1. Homosexuality as a sexual orientation is universal, appearing in all societies.

2. The percentage of homosexuals in all societies seems to be the same and remains stable over time. Everywhere in the world, homosexual populations appear to comprise no more than 5 percent of the total population.

3. Social norms neither impede nor facilitate the emergence of homosexual orientation. Homosexuals appear with equal frequency in societies that are repressive of homosexuality and in societies that are permissive. Repression simply reduces the expression of a homosexual orientation, not its existence.

4. Homosexual subcultures appear in all societies, given sufficient aggregates of people.

5. Homosexuals in different societies resemble each other with respect to certain behavioral interests and occupational choices.

6. All societies produce similar continua from overtly masculine to overtly feminine homosexuals.

Although all six observations are relevant to biologists, to whom they constitute strong evidence of sexual orientation's biological etiology, the third point is most relevant to political debate, specifically the notion that one or the other sexual orientation can be "promoted" or "encouraged." Whitam is careful to note that cultural considerations determine to a degree the way an individual's homosexuality or heterosexuality will be *expressed,* not whether an individual will *be* homosexual or heterosexual. His conclusion: "It does seem clear from the cross-cultural perspective that societies do not create homosexuals. Their emergence appears to be beyond the power of any society to control."

As plates are being cleared from the table at the Yenching, Frank observes, "Steve predicted when we were starting this project that we'd see a reversal of this rough-and-tumble play in the hyenas, higher in females than males. One of our grad students did the study, and sure enough, the female hyenas showed much more vigorous social play than males."

The hyenas (interestingly enough) almost certainly do not serve as an animal model for human sexual orientation per se. Despite their physical and psychological—if it can be called that—masculinization, they display less same-sex mounting than many other species, such as guinea pigs, where females mount frequently. (When they do mount, female hyenas can achieve penetration, which the guinea pigs can't.) What hyena studies do tell us is that hormones we never expected would exist in a given gender are generated in huge amounts by surprising molecular mechanisms in the most unlikely tissues. And this is a discovery with parallels to other species.

This androgen-making hyena placenta is not the strangest item in the world of hormones. In one of the great inconsistencies of biology, it is in fact estrogens, the so-called "female" hormones, not male androgens, that make the brains of rats male. While the male rat brain needs testosterone, a male hormone, to be masculinized, when this male hormone arrives in the brain, it is turned into estradiol, a female hormone, and it is this female hormone, perversely enough, that actually masculinizes.

Why then aren't female rats masculinized, since females are floating in a sea of estradiol? The answer is that they have hormonal armor, something called alphafeto protein which seizes the "masculinizing estradiol"—a bizarre two-word combination that is the quintessence of intuitive contradiction—and stops it from getting into brain cells where it would wreak havoc. Likewise, the vocal control sites in the brains of male frogs have to be masculinized by female estrogens derived from androgens. It may be paradoxical, but this sort of paradox is common in endocrinology and means that hormonal mechanisms may be acting in any of a million ways, some literally unimaginable at present, on human sexual orientation.

Frank notes that "Scientists are becoming more interested in the role of androgens in women. What we don't know yet in humans, on a molecular level, is why you need androgens to be aggressive and tough and tear people limb from limb, but there is preliminary evidence that androgens in women do affect their libidos and their aggressiveness. They just have smaller amounts than us." Women's ovaries, in fact, produce low levels of male hormones like testosterone, androstenedione, dehydroepiandrosterone, "et cetera, et cetera," says Frank. "Testosterone is just the commonest."

Indeed, if, rather astonishingly, the hyena's placenta is producing androgens, human beings have an equally unlikely, and surprisingly common, endocrinological feature called gynocomastia. Virtually everyone at one time or another has seen gynecomastia but most take it as an amusing illusion created by the extra weight that obese men carry. It is, for a number of societal and emotional reasons, difficult to believe that the "breasts" that fat men (and 20 percent of boys while going through puberty) often appear to have are in fact breasts, biologically identical to those of women. Yet they are. Men routinely produce large amounts of testosterone, but fat tissue in human beings converts this male hormone into estradiol, a female hormone; in fact, the principle source of estrogen in male primates is fat, which converts male to female hormones. Sufficiently fat men (and male apes & male gorillas, etc.) make enough estrogen to cause them to develop breasts and are, not infrequently, pseudohermaphrodites. If ovaries and placentas can make androgens, if male hormones can suddenly and subtly be turned female and mere male obesity can give men women's breasts, then endocrinologists like Heino Meyer-Bahlburg who search for hormonal mechanisms creating sexual orientation face fascinating—but almost limitless—possibilities.

∼

In the end, the key to the endocrinological link to sexual orientation may well be hormones in the womb, chemical controls that currently remain so tantalizingly immeasurable. As Mike Baum, an endocrinologist at Boston University who works with ferrets, explained, "There is absolutely *no* data comparing fetal testosterone levels between gay men and straight men. We already know adult levels are identical," he said, dismissing these with a deprecatory motion of his hand, "but *fetal* levels—that would be *very* interesting. You'd have to do the measurements in utero and then follow the subjects through birth, through puberty, and wait until they were certain of their sexual orientation. That's a good decade at shortest, maybe two. And although distinction of human sexual organs is complete by the second trimester, to the extent that the central nervous system is still responding to androgens well after birth, there could still be other effects, so you'd have to try and figure those out." Baum smiled. "It would be a monumental study."

We have yet to figure out how to open safely the human body to examine sexual orientation, both in the womb, where it may be formed, and in the brain, where it lives. To use Evan Balaban's neuroanatomical method of injecting a marked dye to study the brain would mandate either killing the people used in the study or the injection of the dye into a living brain. "You would never get permission from anyone to do this," says Balaban with a raised eyebrow, "and if you *did* get permission, that person would probably be suicidal or certifiable and you couldn't ethically do it."

But if this is so, it points back to the problem faced by biologists who study exclusively human traits and to one of the fundamental questions of biology: What constitutes "proof"? How do you establish it?

In their study of the masculinized female hyenas, the members of the hyena team have a suspect—androstenedione—but they now confront the task of proving a molecular connection. As the endocrinologist Licht states emphatically, "This idea of 'proof' is something a lot of nonscientists get wrong. When you've got a correlation between two things, like this hormone we're trying to track down and a behavior in teenagers, or between the size of some nucleus and sexual orientation, it may support some theory you really like, but a 'correlation' is not 'proof.' Correlation can be excellent science. But it's different from 'proof.' We've assembled an entire picture suggesting that the female hyena fetus is soaked with masculinizing hormones, we think we might know which hormone—androstenedione—and we even think we know the source of these hormones—the placenta. That's our working theory. And everything in the theory makes sense, it all correlates with what we've seen. But none of it 'proves' that hormones are causing this phenomenon to happen. In fact, although we have a hundred experiments that *suggest* this wonderful explanation, if we cannot show the exact biochemical mechanism by which this hormone acts on the genital tissue of these females, no matter how pretty this picture is, at some point we may have to throw it out."

How to show this is the problem. "The only way," says Licht, "that you really prove anything in biology is not correlation but direct manipulation of the organism." He ticks off the steps on his fingers. First, you have to show the hormone is there. "We've done this," says Licht. Next, you have to show there's a receptor on the tissue (in this case, on the genitalia) to receive this hormone, which the team also has done with a traceable radioactive steroid. The last step it has not done. "We have *not* shown," says Licht, "that the androgen binds to those receptors and actually does something. So far we have mostly negative data, which can be interpreted lots of different ways. This is a problem when you look at how people—politicians and the media and such—try to use science. The side that people come down on tends to reflect their bias. If I were very against the androstenedione theory, I would be making a big deal out of the negative data. Given that I am very much in favor of this theory because it makes sense to me that everything else is there, all the parts are there—and because it's my theory," Licht adds with equanimity, "I'm coming down on the side of 'We just haven't done it right yet.' "

He shrugs. "Unfortunately—and this is a huge limitation—when you're doing research in human beings on any trait from handedness to sexual orientation to mental diseases, the vast majority of your data comes only from correlation studies, not direct experimentation. Because you don't experiment on people. That's why all the studies, the LH feedback and the elevation of lesbianism in CAH woman, don't 'prove' anything. They show there's a correlation. That's all."

Neurologists have long confronted it. Any introductory biology book has a map of the brain with different functions carefully ascribed to different areas—math here, language there, hunger or thirst or sex drive—and most of this knowledge comes from natural experiments we call accidents. One of the most famous ones was that of Fineas Gage, a man who, almost a century ago, was planting explosives, forgot to cover them with sand, and set them off when he tamped them down. The explosion blew a rod through the top of the head, destroying part of his brain. (Gage lived.) Seeing what parts of his brain were damaged and what mental functions he did or didn't lose told us much about the brain's function. "That's not a correlation," notes Licht, "that's much more. It's an experiment. You're directly manipulating and altering the brain tissue itself to show cause and effect."

Occasionally the opportunity for direct work on humans arises, as it did with Genie. Neurologists sometimes get permission to do research from surgical patients while, for example, extracting brain tumors. For such procedures, patients must be awake and able to talk so the surgeon does not extract something important, and as they sit in the operating room, their brains exposed, the researcher can stimulate parts of their brain by touching them and ask, "What do you feel?" "I smell berries," the patient will respond, or "I feel hot."

This safe and painless investigation (there are no nerve endings inside the brain) can give us clues as to which brain parts control what, but the fact is that science—including, emphatically, the search for the origin of sexual orientation—is greatly hobbled by ethical considerations. It is a question of efficiency. The natural experiments—the Genies and Victors for human speech, the CAH women and the androgen-insensitive men for human sexual orientation—are interesting, but they are not a particularly efficient way of conducting research. Poking delicately at the brain is fine, but the fact is that the great bulk of our knowledge of brain functioning comes from studies on living monkeys, studies that are a lot more thorough and quite deadly. Studies on humans never achieve the standard of those on lab animals.

Do we really want a cure for Alzheimer's, a solely human disease and one that kills thousands every year? Were we not restrained by the laws of basic humaneness, the study would be obvious. One of the basic questions surrounding Alzheimer's is the origin of the plaques and tangles that seem to strangle the brains of all Alzheimer's patients. To be truly efficient scientifically, we would do a human progressive study, such as those done routinely on monkeys. Select a group of people with Alzheimer's and intentionally allow the disease to progress while keeping track of how much memory they lose. Then kill some of them, slice up their brains, and measure how many plaques and tangles they have. Do the same to a few more people two years later and assess the difference. Then the same two years after that. Such a study would yield spectacularly valuable medical data, but for obvious reasons researchers have never done such human experimentation on Alzheimer's.

They have, however, done human experimentation on homosexuality.

There are more famous examples of human experiments, of course. German doctors performed experiments on human subjects in World War II prisoner-of-war and concentration camps. They kept detailed records of their scientific findings. This was no mere phenotypic research, nor was it tentative correlation work disregarded by the hyena team. This was Licht's "direct examination of the organism itself," the kind that offers proof. The organism in this case was people.

The Germans considered their subjects less than human, which facilitated the research. During the 1940s Dr. Carl Clauberg operated on Jewish women to examine the human female reproductive system and Dr. Sigmund Rascher, in an effort to understand the reaction of the human body to pressure and cold, injected Gypsys with sea water and submerged them in freezing water until they turned blue, lapsed into unconsciousness, and, frequently, died. Measurements were taken of how much heat they could withstand, how much cold, and their tolerance for pain. Men were placed in pressure chambers and blown up, depressurized, and crushed. The data gathered were applied to the construction of better survival clothing and equipment for Nazi pilots flying at extra-high altitudes or downed over water. Dr. Josef Mengele, the most famous of this scientific group, studied genetics, sterilization, and the transmission and progress of infectious diseases, and Dr. August Hirt specialized in human anatomy. They castrated their subjects and forced them to undergo hysterectomies, lobotomies, and injections.[12]

During the war Japanese scientists concentrated on neuroanatomy, performing experiments on living American pilots they had shot down and on Chinese prisoners of war. Among other practices, Japanese doctors removed the tops of their subjects' skulls and observed which of their faculties failed as successive parts of the brain were systematically destroyed. We Americans denounced, and continue to denounce, such practices, but the fact, little discussed in this country, is that we also have conducted human experimentation in our hospitals and research laboratories.

In the 1950s, American scientists performed experiments on homosexuals, viewing them as desperately diseased or simply as people whose lives had little value. In an effort to turn lesbians into heterosexuals, a transformation many psychiatrists then maintained was possible, American doctors forcibly removed the uteri of perfectly healthy homosexual women and cut off their breasts. Based on false endocrinological theories such as the notion that homosexuality was caused by deficient levels of sex hormones, other lesbians were made to undergo estrogen injections. None of these procedures ever altered the women's sexual orientations; the women all remained lesbians. Merely a decade after the Nazis had been defeated in Europe and the German human experiments were revealed to reactions of universal horror, medical authorities in the United States routinely castrated gay men (one particularly enthusiastic doctor is known to have castrated more than one hundred) and lobotomized them. Other gay people were forced to undergo aversion therapy—both men and women were wired to machines that jolted them with electricity every time they were shown an arousing picture of someone of the same sex—and electroshock. One of the attending doctors, curious about the process he was administering, recalled later how he grasped an electrode and turned on the machine. He reported, rather queasily, that the shock threw him across the room. Despite this "direct experimentation on the organism," the gay men's sexual orientations remained universally homosexual.

The procedures continued even after this fact became apparent, however.

One of the better-known series of such treatments was carried out on Alan Turing, the British computer genius and World War II hero who broke the German code for the Allies. When Turing was discovered to be gay, he was arrested and forced to undergo injections of estrogen to control his "deviant" sexual behavior. The injections eventually led him to commit suicide, but they did not change his sexual orientation.

Like the Germans, we were remarkably thorough in recording the process. The recent documentary *Changing Our Minds* about Evelyn Hooker's work contains medical film footage of these experiments. One film from the late 1940s records the lobotomization of a young gay man. His doctor recommended a transorbital lobotomy to alter his homosexuality, and the camera, in close-up, carefully watches as a small device, a kind of double-bladed ice pick with razored edges, is inserted through the man's upper eyelid and skull into the brain. The pick is moved back and forth, reducing his prefrontal lobe, the most reflective and human part of the brain and unnecessary for more basic activity, to a hemorrhaging pulp, although the bleeding, which remains inside his head, is not recorded on film. The young man was successfully lobotomized. His sexual orientation remained homosexual.

The U.S. military also kept careful records. In a grainy black-and-white clip from a 1950s naval research film, a gay man lies in a metal hospital bed. The frames of the film blip and pop in scratchy bits. The man moans softly. Around him, a group of hushed doctors in white work efficiently, attaching electrodes to his head, strapping him down. "We're going to help you get better," says a confident male voice somewhere in the background. The man groans quietly. When they turn on the electricity, his body whiplashes violently and he begins to scream. One of the anonymous doctors surrounding him sticks a tongue depressor into his gaping mouth to stop him from swallowing his tongue during the shock as the camera pulls in, somewhat jerkily, for a close-up.

If scientific and medical interest in sexual orientation had truly been at issue in these studies, in accordance with the most elementary rules of the scientific method, the researchers also would have lobotomized, injected, and electroshocked heterosexuals as a control group. That controls were omitted indicates that the researchers' true motivation was neither changing homosexuality nor researching sexual orientation. Ironically, due to this omission, while the experiments provide overwhelming evidence that homosexuality is an immutable characteristic, they say nothing about whether heterosexuality is also one; although there is much soft correlation data, we still have no hard experimental evidence that heterosexuality cannot be altered.

John Money, the well-known sex researcher at Johns Hopkins University, once remarked that there was, in fact, one easy way to change a homosexual's sexual orientation. This was surprising; Dr. Money, a gruff, no-nonsense scientist who has been active in the field for decades, has amassed a significant body of research, all of which indicates that homosexual or heterosexual orientation is fixed at a very young age, probably by age two or before. "Your sexual orientation is something you've *got*," he has growled, "you've got it for *good*, and you'd better just get *used* to it."

He was asked how, then, could someone's sexual orientation be changed.

"Simple!" he barked. "Remove their brain!"

Chapter Six

GENETIC GRAMMAR 101: A CRASH COURSE

Genetics: the science of heredity, dealing with resemblances and differences of related organisms resulting from their genes.

ON THIS PLANET many languages have been written on the bark of trees or spelled out in dried bits of earth or chiseled in stone. There have been languages written in letters that look like languid circles and letters that look like slashes and daggers of pigment, languages written entirely in numerical signs, languages reserved for the holiest of ideas and those used only for song.

DNA—deoxyribonucleic acid—is an incredibly rich, complex, and ancient language, much older than Greek, Chinese, or Aramaic. It is a language with an elegant, ornate grammar and syntax and a highly sophisticated system of punctuation, although it is written in a surprisingly limited alphabet of just four letters: A, C, G, and T, (Even the Hawaiian alphabet has fourteen, but then binary code has only two, so DNA is twice as letter-rich as that language.) In a book called a genome, each of us is written in the language of DNA.

The alphabets of most languages spoken by humans can be written out in almost any medium—in ink, in liquid crystals frozen on glass, or in glowing ten-foot-tall gas plasma characters. Binary code is written only in numerals (usually represented by numerals) and Morse code in two slightly different units of time, represented any number of ways. DNA is always, and as far as we know can only be, written in chemicals. Its A, C, G, and T stand for adenine, cystosine, guanine, and thymine, substances known as bases. Anyone who has ever plunged a hand into bleach and felt the sliminess knows what a vat of pure bases would feel like; chlorine is also a base. (A base is simply a chemical that has extra electrons and can give some away; an acid has a positive charge, like a hydrogen ion, and can accept electrons.)

Only since very recently—the middle of this century—have we even known what this language looked like, how it was written, and what its basic structure was. Actually becoming fluent in DNA, truly understanding what the language is saying, is, we've discovered since, more difficult than becoming fluent in Japanese. At first glance, Japanese, with its alphabet of forty-six letters and around 5,000 idiograms, makes the A, C, G, T of DNA look absurdly simple, but once a person has mastered a vocabulary of 2,000 characters, he or she can read a Japanese newspaper with ease; DNA's total vocabulary, on the other hand—the number of different genes in the human genome, our genetic book, the volume of information in which we are encoded—is around 70,000, or maybe 150,000 (we don't know yet), and we will need to learn them all before we can read the entire book, the whole human genome.

Hebrew is read right to left, French left to right, and Chinese left to right, right to left, or top to bottom, and all are read linearly. DNA is read linearly as well, yet where French is generally written on paper and Egyptian hieroglyphics were written in soft mud, DNA is spelled out along the surface of two amazingly long and incredibly thin, twin ropelike molecules that twist around each other in a helix. There are two scrolls in this molecule because, unlike Swedish or Urdu where everything is written once, DNA is always written out twice, although not always with exactly the same spelling.

By tradition, scriptural Hebrew always is written out on a particular structure: a long sheet of paper wrapped around two wooden sticks. The doubled DNA molecule also has a very particular structure, which has what is called "handedness." As is the case for many molecules, the two strands of the helix corkscrew around each other as they spiral upward, and the direction in which they corkscrew is their handedness. Hold your hands up in front of you, palms facing away. Twist each hand while moving it upward as if turning a large knob up into the ceiling, and each will twist only in one direction. If a molecule curls upward in the same direction as your right hand, it's right-handed; if it curls upward like the left, it's left-handed. This is the structure of the molecule on which DNA is written:

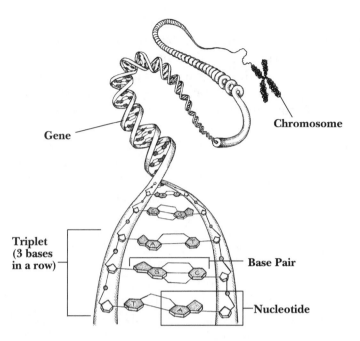

It is a right-handed molecule.

This double helix usually is depicted laid out nicely on the pages of science texts like well-behaved ladders. This is completely unlike reality. DNA, in its natural state, spends 99 percent of its time in a coiled, writhing mess inside the nuclei of cells, tangled up into tight, microscopic balls resembling globes of snarled string. If one were to untwist this long, doubled molecule and stretch it out fully, it would be a sticky, silvery, gossamer thread about a meter long—if, that is, it were actually one long strand. It isn't. In humans, DNA is twenty-three different strands of this doubled molecule, like twenty-three different doubled strands of pearl necklace. Each strand is a chromosome. (In the early nineteenth century when biologists first stumbled on tiny structures in the nuclei of plant and animal cells, they discovered to their delight that these bits and pieces of something—they had no idea what—would turn bright colors when treated with certain dyes for microscopic examination. So they called them "chromosomes," literally "colored bodies.") Chromosomes are numbered according to their lengths, longest to shortest; Chromosome 1 is an elegant expanse of molecule, but Chromosome 21 is short and nubby. Every chromosome, in turn, has two arms, the long arm and short arm (they are called the "p" arm and the "q" arm, "p" because it stands for "petit" and "q" simply because it's the letter after p), and the arms are marked into different sections like the inches on a ruler, and in each section is written out a single set of instructions— almost a recipe—for making enzymes and proteins, the chemicals that build and run the body. These sets of instructions or recipes are genes.

The structure of the language is straightforward: each person's genomic book is divided into twenty-three chapters, but (again) it is a curiosity of DNA that each of these chapters is written out twice, giving every human twenty-three *pairs* of chromosomes. Floating in all our cells are one copy of, say, Chapter 11 (Chromosome 11), and somewhere else in the same cell, another complete Chapter 11. Genes are the individual sentences that, strung together, make up these chapters. (There are, of course, two copies of each sentence, one in both copies of the chromosomes.) Both sentences in a pair may be identical or they may read somewhat differently; one might read ... TTTAGT ... and the other ... TTAAGC. ... Every one of the words that make up the sentences is three letters long, for example, ATC or GGA or TAG. Geneticists call these words, logically if unoriginally, triplets. String together enough triplets (words) and you have a gene (sentence) that tells you how to make one protein. String together enough genes into one complete set, and you have a chromosome. Compile twenty-three chromosomes (chapters 1 through 23, two copies of each), and you have a genome, a book written in a chemical alphabet.

With its alphabet of only four letters and words of only three, the total possible vocabulary of DNA is very limited. Multiply 4 (the number of possible first letters) times 4 (possible second letters) times 4 (possible third letters) and you have only 64 possible triplets, which means DNA is a language with a total vocabulary of 64 words. But the fact is there are only and exactly twenty amino acids total: alanin, arginine, asparagine, aspartic acid, cysteine, glutamine, glutamic acid, glycine, histidine, isoleucine, leucine, lysine, methionine, phenylalanine, proline, serine, threonine, tryptophan, tyrosine, and valine.

Since DNA's vocabulary is 64 words, it is over three times larger than it needs to be, since many triplet words designate the same amino acid. Both CTG and TTA mean leucine, for example. CGA, CGC, CGG, and CGT *all* denote arginine.

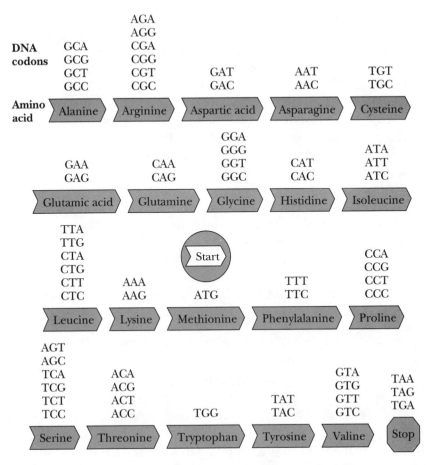

The 64 triplets and which of the 20 enzymes they code for.

The last lesson in "Conversational DNA" is punctuation. Everyone's genome is 3 to 3.5 billion letters long (we don't know exactly) and has between 70,000 and 150,000 genes. (Again, we don't know exactly.) Punctuation helps make sense of it. In English, a period signals STOP the sentence HERE. In DNA, there is a punctuation mark both for stopping and for starting a sentence. The triplet ATG means START HERE when reading this sentence; there are three STOP HERE triplets: TAA, TAG, and TGA.

The chapter (chromosome) is punctuated as well. Its center is marked by the centromere, and its ends are marked by the telomeres.

In the largest (and most impressive-sounding) sense, the human genome is the instruction book for a human being written in the lan-

guage of DNA, a giant blueprint for the complex engineering project that is the human body. In more down-to-earth terms, the instruction book the human genome most resembles is, in fact, a very,
very large book of cocktail recipes. Genes are an extensive string of
instructions for making and mixing together chemicals called proteins, which include both enzymes, the catalysts that direct reactions,
and structural substances, the biochemical equivalents of the steel
beams and cement and copper tubing with which the body is built.
(Enzymes change materials from one thing into another, acting like
little mobile refineries. They swallow the raw material and spit out the
refined product on the other end.)

DNA is a language used for only one purpose: giving orders to
construction workers, groups of molecules called ribosomes, on how
they should assemble enzymes and proteins. If Bechtel, the giant engineering conglomerate, made exclusively for its engineers a highly
technical, precise, specialized language with which to write out technical blueprints, it would be a language like DNA. Computer pro-
gramming languages, languages designed for one specific, stunningly
complex task, also fit this bill.

Genes are simply recipes, a trail of triplets, each of which codes
for a specific amino acid. Read the entire gene sentence, and you have
put together all the ingredients to make an enzyme.

This is how the assembly process works. The DNA text makes a
copy of itself, called RNA (ribonucleic acid), for the assembly-line robots to work with. (It's a very smart system: Transcribe a copy of the
DNA original and keep this original safely tucked inside the nucleus
while sending the copy out into the chaotic cytoplasmic field where
the heavy work is done. If the RNA copy is destroyed, the DNA original can always print out another one.)

The RNA then goes to very large molecules called ribosomes,
which are the assembly-line robots that carefully read these genetic
recipes giving them their instructions, lumber over to the "parts" sup-
pliers, select the amino acids they've been told to, and weld them together to build the enzymes specified by the RNA. The ribosome
takes the RNA copy of the gene and starts reading the letters "AC-
CTGATTC . . ." until it sees the punctuation triplet ATG, the signal
meaning START READING HERE. This is the beginning of the genetic recipe, the start of the gene.

It starts reading triplets.

The first triplet word is, say, CTT, which the ribosome knows
stands for leucine. It deftly grabs a leucine out of the cloud of amino

acids floating around the cytoplasm and holds it ready. The next triplet it reads says ACT, and so the ribosome picks out a threonine, which it connects to the leucine. The next triplet is CTG, another leucine, so the ribosome tacks it on to the end of the growing chain. It reads CCC and adds a proline. TGG is next in line, and the ribosome reaches for a tryptophan, a growing snake of connected amino acids trailing out behind it. The ribosome can continue for hundreds or thousands of triplets, reading, grabbing, attaching, building the amino acid necklace. When it comes to the words TAG, TGA, or TAA, the DNA punctuation that means STOP HERE, the ribosome stops. It's at the end of the gene, the end of the recipe. It fastens the last amino acid to the end, completing the enzyme it has just made, and snips it off, then turns its attention to the next gene—ATG, START HERE—and begins reading a different recipe, making a different enzyme, which will go out into the cell and do something else. By the time ribosomes have read all the genes in the book of the human genome, they have made a human being.

The protein that has just been made, cut loose from the ribosome, moves out into the cell to do its work: ordering more liver cells to grow, directing a lung construction site, plugging a leak in an arterial break. The proteins actin and tubulin build bones, keratin structures and strengthens hair. Proteins are elements of skin and heart and bone and blood, the raw materials that make up the body. They carry on the functions of life, regulating and directing and building.

And these are the elements of genetic grammar, enough to understand a basic conversation in DNA: Read the triplets in a gene, churn out the amino acids the triplets designate, snap them together into an enzyme, dump the enzyme into the cell where it will start work, and go on to the next gene.

There are, of course, complications. The genome has a phenomenal self-maintenance system. Certain genes are devoted entirely to basic construction and keeping the system going. These go about making RNA from DNA, building and maintaining ribosomes (large complexes of molecules built from ribosomal proteins and ribosomal RNA components), giving orders, and supervising the field. Boss genes give orders, "control sequences" direct the work of genes adjacent to them. There are police force genes, road crew genes, and social worker genes that keep mutations in line, suppressing or counteracting them and making enzymes that patrol tirelessly up and down the DNA strand, finding and correcting misspellings and bad punctuation.

And, of course, there are also millions of things we don't understand about DNA. DNA can be packaged in a variety of ways. Human beings have twenty-three pairs of chromosomes, and fruit flies have only four, but pigeons have forty and rhinoceroses have forty-two. Boa constrictors have eighteen, and the Indian fern, *Ophioglossum reticulatum,* has 630 chromosomes. Why the different structures? How do they affect the action of genes? We don't know. And not only has no one read the whole book that is the human genome, we haven't read *most* of it. We are still discovering quite basic things about DNA. For example, you probably remember that in high school you were taught that DNA is a double helix. We now know that DNA also can exist as a triple and quadruple helix. We don't know exactly why or how they work (as far as we know the triple version exists only in the lab, never in nature), but it is suspected that nature uses the quadruple helix in some way to seal the ends of chromosomes. We usually speak of DNA, but in fact there are B-DNA, A-DNA, C-DNA, Z-DNA, all different forms, all with different functions, some of which may remain unknown. Recently it has been discovered that DNA conducts electricity. Although at about one hundred times slower a conductor than copper wire, still it passes electrons along its twisting length. Significance? Unknown.

～

When scientists want to read DNA, they have to start out by freeing it from its prison inside the nucleus. One day, pushing a baby stroller full of vials of human blood out into a corridor at the National Institutes of Health, geneticist Dean Hamer sets off to extract pure DNA.

"There's DNA in every cell in your body," Hamer says, "but it's easier to get it out of certain cells than others." Actually, Hamer isn't pushing a baby stroller (although it looks like one) but a low slung metal lab cart on black rubber wheels. "White blood cells are one of the easiest to get DNA from. For one thing, blood is a piece of you that can be removed relatively easily."

He could, he explains, get DNA out of skin cells or liver cells. "But who wants to cut off a chunk of their skin?" He turns the cart sharply into his lab.

Hamer is looking for the actual gene or genes that create sexual orientation. As a molecular geneticist, he is interested in tracking down exactly which combination of As, Cs, Gs, and Ts on which chromosomes make us homosexual or heterosexual. At first glance his lab gives the impression of crystalline chaos. Shelves upon shelves tower up to very high ceilings, cliffs jammed with glass bottles and plastic containers of glycine and chunks of guanadine hydrochloride. Hamer works rapidly, transferring more vials of blood samples onto the cart. He is wearing latex gloves, and each vial is marked with the name and birth date of its donor. Vials loaded, he expertly wheels the cart around and out of the lab and cheerfully takes off at a brisk clip back up the hall.

"All blood is now handled as if it were pathogen-infected," Hamer says over his shoulder. (Shortly after this conversation, Hamer's lab stopped handling blood altogether. It is now processed at a separate location.) His white lab coat is flapping slightly behind him, and beneath it he is wearing jeans and black Reeboks. The cart and blood suddenly turn a corner, make a quick evasive maneuver around some computer equipment reposing in the hall, and pull into a small room containing rows of glass jars filled with a liquid that looks like vodka and a few contraptions of various sizes. Hamer starts unloading the samples.

Each vial is sealed with a lavender plastic cap. Inside, the blood no longer has a dark-scarlet color but looks instead like a black, evil syrup on top of which is floating a thin, viscous, faintly green pus-colored liquid. "It's congealed," explains Hamer. "The black is the red blood cells. By the way, the red cells have no DNA because they have no nuclei. They just carry oxygen. It's the white cells that have the DNA. The stuff on top is serum, it's just fluid, it carries your antibodies and hormones." In between the red cell syrup and the liquid, he points out a thin layer of powdery white, like a thin snowfall on black earth. "That's your white cells, T-cells, lymphocytes, macrophages. *Those* are the cells that contain DNA. We isolate DNA from the blood gemisch," says Hamer.

He begins the process of extracting genetic material by pouring two plastic vials from the same donor into a larger plastic tube and then adding some of the clear liquid. "These are buffers," he says of the row of bottles. "This is red cell buffer; it keeps the pH and salt concentration right." He adds a detergent, nonionic and relatively mild, which breaks open all the cells and frees the nuclei, inside which is the DNA. After shaking the plastic tube, there is a red froth on top. "Still some red blood cells in there," Hamer says, and slips the tube into one of the contraptions in the room, which turns out to be a centrifuge, and then takes two more vials and repeats the process. He peers at them with concern and sighs. "You're always worried about mixing up the samples," he mutters, and starts checking labels. As the centrifuge, with a high-pitched whine, spins his samples and forces the nucleus sediment to the bottom, he talks about the blood.

"We take five vials of blood from every subject. For every sample, the blood gets divided up into five tubes. Two of those tubes, I extract DNA from. Two I send to a cell-banking facility in Los Alamos, New Mexico, called Vivigen. They take out the white cells and transform them with the Epstein-Barr virus, which immortalizes them. After that, they'll live forever." He grins. "Sounds like a vampire movie. They grow these cells in culture, then extract DNA from that same donor and send it back to me. Then they freeze the immortal cells in a large freezer in Los Alamos so at any time they can grow more cells and extract more DNA, so we have a permanent repository of genetic material. If we discover something and want to go back, we don't have to take more blood. If someone in another lab wants to try to replicate our experiment, all we do is give them a key to the ID numbers at Vivigen, and they can repeat the experiments."

For about ten minutes, the centrifuge howls in the corner.

"The beauty of this is that it's so convenient and so easy to control precisely," Hamer says with satisfaction. "We have infinite samples."

After the vials are centrifuged, Hamer starts removing them. "The fifth vial I send to [geneticist] Jeremy Nathans at Johns Hopkins," he continues. "He's making a DNA bank of people whose sexual histories we know. Say, ten years from now we find the gene, or one of them, and we want to look at one thousand people whose histories we know. There they are."

Hamer takes a seat behind a biocontainment hood, which resembles the plastic shield people see at a salad bar that protects the salad from people. (The hood works on the same principle.) The glass shields his face and also protects the samples from contamination, "which," says a slightly muffled Hamer, who is peering into the hood, "in this case I could care less about. We're just looking at DNA." Under the glass hood, he is spritzing (his word) disintegrated blood from vial to vial. Unseen to the naked eye, inside the vials the outer casings of the cells are being broken apart and shed. But the microscopic nuclei, plumply packed with DNA, are still intact.

In a different buffer, Hamer suspends the intact nuclei, of which there are millions in this thimbleful of fluid. "It's not yet pure nuclei," says Hamer, squinting past the glass hood. "It still contains some other gunk." He is like an obsessive sauce chef working with a very delicate béarnaise. He adds SDS, sodium dodoxyl sulfate, an ionic detergent that bursts open the nuclei, causing them to jettison their innards, which are genes. He adds another detergent—this one frees the DNA from the proteins that usually bind it up and hold it together—and gently sloshes that around in the tube for a moment.

DNA is incredibly sticky and mucuslike, Hamer explains, and that's why the proteins bind it. "Otherwise it would stick all over everything and be a problem."

And all of a sudden, we are looking into a glass tube at pure, raw genetic material.

"Look here." He takes a micropipette and draws up a small amount of the white milky substance. It pulls out in a long, wet, glistening silver strand of bases. "That's DNA," he says. "Pure genes. The Plan." Hamer quickly becomes impatient with the momentousness. "Pure DNA snot," he says proudly, slinging the tube onto a counter and looking around for the tool with which he will extract it. "That's why we're so smart, because we have all this DNA." He rolls his eyes. "I'm joking, of course; there's toads that have more DNA than we do."

After the DNA is extracted, it is reseated in the cart, which he perambulates back down the hall, lab coat flying, on the way to the next stop, where he will read the ACGTs of the areas of the genome that interest him. He'll make copies of these pieces—each about three hundred bases long—then divide the copies up into four groups, each of which he'll spike with a radioactive substance. He'll then race the four groups against each other in four different lanes—a lane for As, one for Cs, one for Gs, one for Ts—in a pool of gelatin with an electrical stream flowing through it, called an electrophoresis gel. Nudged on by the electrical charge, the bases filter through the gelatin sieve, falling into the sequence in the four lanes of the gel that they occupy inside the cell. Then he takes a photograph of the gelatin called an autorad, which comes out on the same gray translucent sheets used for X rays.[1] Then just read up the lanes: ACAGGTTTCCGACTC, and so on.

At this point a scientist will have the DNA sequence, but that is only the start. This is just the order: a G, then two Ts, an A, and so on, and the question becomes: What does this enzyme recipe, written out in tiny chemical blobs, actually do? Answering that question means colliding with an almost universal popular misconception about genes, that genes are "for" things, "for" specific traits. They aren't. Genes are not "for" eye color. Genes are not "for" handedness or height, at least not in the way most of us think about them.

A few years ago *Science* magazine ran a simple one-page genetic chart. It was modestly titled "Map of the Human Y Chromosome," the chromosome that only men have and that contains the gene that turns the rudimentary fetal sex gland into a testis and makes men

men. That gene codes for a protein called Testis Determining Factor; since most genes are named with some combination of up to three or so letters and up to three numbers, its name is TDF. (In fruit flies, the gene whose job is to "transform" an enzyme in the sex pathway is named TRA, and the SDC genes—there are SDC 1 to 3—are named after what they control, *sex* and *dosage* *compensation*. The letter/number combination often has some recognizable link to its function, but sometimes not; some genes are named after their discoverers.)

A reader flipping through the magazine probably wouldn't have noticed what looks like just another chromosomal map. A closer look, however, would have revealed that the ever-serious, ever-sober *Science* had actually published . . . a joke. This map of genes on the male Y was, with the exception of the TDF gene, entirely fictional. It was also rather clearly not written by a man:

MAP OF THE HUMAN Y CHROMOSOME

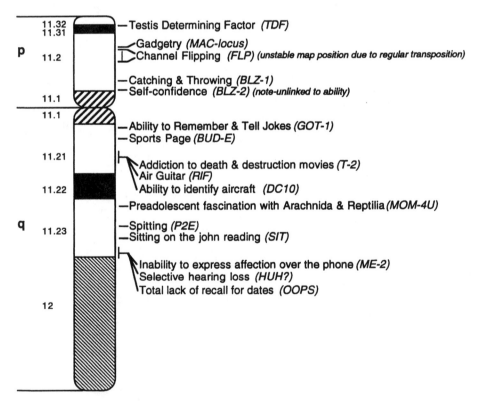

The author of the map[2] had, in fact, two targets. The first is obvious. The second more subtle and serious point concerns the notion that genes are "for" things.

Geneticists, forgetting that the general public has a less sure grasp on their field than they do, frequently use the word "for," talking about, for example, "a gene 'for' Alzheimer's." Journalists writing in *Newsweek* about "a gene for manic depression" only reinforce the wrong idea. Molecular geneticist Jan Witkowski runs the Banbury Center at Cold Spring Harbor Laboratory on Long Island. He is walking one afternoon on one of the vast grass lawns high above what appear to be New England whaling houses across the inlet but are actually genetics labs. On the other side of the bay, Connecticut is a distant line of dark green trees. Witkowski is speaking with his precise and enthusiastic English accent about various aspects of genetics.

"Every living organism has DNA," Witkowski begins, then adds, "except maybe for prions. Disease causing. Thought to be just proteins." He considers prions for an instant. "Anyway, all humans, certainly, have DNA. Our DNA is spelled out in enzyme-creating units known as genes, and we all have the same genes. But what does this mean?" Witkowski smiles. "Huntington's is a terrible and deadly genetic disease caused by a single gene that sits at the very tip of the short arm of your fourth chromosome. You have the gene that gives you Huntington's. So do I."

He pauses a moment after this surprising news, then explains: "Of course we don't all have the same *versions* of these genes. We all have exactly the same set of genes, since we're all human. But only a few of us have this particular gene in the version that gives you Huntington's, while most don't."

What Witkowski is describing is actually quite a simple concept that is contained in one word, a word used everywhere in genetics but—and this is the problem—one that is virtually unknown in popular science reporting: alleles. Alleles are simply different versions of a given gene. Where we talk popularly about "genes" today, in a few years we will be talking about alleles. They are a very handy concept.

Witkowski considers the problem for a moment, looking at the sky, then suggests that the best way to think about alleles is to envision genes as automobile parts.

"Look," he says, "every car has basically the same parts: tires, axles, carburetors, batteries, and so on. And the parts all do the same basic job in each car. A carburetor is a carburetor, no matter what car it's in: It controls the flow of gasoline and air into the engine. But all carburetors aren't identical. They don't all last as long or work with the same efficiency. A Mercedes carburetor is going to perform differently from a Ford or Toyota or Ferrari carburetor. Same part, only a different version, so it does its same job slightly differently.

"Genes are exactly the same way. We *all* have the same genes. (Well—except for the genes on the Y chromosome, which only men have, but we'll come back to that.) Basically, we're like cars: We all have the same parts. Your Gene 1 is the same part as mine, which does basically the same thing, your Gene 2 and mine are the same. But genes come in different versions. The version of Gene 2 that you have might be slightly different from mine. And those are alleles.

"Again, I'm *not* saying the different alleles of a gene all perform equally. They don't. To use the car parts metaphor, your brand of spark plugs, Motorcraft or Champion or whatever—your allele—does the same job as mine, but they might perform differently. You could say that each gene comes in a certain number of brands—alleles. Although they perform the same basic function, if your plugs work better, your car will go faster and run smoother." Witkowski smiles. "Maybe another of my alleles works better than yours, and so I'm better at something else than you are. Maybe at playing basketball. Or seeing at night. Or solving a math problem. Depending on which alleles you have, you'll be different from the next person. And if my allele of this particular gene is a good one and your allele malfunctions and gives you Huntington's, well . . . you're in bad shape."

Although alleles are never mentioned by that name in the press, they are reported on constantly.

"When you see a magazine talking about 'the cystic fibrosis gene,' " says Witkowski, "this does not mean that it's a gene that is there for the sole and express purpose of giving you cystic fibrosis, a gene that some unlucky people have and most don't. We've all got the exact same set of genes, and we've all got this gene—its name is CFTR. What they're talking about is an allele, one particular version of this gene (genes come in all sorts of allelic versions), and this gene in all of us sits on Chromosome 4." He paused for an instant, musing. "I believe the CFTR gene comes in at least five different alleles. I've got one of them. You might have a different one. And each version of the gene produces basically the same protein that does basically the same job.

"But some of these versions don't work. One in twenty-five people has a version—an allele—of the CFTR gene that produces a defective protein. The protein doesn't do the job as it should, or there isn't enough of it. In your car, if you have a bad spark plug, you get a malfunctioning engine with a very specific defect: ignition problems. In you," concludes Witkowski, "if your CFTR gene doesn't work, you get a disease called cystic fibrosis, which, because of the job the protein is supposed to do, produces a specific problem, respiratory problems that are generally fatal."

But we all have a genetic safety system to counter defective alleles. Nature has designed it so that each gene comes with a backup. Like airplanes, equipped with redundant controls—a second complete set to be used if the main set fails—genes come in pairs, and thus all have auxiliaries. If one fails, the other one can take over. Genes are exceptions to Murphy's law, which says that if something can go wrong, it will; if something can go right with genes, it will, and when there are bad mutations, they are usually recessives, weaker alleles that will allow the good backup to turn on and pick up the work. (Strictly speaking, it is not truly correct to describe one gene as the "main" gene and one as a "backup," since most of the time both copies of each gene are functioning, pumping out proteins. At the same time, if one *does* fail, the other does act as its backup, the one providing half the protein that would normally be produced by two. Sometimes this is quite sufficient, and the person carries on utterly oblivious to the fact that they only have one working copy of this gene. Sometimes it is insufficient, leading to illness or even death. It depends on the gene.)

Unfortunately, the backup system doesn't always work. Huntington's disease is actually a dominant, so even one copy of the bad allele produces the disease. The same goes for achondroplasia (a kind of dwarfism) and brachydactyly (very short fingers). Tay-Sachs, on the other hand, is caused by having two bad alleles, as are cystic fibrosis, phenylketonuria (PKU), and 4) albinism. (People are albino because an allele (version) of a certain gene makes a defective protein that fails to convert a chemical called tyrosine into the pigment melanin, which is what gives the body its color.)

Where the popular lack of understanding of the concept of alleles has been particularly troublesome is in the use of DNA fingerprinting in criminal cases. Which allele a person does or does not have can be crucial if the person is the suspect in, say, a murder. When the police find blood at the scene of a crime, lab technicians can amplify it, extract the DNA, and look for specific alleles. They compare the alleles of the genes in the blood evidence to the suspect's. If even four or five genes are clearly of different versions in the suspect and in the blood, this is, statistically, almost absolute proof that, at a minimum, the blood is not the suspect's. If the alleles *do* match, the suspect's guilt will have to be proven by other evidence, but if they don't—and if the police strongly suspect the blood came from the killer—this allelic evidence is virtual absolution.

As Terry McGuire, a geneticist at Rutgers, points out, the statistical frequency of alleles varies hugely between races and even among groups *within* races, a fact that poses a big problem for police, courts, and juries when using DNA evidence. For example, among Cajun French there are DNA sequences that differ dramatically from those of French Canadians, so the data from one group must be carefully evaluated for average allelic frequency. "This can become pretty damn relevant," notes McGuire, "when you're talking about some guy who's a rape suspect. They check his DNA and say, 'Jeez, there's a one-in-two chance it's him by his genes.' And then five years after they've sent him to jail, they realize that one-in-two is the allelic frequency if you're ethnically Mediterranean or French—but he's German or Asian, a totally different frequency of alleles, on average. And all of a sudden—whoops!—you realize the odds are only one in 1,570 that it was him. So anyone doing DNA fingerprinting had better know their allelic frequencies."[3]

It is also alleles that we talk about—and photograph for the covers of movie magazines and envy and coo over and manufacture small disks of tinted polyurethane to alter—when we refer to someone's baby-blue eyes. Everyone is familiar with the concept of recessives and dominants from studying eye color in high school biology, but few people know that eye colors themselves—blue, green, brown, and so on—are reflections of alleles, different versions of the group of genes that create this trait. (Geneticists don't know how many genes are responsible for eye color nor where they are; the question simply hasn't been researched yet.) A gene pair can consist of any possible combination of two alleles. Some genes come in only one or two allelic versions. Others come in hundreds of different flavors.

Alleles that are dominant, such as those that create brown eyes, and alleles that are recessive, such as those creating gray, are, McGuire points out, always defined in terms of their relation to other alleles. "An allele can be dominant in one case and recessive in another," he says. "As far as we can tell, in people's hair color there are probably four alleles: black is dominant to brown, brown is dominant to blond, and blond—which people are used to thinking of as a recessive—is dominant to red. This is the way it works in guinea pigs, more or less. So two things: Blond can be a dominant or a recessive, and it depends on which other allele it's paired with. And if you're a redhead, we know for certain that both of the alleles in your gene that determines hair color are red; if you have brown hair, you've definitely got one brown allele, but we have no idea what the other is. It could be red or blond. In any case, it doesn't matter. It's the brown one, the dominant one, that's going to come through. If you have black hair you could be carrying God knows what." This is, McGuire added, why red hair can show up in anyone's family: Both parents pass the red allele they're secretly carrying on to their child. "Dads who look at Moms a little strangely when the redhead is born and *they* both have jet-black hair—well," he says with a laugh, "you can't tell from looking at the outside when you're dealing with alleles."

Scientists believe there is one major gene that determines hair color, and it works with some modifier genes, but as with eye color—where black is dominant to brown, which is dominant to blue, which is dominant to green, which is dominant to gray, the most recessive in the possible pairings—we don't know where they are found yet.

If allelic differences make us different people externally, they are also medically portentous. Two very important genes named HLA-A and HLA-B determine whether a patient's body will accept a donated organ. These two genes each come in one of several allelic versions, and the alleles in the liver or kidneys that are being donated must match the patient's alleles. If the patient's body identifies them as strangers, it will reject the organ as "foreign," a sort of violent genetic anti-immigrant backlash, and the organ recipient may die. (Incidentally, no one really knows why nature developed these HLA genes. We think now that they aid resistance to certain diseases, but we're not sure. They could also be cancer killers, recognizing cancer cells as foreigners in the body.)

How are new allelic versions created? Alleles are, basically, mutations. They are generated, in essence, when something goes wrong.

Mutation is a fascinating and sometimes terrifying subject. A genetic mutation is simply a change in the ACGT spelling of a gene. Thousands of factors called mutagens can change DNA's spelling, and these can be as common as chemicals in food: corned beef and other preserved lunch meats contain sodium nitrate, which in test tubes is converted into mutagenic compounds such as nitroso amines, which in turn can change a C into a T, although there is fierce debate over whether they actually do so when eaten. One mutagen whose effects we *are* certain of is cigarettes, which contain a huge number of known mutagens that do everything to DNA from misspelling genes to sabotaging enzyme production to reprogramming cells and destroying tissue. Another is radiation, meaning anything from the thermonuclear warmth of a sunny day to X rays and gamma rays—also present in common sunlight—filled with ions that crash into tissue like a deadly shower of meteors.

The basic, standard spelling of a gene—the one that exists in the wild—is called, logically, wild-type, and a mutation can change a wild-type to something strange and new ("forward mutation") or *from* something strange and new back to the standard wild-type spelling ("reverse mutation"). Mutations can occur both in somatic cells, the cells that make up the parts of the body (*soma* is Greek for "body"), or in germinal or sex cells, the ones that become sperm and eggs.

Constant mutation in our genes means that new genes are constantly being created and added to our total genetic material. But the old genes are, over millions of years of evolution, continuously retired from active service. They don't go anywhere, however; they simply go to sleep. This is partly why only 3% of our DNA actually makes proteins and the other 97% is "junk," making as far as we can tell nothing at all, genes that simply lie in our helices, dormant, although the expression "junk DNA" is now considered politically incorrect in many genetics circles. This non-coding DNA contains, we speculate, rather amazing genes called atavistic mutations, ancient codes for primeval traits we carried over 25 million years ago, even before we were homonid apes. These deeply buried genes re-create complete coats of fur in humans—members of one large Mexican family had everywhere except their palms and the soles of their feet a thick, fine coat of hair attributed to these genes—and, in some whales, the vestigial hind legs their landed ancestors used; the limbs were discarded over forty million years ago, but they remain coded silently in the genes.

Atavisms are driven by "phylogenetic inertia," and can also be traits that are occasionally harmful to individuals but that evolution hasn't gotten around to eliminating, probably because they have little overall effect on reproduction. In human beings these include the appendix, an organ that does little of anything, and the structure of the lower human back, which is somewhat unsuited to walking upright. We still have the vestige of our tails (the coccyx) and, for men, those purely ornamental nipples. Atavisms frequently appear in many human embryos, vanishing by the time we emerge from the womb, although sometimes they remain, as with extra sets of nipples arranged, dog-like, down a human chest. This is my case; I have an atavistic nipple from some gene millions and millions of years old.

In her dark, mesmerizing novel *Geek Love,* Katherine Dunn tells the story of a carnival family led by Al and Lil Binewski of the traveling Binewski Fabulon who, as a cost-saving measure, hit upon the idea of breeding their own freak show of dwarfish, albino, web-footed children. Dunn, in a fascinating literary passage combining biology and art, harnesses genetic mutagens into the service of her plot:

> The resourceful pair began experimenting with illicit and prescription drugs, insecticides, and eventually radioisotopes. My mother developed a complex dependency on various drugs during this process, but she didn't mind. . . . Their first-born was my brother Arturo, usually known as Aqua Boy. His hands and feet were in the form of flippers that sprouted directly from his torso without intervening arms or legs. . . . I was born three years after. . . . My father spared no expense in these experiments. My mother had been liberally dosed with cocaine, amphetamines, and arsenic during her ovulation and throughout her pregnancy with me. It was a disappointment when I emerged with such commonplace deformities. My albinism is the regular pink-eyed variety and my hump, though pronounced, is not remarkable in size or shape as humps go. . . . The drawfism, which was very apparent by my third birthday, came as a pleasant surprise to the patient pair.

The substances Dunn describes may, alternatively, have been not mutagens—from the Greek, "makes changes"—but teratogens—"makes monsters," denoting a factor causing malformation of the embryo. Thalidomide is the best-known example of a monster-maker.

Not all mutations produce impairment. They can create statuesque bodies and superstrong muscles as well as grotesque physical deformities and death. There are morphological mutations that change the shape of the body; lethal mutations that kill; conditional mutations, (which act only under certain conditions but not others); biochemical mutations, which cause the loss of or a change in some biochemical function of the cell (in many cases, these can be repaired by taking some specific nutrient); and resistant mutations, a sort of monster mutation that permits the mutant cell to grow and multiply wildly despite inhibitors that are supposed to turn it off.

The actual changes in mutated genes vary. There are chromosomal mutations, where huge chunks of chromosome swap places with each other, and then there are mutations so subtle that hypersensitive biochemical techniques must be used to detect any difference between them and the wild-type. Point mutations, changes at a single point in a gene, can mean simply changing a C to a T at one point in some gene. (Sickle cell anemia is caused by a change in one specific base in the globin gene.) A point mutation looks like this. Start with the genetic code of triplet words, each word coding for one amino acid:

THE FAT CAT ATE THE BIG RAT

Mutate a single letter, one point, and the result is:

THE FAT CTT ATE THE BIG RAT

Delete just one letter, and you get:

THE ATC ATA TET HEB IGR AT

Imagine the strange enzyme the ribosome makes when it tries to read this. We see the results: dwarfism, mental retardation, albinism, cancer, death. But the fact is, in the end, despite all the problems they cause, mutation is rare. The tremendous stability and constancy of species from generation to generation shows this. The great irony is that we need mutation to evolve. Evolution—adaptation to the environment and the development of all sorts of mental and physical faculties—is caused by the natural selection of genetic mutations that allow us to endure and thrive.

∽

Genes are the place that Bill Byne's stern warning against extrapolating from other species to human beings begins to break down.

Jasper Rine, a molecular geneticist at UC Berkeley, is fascinated by the similarities between genes in all forms of life or, as he expresses it, sitting in his office in the Berkeley genetics department, by "so much demonstration of underlying unity in biology and biochemistry." Consider, for all the differences between us *Homo sapiens* and *Saccharomyces cerevisiae,* how deeply similar we are. Consider, despite how we look down on them, that many of their genes are so similar to ours that we could remove theirs from their measly sixteen chromosomes and plug them directly into our twenty-three with virtually no loss of genetic function, no difference between the enzymes they make and the ones we make. Organ donations sometimes don't even work between family members, and yet, amazingly, these lowly creatures could lend us many of their genes wholesale. As sophisticated and intelligent as we are, as simple and modest as they appear, many of our genes are exactly the same as those of . . . yeast.

"From yeast to humans," notes Rine, "genes are interchangeable in their biochemical pathways. Yeast and thousands of other species from mosquitoes to penguins to armadillos have genes that are 'homologues,' interchangeable parts in you and me. Example? The RAS oncogene in yeast—and elephants and pandas and whatever else you want—you can take and stick it in a human and it will function pretty normally." Rine proffers the gene that makes a protein called tubulin, which is used in the cytoskeleton of cells. Yeast tubulin and human tubulin are 90 percent identical, and the yeast and human tubulin genes—or the tubulin genes of a rhinoceros and a muskrat—are interchangeable, which, given a billion years of evolutionary distance, is remarkable.

The similarity between yeast's genes and ours allows us to use yeast as test sites for drugs meant for human beings. You'd think this would be impossible, says Rine. "How could you give a drug to an organism like that and have it work on us?" In fact, there is a gene that makes the protein HMG Coereductase, the major component in one of the most famous drugs in the world, Mevacor. The drug lowers hypercholesterol levels by slowing down one of the steps on cholesterol's biochemical assembly line. Merck produces this drug (its generic name is lovastatin), which is worth around one billion dollars a year in retail value. Mevacor works in yeast just as it does in humans. Extract the human gene and put it in, and it works; in humans it lowers cholesterol, and despite the fact that yeast makes a slightly different steroid called argosterol, it lowers that as well.[4]

This genetic interchangeability between species is fundamental to the current focus of Rine's professional life. He is working on a project rather different from any other. A geneticist, he is content to leave the Human Genome Project, the great effort to map all the genes in the human genome, to others. "It so happens," Rine says, his voice becoming even more animated, "that I am working on the *most* interesting genetic problem in the biosphere. Now, any geneticist will lead with that statement. But in *my* case, it's true. This is how it works."

The major lessons in biology, says Rine, settling into his chair, have come to us from two sources. One was Charles Darwin (1809–1882), who taught us that gene selection, either in the wild *or* human-directed (animal breeding), builds fantastically different creatures. "Better to fit the niche in the ecosystem," Rine says, "or better to respond to some breeder's desire for a particular trait. So Darwin taught us how to make creatures different." The other was Gregor Mendel (1822–1884), an Augustinian monk in Austria, who taught us that we could find the genes that determined these different traits by looking at how these traits segregate down through generations. Watch the traits filtering down from parent to child, and that pattern gives you clues to where the genes are and how they are distributed in this cascade of heredity.

"If you wanted to do a really interesting experiment," Rine adds, "what you'd want to do is introduce Mr. Darwin to Brother Mendel, who lived at the same time but never met. Those of us working on this project are using our twentieth-century technology to make that introduction." For example, if you wanted to study the genes that are responsible for the length of the neck, ideally you'd cross a giraffe with a warthog—Darwin's contribution—and then watch the offspring, and their offspring, and *their* offspring after that, to see how the trait neck length segregates down through the generations—Mendel's contribution. And this would be a very powerful way of tracking down the genes that evolution has acted on to control the length of the neck. The problem, of course, is that this is impossible. Different species can't interbreed. Actually, that's the definition of a species, that members of different species can't breed with each other. The sperm of one species won't penetrate the egg of another. So we can't do our potentially Really Interesting Experiment.

Rine shrugs darkly, then brightens. "So what do you do?" he asks, leaning forward, eyebrows raised. "What do you do?"

What you do, if you're Jasper Rine, is come up with another way to present Darwin to Mendel: You come up with the Dog Genome Project.

"Dogs," Rine explains enthusiastically, "allow us to make this introduction. For 150 years breeders have been doing what natural selection has been doing, but much more quickly. They've been purposely cross-breeding to get dogs that either *look* the way they want or—and this is the part that everyone fights about when it comes to humans—*behave* the way they want. That each breed has a characteristic body, coat, and size indicates that these traits come from genes. That's completely uncontroversial. That each breed of dog has a characteristic behavior, however, says that behavior too is rooted in genes.

And this is where people start to get uncomfortable." Take a Chihuahua, says Rine, and raise it in Mexico City or in Shanghai, two completely different environments, and it still behaves like a Chihuahua. Take a Tibetan Lhasa, raise it in Chicago and it is still a Lhasa, *not* just in the way it looks but—"and this is the point," says Rine—in the way it behaves, which says that these behaviors are not influenced by environment. "So our idea is to build a genetic map of dogs who, through Darwin's selection, have become different breeds. Once we have that map, we cross the dogs according to Mendel, and tease out the genes that control the physical *and* behavioral differences for the different breeds: the Dog Genome Project."

Most of us think of behavioral genetics as the province of human-subject research, but Rine is on the front lines of the pursuit of genes that control behavior. "We want to find the genes that make border collies herders—a stable, consistent, characteristic set of behavioral traits in this breed that makes them perfect for performing the herding task. What makes Newfoundlands behave generation after generation like swimmers and rescuers? In the days of sailing, ships used to carry Newfoundlands to rescue sailors who were swept off the deck in stormy seas. You ask why you don't see stories of dramatic sea rescues by Chihuahuas or German shepherds. With Chihuahuas, you could argue it's a matter of size, given that they'd be better for rescuing ladies' handbags than the ladies themselves. But German shepherds are, if anything, larger than Newfoundlands; what distinguishes them is their *behavior*. And it's genetic."

Rine is looking for the genes that make pointers point. "The approach to hunting is a behavior which is bred in the genes." Salukis and greyhounds and so forth maintain contact with their prey by sight, he points out—basset hounds and beagles primarily by scent. And basenjis, the barkless African dogs, can bark, but they don't. When they do vocalize, instead of a sharp barking sound they make a sort of eerie, diffuse wailing, the argument being that it's harder to figure out where they're coming from, which protected them against the large cats in the African plain. So they've selected for this different behavior.

"One of the things that is clear," says Rine, "is that breeders can select in a very short number of generations for substantially recognizable behavioral traits. Like courage. Sheepdog breeders, for example, can breed dogs that have more or less courage. Operationally, courage is the willingness of the dog to work in close next to very large, dangerous rams, and what we'd call courage is in the genes. Mothering behavior is clearly genetic, and you can breed for it. Loyalty is under genetic control in dogs. The dogs people love as family pets, like golden retrievers, have almost no genes for loyalty—they'll happily play with anyone. Fiercely loyal dogs like bulldogs are poor family pets in general because they develop strong attachments to one person in a family and everybody else is the outsider. And by sorting out the genetics of dog behavior and personality differences, we will gain a lot of insight into how genes control both."

Apply the idea that genes control behavior to humans, however, and chaos ensues. Rine has spent a good amount of time pondering the reaction. "I can't figure it out," he says. "Suggest that our behaviors are influenced by our genes, and people get whacko. Yet show them two breeds of dog, say this breed behaves like this and that breed behaves like that, and they sort of look at you like you're an idiot and say, 'Well, I *know* that, my cousin has one of those.' "

He sighs. "It makes no sense at all. Genes control behaviors. I'm sorry, but they do. I think the problem may be that the observation is so obvious that people take it to mean more than it does. Giraffes behave like giraffes and not like horses because they have the genes for being giraffes. They didn't read a manual on 'How to Behave Like a Giraffe.' " Dog behavior—and hyena and shark and ant behavior—is as particular to its species as is the look and build of its body. It was this observation that led the famous naturalist E. O. Wilson to formulate his theories of sociobiology, the science of how biology creates behavior. Wilson is an ant expert who realized that the intense social ability of these creatures must be wired in, evolved, and entirely genetically determined. Their tiny brains can't be expected to learn much, yet ant behavior is amazingly complex. If social behaviors can be evolved in ants, termites, bees, and dogs, why should the same not be true of humans?

But Rine faces a methodological problem: how to separate the genetic from the nongenetic influences on behavior. "Say there's a species of animal—like a sea turtle—that is always born from an egg that is abandoned. Its parents go off to other pursuits and the turtle never knows them. And yet the orphaned turtle grows up to behave just like them. How is this? Clearly it didn't get its behavioral cues from a note taped to the refrigerator. Its behavior, which is identical to that of its parents, has to be in the genes. So clearly much of this particular animal's behavior is genetic. But this is a sea turtle. What about when you get to the social animals, to species with more complicated social structures? For instance, what about our giraffe?

"Our giraffe is raised by his mom, who could be whispering in his ear as he's sleeping on how to behave like a giraffe. What do you do?

You raise him in the absence of Mom. In the case of monkeys, some famous experiments were done by the Harlows, who raised infant monkeys without contact with their parents. The effect was clear: The monkeys suffered tremendous behavioral deficits from the absence of parents, so undoubtedly many primate behaviors are influenced by socialization. And yet these monkeys were still clearly monkeys, behaviorally. They hung by their hands in their cages, they treated each other in monkey ways, they had monkey reactions. Not cat reactions or even ape reactions. Monkey reactions."

Rine notes a similar experiment involving dogs done at Jackson Labs in the 1960s by Scott and Fuller. The researchers took newborn puppies from six different breeds and nursed them on mothers of very different breeds. In adulthood, they ran a series of nine quantitative behavior tests—ability to run mazes, to solve problems, to respond aggressively or passively to being on their backs (all of which were different from responses to commands that would involve training, such as behavior on a leash). They then asked how much horizontal transmission of behavior there was from being raised by a mother of a different breed, entirely out of contact with their own.

"And they found none," says Rine. "Zero." The dogs were socialized, happy, healthy, nursed and cared for by their adoptive mothers, but their behaviors in the nine tests showed no influence of learning from their adopted breed but a great deal of influence of learning from their genes. "What that says is that much of the behavioral differences in dogs are innate. They don't pick up the behavioral characteristics of the other breed. Many of these cases, even highly social pack animals like dogs, which you'd expect to pick up some knowledge from their culture, are really picking up very little of their behaviors from their culture.

"So now," says Rine, "let's go back to gene homologues." The existence of gene homologues, such as the HMG coereductase gene that's pretty much identical in yeast and in humans, makes it a molecular given that when we map those genes that create behaviors in dogs, humans will have the human form of the exact same genes. What is not at all certain is whether these genes will play the same role in humans. Rine spreads his hands. "But when we find the genes, we will begin answering that question. And only time will tell." He adds pointedly, "But time *will* tell. Even if the behavior genes in dogs have different functions in us, what we learn about the ways genes create behavior in dogs we'll be able to extrapolate to the genetics of creating behavior in human beings."

But this fact meets head-on the widespread fear that knowledge of behavioral genes will somehow choke off human free will. Rine finds this silly. If one day, for instance, we find the genes for right- and left-handedness, will that limit our ability to choose the handed- ness we want? "That would," Rine says tightly, "be the logical impli- cation." He sighs. "People don't think clearly about this. It's wrapped up in so much emotion it leads to sloppy logic. People want to have free will, and they equate—I assume because of a panic that overrides logic—the presence of a gene as a factor in a given behavior as the loss of free will in *all* behaviors.

"If you have a Y chromosome with a Testis Determining Factor gene on it that turns your gonad into a testis, you do not have free will to be male or female. You're male. You don't have a choice. Do we know where the gene is? We sure do. Before we knew where the gene was, did you have a choice to be male or female? Nope. So find- ing the gene has nothing to do with having free will. If you don't have the gene for being right-handed, you still can't choose to be naturally left-handed.

"I happen to have genes that make it easy for me to gain weight," continues Rine, "and although I can choose to eat moderately and stay thinner, or I can choose to eat a lot and get fat, I can't, no mat- ter how much I might want to, choose to eat a lot and stay thin. My genes limit my choices. But among the choices I have, I have free will. I can't choose to be Asian or not, but," he says testily, "I just *really* find it hard to get too worked up about the fact that my genes took away my choice to be Asian." He considers, then adds, "People get upset at talk about genetic differences between races. Of course there are genetic differences between races! That's the very definition of race."

Instincts are genetic. When we hear a sudden loud noise behind us, we jump. There is no choice in that reaction: It's genetic. We get hungry—which is genetic—and thirsty and tired and sleepy, all of which are genetically programmed internal feelings and desires that cannot be chosen, although what to do about them can be. "Sex," says Rine bluntly, "we don't even need to discuss." An aggressive con- frontation meets with an emotional reaction, and the rise in heart rate, the flow of adrenaline, the anger and fear, are biochemical changes created by genes. Yet unless there is an overpowering stim- ulus that causes us to react immediately, we can fight off our initial reactions. We can, notes Rine, hit the person, or run and hide, or choose a middle course of reaction. We can train ourselves to react one way or another.

"And that's free will," Rine says. "Which is, incidentally, genetic.

Where did you think we got free will?" He smiles. People want something mystical and spiritual in their lives, he notes, and then they think that if genes create free will or the capacity for a spiritual sense of life, then free will is somehow "less free" than it would be if it were created by something else. But genes and free will are not mutually exclusive. Remember, he points out, conscious of the irony, as a human being you have free will, but you don't have free will to decide whether you have free will or not. "And if a person has a homosexual orientation, they are sexually and emotionally attracted to people of the same sex. But that's all it means. Whether they act on their desire is a matter of free will." He pauses. "And their hormones. Got to remember those."

Rine looks out the window at the Berkeley campus. "The word 'genes' does not equal destiny. Genes equals options. Maybe, in the end, the reason people get freaked out is because they take obvious facts to mean things they don't."

⁓

Genetics—as a labor-intensive, painstaking art practiced in a laboratory—is an exercise in chemical sleuthing, a discovery process, and as in all such sleuthing, it is clues that guide the gene hunters. There are all sorts of clues and tricks of the trade, oddities to notice, such as biological disasters like CAH, the masculinization of female genitalia that constitutes a seam of hormonal information waiting to be mined. In biology, aberrations are invariably information; any deviation from the norm must have a cause, and the cause, if its secret is cracked, will reveal a hormonal function, a molecular mechanism, a neurological effect. But "aberration" doesn't always mean "disease." Men are one of these aberrations. There is a one little corner of men's genomes that holds an exquisitely clear look into the spectacular effects of a single gene, and this corner is the male sex chromosome.

The sex chromosomes, the twenty-third pair, are unique because in men they're mismatched. They are the only place a genome structurally differs from that of someone of the opposite sex. Where women's sex chromosomes are XX, always giving them a gene *and* a backup gene in nature's usual safety-conscious way, men have only one X—there is no backup on this one chromosome, no safety net, no built-in redundancy—and one Y.

What this means to the bottom line of the genetic manufacturing of proteins is simply, and harshly, that in women, if a gene on the X chromosome breaks down, its backup will always kick in. And so it is impossible to tell just from looking at her that a woman has a defective allele. Her backup masks it and compensates for it.[5] But if in a man the malfunctioning allele is on the X or Y—too bad. On their X, men get only one shot for each gene. What this means for scientists like Witkowski toiling in genetics labs is simple: Whatever genes a man has on his X or Y are what he shows, *including* recessives. Where women have a backup system that masks the blemishes, the cracks, the problems, men are exposed. What you see is what you get. Anytime a man has a defective or mutated allele on his X, he shows it. A man's X is thus a goldmine of genetic information.

Y—"the male chromosome"—is a tiny chromosome that seems so far to be a somewhat barren genetic landscape containing the gene that makes an embryo male, some other indistinct genetic scrub brush, and very little else. (This is something of an exaggeration, but not a great one; the Y in addition boasts a few genes directing sperm production and a gene that makes men's ears hairy when they reach middle age.) But X is a full-fledged chromosome carrying a great number of genes responsible for a wide range of traits, and in men, every one of these genes—all the mistakes and all the recessives—shows up. It is for this reason we were able to track down the gene allele causing hemophilia. Virtually all hemophiliacs are male, which was more than a clue; it was a neon sign indicating the gene was on the X, just as it proved to be. If you glance at a chart of all the disease genes that have been identified on the average autosome (all the chromosomes *except* the sex chromosome are called autosomes), you will find that the extent of our knowledge is still somewhat limited. Here, for example, are the disease genes that have been located on Chromosome 18:

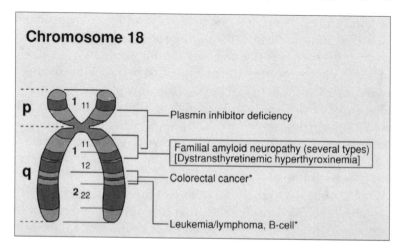

Chromosome 18

p

1 11

q

1 11

12

2 22

Plasmin inhibitor deficiency

Familial amyloid neuropathy (several types)
[Dystransthyretinemic hyperthyroxinemia]

Colorectal cancer*

Leukemia/lymphoma, B-cell*

A map of the Y chromosome, incidentally, is even more barren. Y currently looks like this:

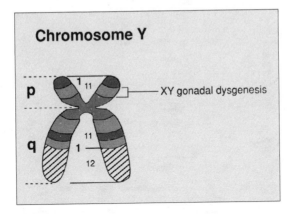

But our knowledge of X is approximately the same size as Chromosome 18. See the illustration on the facing page.

Duchenne muscular dystrophy was found on the X, and Fragile-X syndrome, and Kallmann syndrome, and Charcot-Marie Tooth disease. In fact, it was not until 1968, Jan Witkowski points out, that the first gene was mapped to an autosome, a nonsex chromosome. (He reflected a moment, smiled, and sighed. "I took my BSC in 1968 at the University of Southampton in the UK. It's absolutely amazing to look back on it from what we know today.") Our relatively vast knowledge of the X is due overwhelmingly to research in men, not women, simply because men are genetically simpler for studying the X. Missing their backups in this one spot, men have a genetic window of vulnerability through which geneticists can peer.[6]

Identifying abberations is not the only strategy in the hunt for genes. There is also logic. "If you know the biochemistry of what's happening in a particular disorder," Witkowski points out, "that gives you a clue as to where to start, a candidate gene as it's called. Sickle cell anemia, which is a disorder of the blood, logically leads you to look at the hemoglobin gene because it carries oxygen in red blood cells." Statistics are also a standard weapon in the geneticist's arsenal. "Huntington's," Witkowski explains, "a very rare genetic disease, is caused by a dominant. All a person needs is one copy, and he or she's got it. But statistically, there's a stark picture of carriers and noncarriers. If one man has it, take someone else at random from the population and you've got a one in 10,000 chance of his having Huntington's;

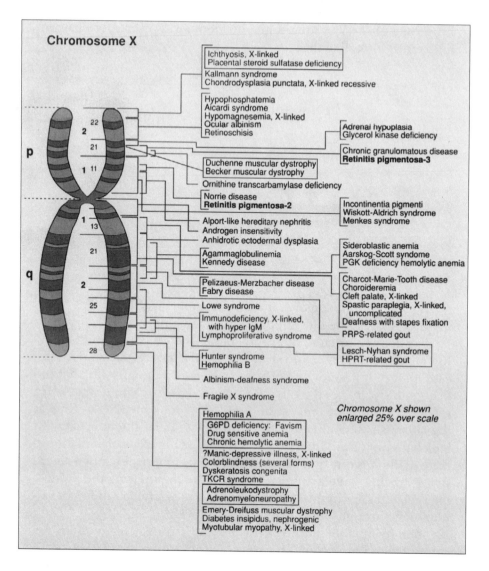

Chromosome X

Ichthyosis, X-linked
Placental steroid sulfatase deficiency
Kallmann syndrome
Chondrodysplasia punctata, X-linked recessive

Hypophosphatemia
Aicardi syndrome
Hypomagnesemia, X-linked
Ocular albinism
Retinoschisis

Adrenal hypoplasia
Glycerol kinase deficiency
Chronic granulomatous disease
Retinitis pigmentosa-3

Duchenne muscular dystrophy
Becker muscular dystrophy
Ornithine transcarbamylase deficiency

Norrie disease
Retinitis pigmentosa-2

Incontinentia pigmenti
Wiskott-Aldrich syndrome
Menkes syndrome

Alport-like hereditary nephritis
Androgen insensitivity
Anhidrotic ectodermal dysplasia
Agammaglobulinemia
Kennedy disease

Sideroblastic anemia
Aarskog-Scott syndome
PGK deficiency hemolytic anemia

Pelizaeus-Merzbacher disease
Fabry disease
Lowe syndrome

Charcot-Marie-Tooth disease
Choroideremia
Cleft palate, X-linked
Spastic paraplegia, X-linked,
   uncomplicated
Deafness with stapes fixation

Immunodeficiency, X-linked,
   with hyper IgM
Lymphoproliferative syndrome

PRPS-related gout

Hunter syndrome
Hemophilia B

Lesch-Nyhan syndrome
HPRT-related gout

Albinism-deafness syndrome

Fragile X syndrome

*Chromosome X shown
enlarged 25% over scale*

Hemophilia A
G6PD deficiency: Favism
Drug sensitive anemia
Chronic hemolytic anemia
?Manic-depressive illness, X-linked
Colorblindness (several forms)
Dyskeratosis congenita
TKCR syndrome
Adrenoleukodystrophy
Adrenomyeloneuropathy
Emery-Dreifuss muscular dystrophy
Diabetes insipidus, nephrogenic
Myotubular myopathy, X-linked

take his brother, it's one in two, a 50 percent chance. A staggering difference."

Alzheimer's researcher Alan Roses, discoverer of a link between APO-E4, an allele of the APO-E gene, and Alzheimer's, took sets of people with the disease and looked to see what version of APO-E they were carrying. In an average population, 16 percent of the people have APO-E4. But Roses found that in people with Alzheimer's, the percentage was 52 percent. This method is called shared-trait analysis. A researcher takes people who share the trait (like Alzheimer's) and looks for the statistical incidence of some allele, which he or she then compares to its incidence among people who don't have the trait. "Roses found two dramatically different numbers, 16 percent and 52 percent," notes Witkowski. "When you look at those, immediately the light bulb goes off: something is up here."

We rarely really know what we have hidden away in our genes, Witkowski stresses. Our genomes are packed with untold stories, dormant alleles shielded from view, and masked potential traits that nature carefully overrides or suspends or bypasses. "Something like 5 percent of all people carry the cystic fibrosis gene, one in twenty people," Witkowski notes. He gives me a level look, up and down, and says, "You don't have cystic fibrosis, but you absolutely could be a carrier of one CF copy. Without looking in your genes, we can't say. If carriers of one copy are one in twenty, the number of people who carry two copies and actually express the trait is only one in two thousand. If there is a homosexuality gene, it is not a dominant—and we know it's not because of its inheritance pattern—but it could be a recessive, and thus there could easily be tens of millions of people who are carriers. If you're heterosexual and your husband or wife and your children are heterosexual, that doesn't necessarily mean a thing genetically. If there's a gay gene, one or all of you could be carrying it. Your next child, or perhaps your grandchildren, may by chance get the genetic combination, and suddenly you've got someone gay in your family."

We have found recessives like cystic fibrosis, tracked down all sorts of elusive necklaces of bases that create a variety of traits. What if there is a gay gene? How would you find it?

Dr. Dean Hamer may already have found it.

Chapter Seven

THE GAY GENE: THE DISCOVERY
OF XQ28

A SHORT WALK from the Bethesda-Medical station of Washington, D.C.'s sleek Metro system and in the direction of the hum of traffic on Old Georgetown Road stands Building 37 of the National Institutes of Health, a large glass-and-concrete box that houses some of the labs of the National Cancer Institute (NCI to its familiars). The NIH itself is a gigantic complex that sprawls across acres of gently rolling grassy hills just outside the capital, and it has the look and somewhat removed, contemplative feel of a large university campus. It is a world of pure science. In Building 37, researchers gleefully tape Far Side comics about mad scientists to the doors of their labs, rewriting the balloons with references to their own experiments. Bright advertisements in neon pink and green are taped to the walls, reinforcing the college dormitory feel, although rather than offering word processing for English literature papers or off-campus housing shares, these ads read *"DNA SYNTHESIS—*new low prices! $2.33 per base. OPC PURIFICATION: $25.00 per oligo. Free delivery to your bench! *Overnight* econoligo™!" Each price is printed with a little smiley face next to it.

On the cinder-block walls, painted a soft mauve, hang signs of personal lives, photos of someone's soapy kids in a bathtub and of a recent trip to Peru, a reproduction of a Hopper painting of storefronts at late evening, and a large *Jurassic Park* poster featuring a rampant T. Rex taped to a door marked DEVELOPMENTAL GENETICS. The little black nameplates outside each scientist's lab seem to feature every conceivable race and nation. On one door is a chart in Korean, on another a poster of the Taj Mahal, jokes in German and Italian and scientific philosophy in Chinese, and on another a sign that says PARKING FOR ARGENTINOS ONLY.

Down the hall, four huge, ugly metal tanks of liquid nitrogen are cordoned off with DANGER! in neon orange plastered on them. Small diamonds color-code doors to warn firefighters what chemicals are inside: black for flammable, dark green for poisons and toxins, orange for radioactivity. Next to the ubiquitous biochemical waste instructions plastered everywhere and the HAZARD! NUCLEAR WASTE! sign someone has Scotch-taped a *Wall Street Journal* cartoon of a rat warning its child "And stay away from scientists! They cause cancer." A plastic NIH sign identifies the corridor as NIH Cancer Research. Below is a cartoon, in unmistakable *New Yorker* style, of a rat lounging in a cloud of cigarette smoke saying amiably to the rat in the cage above him "Good news, I'm down to eighteen packs a day." In the hall whose large windows look south over a visitors' parking lot and toward Washington, a cartoon on a door features two rats in a sterile, numbered cage very irritably blow-drying their hair. One has curlers, one doesn't. The one with curlers is griping, "Remember when lab work was easy and all we had to do was eat stuff?"

This corridor is filled with light. At about its midpoint is the door to the lab of one of the best-known members of the NIH. Dean Hamer is a molecular geneticist, a geneticist who works directly with genes on the level of molecules, as opposed to working indirectly with them on the level of their ultimate products: people. "I study how genes affect the phenotypes of organisms," he explains. Hamer looks about thirty-five (he's older) and radiates a constant, controlled, chipper energy. He was described in a magazine article as looking like "a young Harrison Ford," a description that made him very happy. His colleague Maxine Singer objected that she always thought he looked more like Paul Newman. *"Whatever,"* says Hamer. "I accept." He is also formidably intelligent, a trait that becomes apparent quickly when he begins talking about genetics.

Hamer works in a division of the NCI, where he is chief of the Gene Structure and Regulation section of the Laboratory of Biochemistry. He also just so happens to be continually in the process of quitting his smoking habit, sometimes via nicotine patches, sometimes by switching brands, sometimes cold turkey, all of them agonizing. The lab is part of the Division of Cancer Biology and Diagnosis. After graduating from Harvard with a Ph.D. in genetics—he was enrolled at the same time in Harvard Medical School, doing a joint Ph.D./M.D., "but the M.D.'s kind of been on hold"—Hamer moved to the NIH for a postdoctoral fellowship. There he was one of the first geneticists to transfer genes into animal cells, a methodology that is at the heart of modern molecular biology and the biotechnology industry. After he got his own lab, Hamer turned to gene regulation—how organisms turn their genes on or off—and established a very solid reputation working on metallothionein, a sort of genetic safety mechanism that everything, from yeast to humans, uses to wash out heavy metals when they pollute cells.

In 1992, when the metallothionein puzzle had been largely worked out, Hamer began thinking about the role of genes in "complex traits," a broad category that can cover just about every aspect of human health and disease. His laboratory at the National Cancer Institute was especially interested in Kaposi's sarcoma (KS), a cancer of the skin cells that appears most frequently in Mediterranean men—Greeks and Italians—and in gay men with AIDS. As he went about setting up a family genetics study for Kaposi's, Hamer realized that the population he'd be assembling would represent a unique resource to study a completely different but no less interesting question, the possible role of genes in sexual orientation.

For this part of his research he began assembling a team that he would lead in a search for the gay gene, something he did with the care, and nearly the attendant anxiety, of the head of a difficult search and rescue operation. They would be attempting to storm an interior wilderness of deoxyribonucleic acid, which meant that each team member would have to excel at his or her specialization to get them where they wanted to go and yet at the same time be able to work as part of a group with precision and dedication, relying on the others. Hamer knew that they would have to get along if they were going to locate their quarry. He was also frankly hoping for a little luck. As he went about preparations for the study, Hamer sent out word that he was searching for collaborators on this significant task. In an odd twist of fate, *Newsweek* magazine did the first part of his search for him.

Angela Pattatucci, a young Ph.D. in molecular genetics three years out of graduate school at the University of Illinois, had come to the NIH to work on fruit flies. Once she arrived, however, she found she had made "a terrible mistake." Pattatucci is a serious, quiet woman with dark blond hair and a thoughtful manner. "I was researching with another scientist," she recalls, "but it almost immediately became obvious that the personalities, and thus the research, weren't going to work. I had to cut my losses, get out, maybe teach somewhere. Then the *Newsweek* cover story 'Is This Child Gay?' came out, and they said *someone* at NIH was doing genetic work on sexual orientation. I was fascinated. I was hooked. And I had absolutely no idea how to find this person. There're 15,000 researchers here, and that was too small a needle in too big a haystack. One day I noticed on a list of NIH lectures that Roger Gorski was going to be lecturing in Building 37. I knew who Gorski was because I had been interested in his work. So I wandered in, sat down in an empty seat next to some guy, and after five minutes—because Gorski made some references to him during his talk—realized that this person sitting next to me was the guy doing the research. Gorski happened to be working with him. I pulled out of my other position and started working with Dean, who was happy to have me."

Hamer already had a lab technician for the team, Stella Hu, a small, stern Chinese woman born in 1932 in Nanjing, to do the bench work, the tedious, meticulous work of analyzing genetic markers in hundreds of samples. Now he had Pattatucci. The next person added was Nan Hu (who was also Chinese but no relation to Stella), an expert in cancer cytogenetics, the study of chromosome morphology. She would spend her time peering through a high-power microscope, searching for any minute changes in the chromosomes of the KS patients or of perfectly healthy men who just happened to be gay. The final member of the team was Victoria Magnuson, a recently graduated Ph.D. molecular biologist who would help out with some of the DNA assays in the later stages of the projects.

~

At the outset, Hamer faced three preliminary issues.

The first was that he had to confirm that from all the phenotypic clues—clues on the trait level, ones he could see—it was likely that a gene hidden in the depths of human DNA was influencing sexual orientation. Did the trait they were hunting with genetic techniques appear to have real genetic roots, or was it more likely merely an environmentally or culturally created mirage, something that looked genetic but wasn't? The second issue was a conceptual one: how to narrow down the trait to its essence to get the clearest, most solid target for the team. And third, on a practical level, the team had to answer the basic question of just how they would go about the search, a question of methodology.

The first task on the list was filling in as exactly as possible the trait profile, defining this black box that was sexual orientation. "As you live in the world, you notice that people have lots of different traits," Hamer says, recalling the pitfalls the geneticists faced at this initial stage. "You might notice, for instance, that lots of people speak English and lots speak Chinese." (Hamer tends to use Chinese in his examples. He speaks the language, conversing in serviceable Mandarin with Stella Hu, and covering his tiny private cubicle in his lab are photographs of Shanghai, photos he took while teaching a course there in molecular genetics.) "So if you're a geneticist and you're kind of already professionally obsessed with this stuff, you might say to yourself 'My gosh, they're *everywhere,* people speaking one of these two languages. There must be a gene!' If you then were to say 'I'm going to study the genetic basis of speaking English or Chinese,' the odds are that the first one hundred Chinese speakers you found would be Asians and the first one hundred English speakers you got would be Caucasians. And you would hurriedly look in their DNA and say, 'Look! White guys and Asian guys have a genetic difference right here in this gene that makes them speak these two different languages. And then you'd spend years writing your grant proposals and figuring out your methodology and conducting interviews and running gels to look at the genes. At the end, you'd announce, 'I've studied one million people, and 99 percent of the Chinese speakers have this version of the gene and 99 percent of English speakers have this one. I've found the language gene which makes some people speak English and some Chinese.' And you'd sit back and wait for the phone call from Sweden.

"And of course you'd be *totally* wrong, because obviously there are lots of genetic differences between Asians and Caucasians. You can see this whenever you compare them. The reason our eyes are different shapes and our hair is different colors is because of these genetic differences. What *you* would have done is taken the genes that form these traits and misassigned them to language. Obviously the ability to speak language is genetic, but the language you speak, Chinese or Ukrainian or Farci or whatever, is not. And you'd see this the moment you raised an Asian genotype—an Asian kid—in an English-speaking environment and found you'd gotten an English-speaking person who was genetically Asian. It's easy to misassign genes, which is why you have to be very careful when you do genetics that your work is based on more than just associations because you can confound one trait you see in a person with another." He added, "Whenever you have a gene that you *think* is associated with a trait—any trait from hair color to speaking Chinese to sexual orientation—you'd better be able to prove that in some scientifically reasonable way."

One of the first things Hamer's team had to determine about sexual orientation was whether it was continuous, expressed across a range, or bimodal, divided into two expressions with few or no fine gradations in between. "A good example of a continuous trait is height," Hamer explains. "People have all different sorts of height from very short to very tall, so you know there are a lot of genes affecting this trait, and that it's genetically very complex. Which makes studying the genetics of height—and all continuous traits—really tough. You can draw a picture of this idea, by the way." It's called a distribution curve.

He takes out a piece of paper and a pencil and draws this:

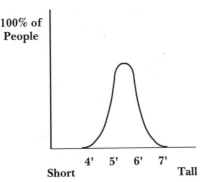

"This," he says, pointing at it with his pencil, "is a classic bell curve. Say you pick one hundred women at random. The women would be all sorts of heights, which would be roughly distributed along this curve, most of them in the middle here where it's highest, but some on the ends here where there are a few very tall women and a few very short."

Bimodal traits, on the other hand, are either/or propositions. Diabetes in rats, Hamer notes, is controlled by around four or five genes, a relatively simple genetic etiology, and it's pretty much a simple bimodal trait: either the rat has diabetes or it doesn't. Albinism is controlled by only one. Again, you have albinism or you don't, bimodal. "The distribution of handedness," says Hamer, "is basically bimodal, right or left, although we haven't found the genes yet."

He picks up his pencil and bends over the paper again:

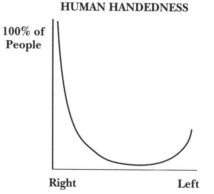

HUMAN HANDEDNESS

100% of
People

Right   Left

"This is a J curve, and any geneticist is going to recognize it instantly. It's pretty much the opposite of the curve you get for a continuous trait. Here the trait is either one thing or the other." He stands, looking down on the two different distribution curves. "If I were working on handedness," he says, "I would say hey, this is a good sign."

Why "good"?

He laughs and looks almost sheepish. "Um, I guess I'm speaking as the guy who has to do the work. When I say 'good' I mean it makes things experimentally attractive. If you want to try to study a trait, isolate individual genes and figure out how they work, which is really what genetics is about, after all, then purely and only as a matter of practicality—practicality being how many headaches you want to give yourself—you'd prefer to be researching a trait that's bimodal, like handedness is, because it's probably caused by one or a small number of genes, each with a decent share of influence that you can actually measure. A continuous trait like height is probably controlled by a whole bunch of genes, each with tiny, difficult-to-detect influences that your lab can go off on some frustrating wild goose chase trying to track down. Take on height, and even though it's pretty clear height is overwhelmingly due to genes, you're going to have a major headache figuring out how they work." To tackle height, says Hamer, you'd have to be very gutsy.

"We were sort of holding our breath at that point," Hamer would say later, "because the distribution curve we would get for sexual orientation would give us a big clue up front as to whether or not we'd be able to pull this off."

Hamer and Pattatucci began the painstaking job of sketching sexual orientation's distribution curve—a process Pattatucci poetically likens to "painting in strokes of empirical observation the portrait of the trait" ("Yeah," Hamer grumbles, "more like a mugshot")—by finding subjects, gay and straight, and asking questions: Was the person sexually attracted or oriented to men or women or both? Had they always felt this way? How consistently had they felt this orientation? Along with distribution, they established seven indices to measure sexual orientation: self-identification, fantasy, sexual behavior, sexual attraction, romantic/emotional attraction, social attraction ("With whom do you prefer to socialize?"), and lifestyle. Questions designed to assess these indices were carefully drawn up and put to the subjects.

Part of the challenge in establishing the distribution curve for sexual orientation, Pattatucci and Hamer knew, was a well-known genetic consideration that complicates all such studies. It is called expressivity, and it affects the shape and behavior of the curve. This factor surfaced, for example, in the work of Dr. Rob Collins, a researcher at the Jackson Labs in Maine who studies paw preference in mice.

Scientists had noticed that mice seemed to be pawed, right and left, just as we are handed, and the trait appeared to be inherited from parent to child, as is human handedness, implying there is a genetic component to it. It was easy for Collins to decide he wanted to research the trait; the problem with actually doing it was untangling the various, complex ways genes can act. The "distribution" profile of the trait pawedness in mice is not merely a matter of right or left. Geneticists also need to account for "expressivity"—how clearly the trait is expressed in the animal. Collins tackled the problem of separating these aspects of genes and came up with an ingenious method.

First, he nailed down distribution. Collins created a Plexiglas box and in the center of one of its walls drilled a short tube. Into the tube he stuck a small food pellet. Put the mouse in the box, and, with its preferred paw, it would simply reach into the tube to get the food.

Collins found that the mice consistently used either their right or left paws, which showed a bimodal trait, but the numbers of right- and left-pawed mice were also interesting. In any general mouse population, the two orientations of the trait were evenly distributed, 50 percent right-pawed and 50 percent left-pawed. Graphing them, he got a characteristic distribution: a U curve, a variation of the standard bimodal J.

This was the distribution of the trait, 50/50. Now for expression.

You could say that the expression question asks "If a mouse is right-pawed, just *how* right-pawed is it?" To measure this, Collins put his mice into the same Plexiglas box but created a "right-pawed"

MOUSE PAWEDNESS

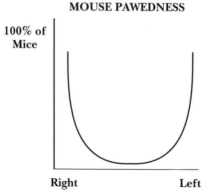

world. This time, the tube was drilled not into the middle of one side, with easy access, but squeezed into a corner of the box, running at a 90-degree angle to the right wall. With this placement, a mouse can easily reach its right paw in for the pellet, but it's a very difficult maneuver to reach in with its left. Collins found, unsurprisingly, that 100 percent of the right-pawed mice in the right-pawed world reached for the pellet with their right paws. The surprise was that most of the left-pawed mice, though some were hesitant about it, also reached for the pellet with their right paws. However, 10 percent of the left-pawers *would not* reach with their right paws. "You cannot make them, period," Collins says. "These left-pawed mice are so left-pawed—that is, the left-pawedness trait is expressed in them so completely—that they will, literally, stand on their heads in the corner of the box in order to be able to reach into that hole with their left paws, rather than use their rights. And the same is true for 10 percent of the right-pawed mice in the left-pawed world." The answer to the expression question was that 90 percent of right-pawed mice are only moderately right-pawed, but 10 percent are absolutely, unalterably right-pawed.

The basic distribution of the pawedness trait in mice, Collins thus established, is a bimodal U curve distribution. The trait of handedness in people, while a slightly different distribution (a swooping, clear-cut J), is also bimodal. The distribution curve of height is continuous, arching across in a bell curve. Hamer and Pattatucci, applying their questions and their measurements week after week to hundreds of subjects, collected their data in what seemed like an unending series of interviews, hoping for a clearly defined trait. Gradually, meticulously, they put it together, tabulated it, and derived the distribution curve for human sexual orientation.

So which is it? Continuous or bimodal?
Hamer grins and draws this:

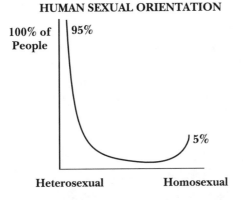

HUMAN SEXUAL ORIENTATION

"It's bimodal," he says. "That's what our data and the data of others tells us. This was a major debate in sexology that began with the creation of the field. Kinsey's rather vague notion was that everyone is born essentially bisexual, and if that were true, human sexual orientation should be a bell curve, a continuously distributed trait. He created a scale from 0 to 6, heterosexual to homosexual, and assigned the population to it in a bell-shaped distribution. This was incorrect. In fact, we now know that the Kinsey scale—at least as far as it decrees a bell curve distribution—is basically a dead model because it doesn't reflect reality. Sexual orientation isn't a continuous bell. It's a bimodal J."

This J curve was just a rough first assessment. When they completed analyzing the data, they found the trait actually had not one but *two* markedly different curves, yet these curves were not, as many would expect, gays on one side and straights on the other. Actually, the two distribution curves for sexual orientation are divided between men and women, and they differed in three significant ways.

The first way was the shape of the curves.

"Sexual orientation in men," human geneticist Richard Pillard once joked, "isn't really a J curve. It's more like an 'L.' " The distribution of sexual orientation among men, Hamer and others have found, is essentially a standard J curve, like handedness, but with such an extreme distribution that it is almost completely bimodal, either/or, gay or straight. Women, on the other hand, have, as Hamer put it, "a real J." Although sexual orientation is clearly bimodal among women as well, there are more bisexual women than there are men—

more 2s, 3s, and 4s—so the curve has more of a curve. Side by side, they look like this:

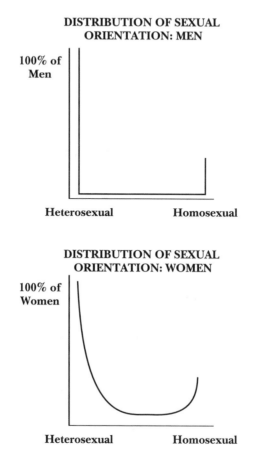

The second way the curves differed was in the percentage of gays to straights. Hamer and Pattatucci found that approximately 2 percent of all men were gay (and 98 percent straight), but the percentage of lesbians was roughly half that, around 1 percent. (These figures are almost certainly low. Research on the population percentages for homosexuality has begun only recently, and the figures are still extremely inexact since political, social, and religious pressure has always dissuaded homosexuals from identifying their sexual orientation. In addition, the low figures may result in part from scientific stringency in assessing the data, and Hamer and Pattatucci themselves believe that the numbers are actually around 4 to 5 percent for men, 2 to 2.5 percent for women, or perhaps a point higher, although these too are only estimates. But the 1992 presidential elections provided a fascinating illustration of the social change that is giving us an increasingly accurate look at the true ratio of homosexuals to heterosexuals. This change shows up, as so many do, between generations. A consortium of CBS, ABC, NBC, and CNN did a nationwide exit poll, conducted by Dr. Murray Edelman of Voter News Service in New York, and found that only 3 percent of all voters identified themselves as gay and lesbian. The story, however, lay beneath the 3 percent figure: for voters older than 60, only 1 percent self-identified as gay, but for those between ages 18 and 29, more than 5 percent did so. Pattatucci and Hamer, in trying to measure scientifically a trait whose expression is subject to social censure, are made constantly aware of the problem (one that faces all biologists to one degree or another) of establishing an accurate phenotypic profile. But as social pressures recede and more accurate data comes forward, that profile will become increasingly reliable.)

But the third difference between the sexes was perhaps the most striking, and this was a difference of expression. Pattatucci conducted the interviews with the women in the study, evaluating, as she said, "tons of letters, tons of conversations on the phone, tons of interviews." She confirmed in the end that women do something men virtually never do: They move among straight, bisexual, and lesbian.

Again, the genetic concept of expression for the trait sexual orientation could be posed as the following question: If a homosexual is homosexual, just how homosexual is that homosexual? It turns out that the answer is different for men and women. For men, the answer is usually: completely. If a man is homosexual (or heterosexual), he expresses that version of the sexual orientation trait 100 percent. For women the answer is: sometimes not as homosexual as homosexual men. And straight women are not as straight as straight men either.

While most women are stable in their orientation, Pattatucci determined that it is not unusual for a small portion of women to report feeling themselves to be straight at age sixteen, perhaps lesbian at age twenty-four, maybe bisexual at age thirty-eight, and straight again at fifty-five. With men, Hamer says flatly, "This sort of movement is very, very rare. It's pretty much a phenomenon we see exclusively in women." Hamer tells of an interview he conducted with a woman in her sixties. "She's been happily married to a man, married her whole adult life, but he had died. Her answers to all the questions are down-the-line straight: never slept with a woman, never fantasized about sleeping with a woman, never *thought* about sleeping with a woman, always attracted to men, never thought of herself as lesbian, et cetera. So the interview's over, I've marked her down-the-line heterosexual, and I ask her if she has any questions about the study. She looks at me and comments matter-of-factly, 'You didn't ask me about the future.' My very eloquent response was, basically, 'Huh? What are you talking about?' She says calmly, 'Well, you know, I'm only in my sixties. My sex life isn't over, and who knows if in the future I won't be with a woman?' " Hamer smiled and jabbed a finger on the table for emphasis. "Erotically, men do not experience even having the interest. You never, ever get a sixty-year-old straight guy who reports, 'I've slept with women all my life, was married happily to my wife forty years, never had the slightest sexual attraction to any man or even *thought* about a guy romantically, but hey, who knows! Maybe next month I'll start doing it with men!' "

Pattatucci concurs. "If you track one hundred men, after a few years, maybe—*maybe*—one will have moved. Just a little. Maybe. With one hundred women, it's almost certain a number will move. Although," she adds, "women are actually more stable than we thought they would be." In fact, Pattatucci found that bisexual women are just as anchored in their bisexuality as straight and gay women in theirs. "More often than not, women identifying as 0 or 1 or 5 or 6 say they are fixed in that position," she said, "but even bisexual women at 2, 3, or 4 responded at a fairly high rate that their sexual orientations were fixed at 2, 3, or 4. I've found virtually no evidence that 2s, 3s, and 4s are any less stable in their categories than 0s or 6s. When women *do* move, they tend not to go drastically from 6 to 0 but maybe from 6 to 3 or 2 to 4. They don't make large movements. And I saw just as much movement from outside of the scale to inside as vice versa." Pattatucci learned that there is less to movement than meets the eye. "Sometimes a woman will identify as a 2 and then a year later will say she's a 4. At first I thought, 'Ah! Her orientation has shifted. Now let's look into this.' Well, then I find that when I ask why she's telling me she's now a 4, she'll respond that in fact *nothing* has changed for her as far as her desires and feelings—the biological definition of sexual orientation—go. What has changed is the way other people react to her. Last year she had a boyfriend and went more to straight bars and places, but they broke up and so this year she has more access to the lesbian community. The orientation hasn't changed, just the social situation, and thus, the way she identifies herself. So in fact, her orientation is quite stable."

Aside from the complications of movement with women, Pattatucci found a number of women identifying in the 0 and 1 and 5 and 6 range, and quite a few as 3, but less as 4 and 2, which gives this subtle variation in distribution:

MORE EXACT DISTRIBUTION OF SEXUAL ORIENTATION: WOMEN

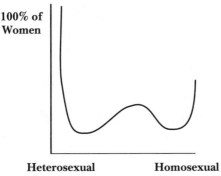

"It's more intellectually arresting," she says, "but more methodologically problematic. It's basically a J, but with these variations." She sits looking at the curve she'd drawn on a white piece of paper on a lab countertop and smiles, shaking her head. "It's sort of a strange distribution," she says intently. "But interesting, don't you think?"

Between men and women, she observed another fairly spectacular difference. "When," she says, "you ask men, gay or straight, if they have ever wondered about their sexual orientation—if, for example, a straight guy has wondered, after a breakup with a girlfriend, if he's gay or a gay guy, if he meets a beautiful woman, if he's straight—they almost invariably say no. It's extremely rare to hear that it has ever occurred to them, that they've ever felt this, and in the few instances when you do hear it it's usually the result of some major event and goes away the instant the event is resolved. Well, it's not at all rare for us to have women say they've wondered. A large number of straight women and almost as large a percentage of lesbians have wondered about their sexual orientations." She adds quietly, "I think few people want to say what the real answers are because it's too politically sensitive."

Pattatucci found the phenotypic part of the study fascinating, and it struck her that she was engaged in research that revealed, it sometimes seemed, more about the sexes than about sexual orientation. Molecular geneticists, she and Hamer found themselves doing more than a little male/female sociological research in the process of nailing down the phenotype of sexual orientation. After the first, phenotypic half of the study, Pattatucci observed, "Men seem to be able to average their experiences—any experiences, not just sexual—with ease. You ask them about something, and they are able without pondering it too much to say decisively 'I guess that overall, although I'd say no in a few instances, in general the answer is yes.' For example, when you ask men about their lives—'Were you a good student in high school?'—you find they'll either give a yes or no. Experience is evaluated bimodally without undue effort or concern. You ask women, and they'll say, 'Well, it depends.' Women look more at the details: 'Do you mean English or biology or algebra?' They separate it into categories and they focus on each category instead of giving you a clear general answer—and are, in fact, generally averse to yes-or-no division of experience. As a clinical researcher, I'm not able to pull a general answer out of that.

"If I ask, 'Did you have sex in high school?' women will respond, 'Well, what do you mean by "sex"? Petting? Making out? One time or five times?' If I ask a man, he'll give me an almost immediate response, yes or no. I think that for men, having sex is a more narrowly defined phenomenon than for women. That is a striking difference, in my opinion, and it speaks to a difference people have been talking about for a long time, that women's sexuality is expansive, that we need to have more things fulfilled than just erotic desire. On average, men get satisfaction purely from eroticism. If you ask women about their fantasies, they typically are very situationally oriented. It's not just sex, it's sex culminating from a series of attractive romantic but not necessarily sexual circumstances. That makes asking the same question for men and women a different enterprise. I've found women more informative if you're talking about sheer volume of information, but women's responses can therefore be frustrating for me as a researcher because there are questions to which I need a general answer. It's easier to get to the core of a man since they tend to define things more narrowly. If I ask men a specific question, they'll give me a specific answer, and if I ask a general question, they'll give me a general answer. On the other hand, if I ask women a general question, they will often give me a specific answer, filled with detail, interspersed with political views and personal experiences, and contingent on conceptual arguments. And within women, there is a subset who seem to have so blended their uncontemplated desires and emotions with their politics and their feminist consciousness that I'm not able from their answers to separate the two."

This is, Pattatucci notes, an example of a fascinating, frustrating process crucial to the team's research: the separation of the biological definition of sexual orientation from the cultural definitions. It is a process common to all biological research of traits expressed behaviorally. Pattatucci faced it in one of the measures that she and Hamer were using to ascertain sexual orientation: "lifestyle."

She grimaces. "I hate the word 'lifestyle.' It's idiotic when applied to sexual orientation—would you refer to left-handedness as an 'alternative lifestyle?—but the problem is that through misuse by the media and in political rhetoric it's become ubiquitous now, and there's nothing we can do about that. When reporters use it, it is simply intellectual laziness. But some people adore that word, and the reason is probably in many cases, I'm very sorry to say, that it is such an inaccurate description of homosexuality, implying that sexual orientation is something one chooses, something frivolous or faddish, determined by what you do as opposed to an internal orientation that is a component of what you are. It's gotten to the point that when your proposed study protocol gets reviewed, they nitpick you to death. 'Why didn't you include "lifestyle" as one of your criteria?' " She shrugs and rolls her eyes slightly. "So we put it in because of politically formed expectations and we don't include it in linkage analysis because it's substantively meaningless for sexual orientation in a biological context."

In fact, in contrast to the popular misuse of "lifestyle" as synonymous with "homosexuality," Pattatucci and the NIH team use the word "lifestyle" for the express purpose of identifying and weeding out from the study those who are *not* homosexual. "A relatively small number of women will say in the interview, 'I'm not a lesbian, I just fell in love with this one woman,' " says Pattatucci, "and it's apparent that their feelings are, by pretty much every measurement, basically heterosexual. They perceive themselves as having had a serendipitous experience. They fell in love and committed to this particular woman, and sex became part of the relationship. We've noticed that one of the consistent identifying characteristics of this phenomenon is that these women will describe their relationship as living a *lifestyle*. They're quite comfortably certain that they are attracted to men *except* for this one woman."

Pattatucci does not include these women in the study. "Simply as a matter of good methodology and effective research, we have to study people most likely to have a genetic influence, and if you're assuming there's a genetic influence for homosexuality, then it stands

to reason that it behaves like a lot of other traits. The way traits behave is that the same trait might show a strong genetic influence in
some people, some weak, and some virtually none. Obviously the
people who are in this last group won't be good candidates for a genetic study. That's axiomatic. We only have so many dollars to allo-
cate, and we put our resources into clearer possibilities. People in the
first group are the best candidates. We're not doing anything different from what biologists do for any other trait. Obviously if you're
looking for a needle in a haystack and there's twenty-three haystacks,
you'd better narrow your field as much as you can. The best place to
look is at people who show the greatest amount of expression."

(A short time after beginning the study, Hamer and Pattatucci
dropped "lifestyle" and settled on four criteria: self-identification, behavior, and, most important, sexual fantasy ["When you masturbate,
do you think about men or women?"] and attraction ["When you walk
down the street and pass people on street, do your eyes jump automatically to men or women?"].)

Another category of subjects that Pattatucci excludes from genetic
analysis is what she genially refers to as "political lesbians." "These are
women," she says, "who identify themselves as lesbian for political and
ideological reasons with little or no evidence of a romantic or sexual
attraction to women. If we had set up a large study of the genetics of
left-handedness, got it funded, took blood, done the gene assays, and
found that we were working on subjects who had identified themselves
as left-handed simply in solidarity with the political and social op-
pression of left-handers but were actually right-handed, it would be
scientifically pointless; *of course* you're not going to find a gene for left-
handedness in these people's DNA. You're looking at right-handed
people who don't have the trait you're researching."

Pattatucci adds, "One of the basic political positions of these
women is opposing discrimination against people who are gay and
support for ideas like women's equality. Am I sympathetic to those
goals on a purely political level? I absolutely am. I'm a woman, and
I'm a lesbian. But I am also a geneticist, and *that* is the context in
which I operate here. I'm researching a scientific question, not a political one." Pattatucci pauses to stress her point. "I'm making no
judgment of their politics. But politics and science are not the same."

Pattatucci and Hamer both note that this is one more illustration
of the difference between men's and women's sexuality. "With
women," says Pattatucci, "we ask the question: Is this person a lesbian
for political reasons? This is a question we *never* ask men." Hamer con-

curs. "There are no 'political gay men,' and you virtually never see a straight man fall in love and have sexual relationships with 'one special guy.' "

For centuries scientists have observed the profound differences between male and female sexuality, of which sexual orientation is only one component. The most fundamental dividing line runs not between gays and straights but between men—both straight and gay—and women, both straight and lesbian. As Simon LeVay commented, "A major question that's been raging for years is 'Is homosexuality really just "reversed sexuality"?' 'Are lesbians just straight men in women's bodies?' In a word, no. The fact is that gay people are sex reversed in some ways; but in *most* ways, lesbians and straight women are, in the ways they view and experience sex, fundamentally the same. And gay men and straight men, who in just about all the various aspects of their sexuality resemble each other almost exactly, are most appropriately grouped not in opposition but together. There is the one obvious reversal everyone concentrates on and makes a fuss about: Gay men like men, straight men like women. But—and this is in a sense even more interesting—*aside from* the gender to which they are attracted, virtually all the typical sex behaviors, responses, desires, and fantasies of straight men are identical to those of gay men."

There is the well-known story about Mrs. Coolidge who, on a presidential tour of a chicken farm, noted approvingly a rooster copulating with a hen. "Does he do that often?" she asked. "Oh, several times a day," said the farmer. "Please tell that to the President," she directed pointedly. Coolidge turned to the farmer. "Always with the same hen?" he asked. "No, all different," was the response. "Please tell *that,*" Coolidge said, "to the First Lady."

The male-typical trait of being sexually aroused by diversity can be seen in the reproductive products we manufacture. Having two sexes is a popular, though by no means universal, mode of reproducing. Some lower plants have only one sex, making reproductive cells and mixing them up at random to create the genetic blend that results naturally from the two-sex method. Fungi have hundreds of sexes. Humans, with two, are specialized into making eggs or sperm, and our sexual behaviors originate from our respective specializations. Form and behavior, in a sense, follow function.

While all other human cells have a full complement of the twenty-three pairs of chromosomes, sperm and eggs are haploid, carrying exactly *half* a full complement. Each sperm only gets half of a man's chromosome pairs, and the same is true for a woman's eggs. As Hamer put it, "Basically sperm and egg both carry twenty-three single socks out of twenty-three pairs of socks in the wardrobe." But the similarity between sperm and egg ends there.

A man releases roughly 100 million sperm each time he ejaculates. Sperm are marvelous studies in minimalist design. Tiny even by cellular standards, they are the most stripped-down, quick and dirty biological contraptions imaginable, purely functional delivery systems. Sperm are essentially just flimsy warheads packed with genetic material. The DNA payload is strapped to an expendable engine (the tail) that contains only enough stored-up energy to reach the target. (A notable exception to this is certain species of *Drosophila* fruit flies, such as *D. hydei* and *D. littoralis*. They produce the longest sperm in nature, at around two-thirds of an inch, 300 times longer than that of humans and 600 times longer than the sperm of hippos, who make the world's shortest. The fly's sperm are also six times longer than the fly itself; if men's sperm were proportionally sized, they would be forty feet long. And these fruit fly sperm do not swim. The male deposits them in the female's uterus, a sort of waiting room where they linger until an egg comes by to claim one of them. The male flies make far fewer sperm than do human males—only around 250 per deposit—but they are extremely effective. Anywhere from around 50 to 80 percent wind up fertilizing eggs, whereas men generally ejaculate a hundred million tiny sperm, usually all of them unsuccessful.) By contrast, a woman's eggs, 85,000 times larger than the sperm, are large, solid machines. They are resistant to drying and, with their own emergency food ration of yolk, self-sufficient. A woman puts a great deal of energy into making each egg. She will produce only about 400 in her lifetime, and at most only around twenty of those will become children. (A man could easily fertilize a lifetime's worth of 400 eggs with a single ejaculation.)

Given what they have to work with, men and women have evolved conflicting sexual modus operandi, which operate on the instinctual level. As sociobiologist E. O. Wilson has said of mating, "It pays for males to be aggressive, hasty, fickle, and undiscriminating [while] it is more profitable for females to be coy, to hold back until they can identify males with the best genes."[1]

"How many sex partners a person wants is clearly different not between gays and straights but between men and women," LeVay says. "Men, both straight and gay, and all male animals want to have more sex partners than females do. Women, gay and straight, and all female animals want fewer sex partners. This is rooted relatively simply in biology and has to do with strategies for reproduction. Males try to have more sex partners than females because it's so 'cheap,' in biological terms, for males to inseminate females. Sperm is *not,* relatively speaking, a scarce resource. By contrast, it's very 'expensive' for females to get inseminated because then they have to go through pregnancy, childbirth, nursing, and caretaking, and they can only have a limited number of offspring. If you're a male you are free, if you're powerful enough, to have a virtually unlimited number of partners and offspring, which is great for you. If you're female, your interests are very different, and you're going to be as choosy as you possibly can. This biological phenomenon is clearly visible in our behavior."

LeVay points out that in one study, researchers Bell and Weinberg found that the number of sex partners reported by gay men averaged thirty-three times what lesbians reported. Straight males report a number of partners lower than gay men and higher than lesbians. "But," LeVay asks, "is that what straight men and women really *wanted?*" Not really. As is frequently the case, behavior taken at face value can be misleading. "Lesbians present you with an opportunity for studying female sexuality in its purest, most undiluted form, and gay men for studying pure male sexuality. With heterosexuals, if you're trying to distinguish how men feel and how women feel sexually, the waters are muddied because perforce you're mixing the two. There's a sort of battle of the sexes there because of the clash of biological interests and strategies. Straight men are limited in their sexual activity by the number of straight women who want to have sex with them. So with heterosexuals, behavior misleads. With gay people, however, the drive is the same in both partners, so they have a much better chance of finding partners interested in the same frequency of sex and the same variety of partners as themselves, and the data are clearer. This is why lesbians report fewer sexual partners than straight women and gay men report higher numbers than straight men. Of course," LeVay adds, "as we've seen in the age of AIDS, the male sex drive can lead to an epidemic."

Pattatucci comments, "In our studies we're showing that when you ask people how many sexual partners they have over their lives, most lesbian women will say one to five or six to twenty and less will say twenty-one to forty and almost none forty to one hundred. But when you ask gay men how many partners they've had over their lives, it's reversed: the numbers are typically high, forty to one hundred is not rare, and the atypical ones are one to five. But what is interesting is that when you pose the same questions to straight men and straight

women, you get similar answers: The straight men want more variety—
men like big numbers—and seek that, and the straight women, like
gay women, want less." Aside from all that it says about disease and
infection, as LeVay points out, the epidemiological data from the
AIDS epidemic and the infection pattern of the disease, particularly
in Africa, demonstrate with cruel clarity the biological point that
males desire a far greater number of sex partners than women. Male-
typical sexual desire added to a common medical observation—that
because the lining of the anus is torn more easily than that of the
vagina or mouth, giving viruses easier access and resulting in higher
viral infection rates from anal sex—provides the simplest explanation
of the epidemiological demographics of the AIDS epidemic, which
has spread primarily through men who are also homosexual.

Another important sex-related trait that is not reversed in homosexuals is gender identity. If gay men are the opposite of straight men
in terms of the gender to which they're sexually attracted, they are
completely like them in terms of their sense of themselves *as* men.
"The feeling of what sex you are, in my view, has deep biological
roots," says LeVay. "It's not just that you look down and see you have
a penis, and your parents tell you you're a boy, and you say, 'Oh, I'm
a boy, okay, great, that's my identity.' In fact, I think there must be
some internal representation of what sex you are independent of
things like the appearance of your body. My evidence for saying that
is simply the existence of transsexuals, people, male and female, who,
in the face of every possible evidence of their anatomy and what society is telling them, are absolutely, irrevocably convinced that they
are of the opposite sex."

LeVay raises his eyebrows thoughtfully. "Transsexuals attest to
gender orientation being as real a phenomenon as sexual orientation.
What's interesting is that homosexuality and transsexuality are distinct
but perhaps biologically parallel phenomena, one a reversal of the
gender of the people to whom you're sexually attracted, the other a
reversal of the gender that you feel you are. Gay men are aware of
some degree of femininity in themselves, and gay women some degree of masculinity, yet there is no reversal of gender identity. Drag
demonstrates this. There's a thousand jokes about this, which gay people enjoy as much as—actually, I think, more than—anyone. But this
is completely and notably different from the sense that you *are* a
woman or a man. Even for gay men who dress up in the wildest
women's clothes, the *point* of the whole thing is that it is parody,
which is why it's funny. Gay men are absolutely convinced they are

men, lesbian women are absolutely convinced they are women, and if you ask gay men, they respond, 'I have a firm, inner sense that I am a man. And *this* is a show.' Frankly, the very essence of what makes camp camp is the conscious tongue-in-cheek aspect. At the point at which it is taken seriously—at which one says that this man in a dress truly thinks he is a woman—it stops, by definition, being drag."

The distribution curve dictated by Hamer and Pattatucci's data raised some eyebrows among scientists. The fact that sexual orientation is clearly "either/or" for men is still a relatively new concept to many scientists, several of whom expressed initial doubts about the data. One college genetics professor insisted that the trait sexual orientation was "continuous"—that is, a bell curve—adding "I don't really have any evidence for it, but it just makes sense."

Hamer points out, somewhat testily, that his distribution curve has been confirmed by several studies. ("I didn't *tell* these men to answer 0 or 6," he mutters, "it's just that almost all of them did. Am I supposed to *pretend* the trait is continuous?") One of these studies, he notes, was conducted at a military hospital where the military's strict ban on homosexuality biases strongly against a response of 6, or homosexual. "You'd expect to get a large number of 'bisexual' responses in the military," explains Hamer, "2s, 3s, and 4s, because bisexuality would be a convenient way of shading the answer and protecting yourself. But again almost all of the men answered 6 or 0." The latest such study was conducted in Australia by Michael Bailey and Nick Martin, and had a sample size of more than 2,000 respondents.

Hamer takes out a paper the dubious genetics professor had published that criticized Hamer's finding. The professor had written, "Although the probands reported a wide range of sexual behaviors, identities, and fantasies, [Hamer and Pattatucci divided] the men into homosexual and heterosexual." "The 'wide range,' " Hamer responds sarcastically, "is here." He snaps open a copy of his study, pins it to his desk, and points briskly to the four raw-data charts on page 1 from which the distribution curve was derived. The numbers clearly indicate bimodal distribution in men, and the two most accurate measurements for defining sexual orientation—Fantasy and Attraction—are the most bimodal.

Asked if he had anticipated this striking bimodality for male sexual orientation, Hamer says, "Well, how many truly bisexual men have you ever met? I have no theoretic argument with bisexuality. It's just that before I started doing research, I'd never met any. Of the men we've interviewed, most identify themselves as either gay or straight.

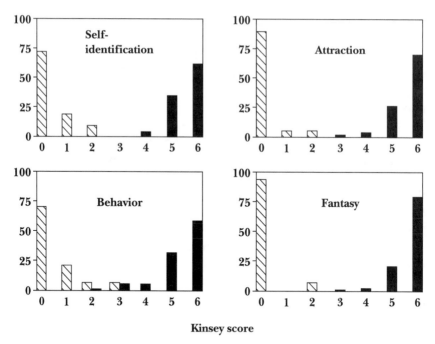

Distributions of Kinsey scores in study participants. (■) Homosexual probands and relatives ($n = 144$). (◨) Nonhomosexual relatives ($n = 22$).

A handful identified themselves as bisexual, and we did not include them in the DNA analysis because of the possible genetic complexity in their cases and our need, at this stage, for simplicity. But of the few who said—even insisted—they were bisexual and made their case with the fact that they were also sleeping with women, it would become clear with most of them after just a couple of casual questions that they were really only attracted to men but were in the process of coming out and felt more at ease at that point calling themselves 'bisexual' than 'gay,' which was a more radical term for them. They were still sleeping with increasingly fewer women for the same reason. Besides, as a geneticist, to be blunt about it, I don't really give a damn what label anyone uses, or even what they do, or with whom. I care about what they feel inside. When I started researching this question systematically, I was asking 'continuous' type questions—that is, instead of asking 'Are you gay or straight?' which would bias you toward a bimodal model, I asked, 'How do you ID yourself on this scale, who do you fantasize about, who do you have sex with?' and so on. That type of questioning biases toward a bell-shaped distribution. We even based our data charts on Kinsey's four measurements—Kinsey, the king of bisexuality. If *anything*, that should bias the answers in favor of a continuous distribution. And still, after all that, when we looked at the responses to these continuous-type questions, this is what they told us."

～

After distribution and expression were nailed down, Hamer and his team had three more crucial aspects of the sexual orientation black box to determine before he could begin looking for a gene. During the 1980s, researchers had been turning up data suggesting that homosexuality ran in families like any genetic trait. "We knew that, if there were a gene that influences sexual orientation in men," Hamer explains, "then male homosexuality should run in families, and gay men should have more gay men in their families than on average. Genes go from parent to child, so you'd see familial aggregation of the trait. Of course," he adds, "having something run in a family doesn't *prove* that it's genetic."

Jeff Hall expressed it as: "If you're a geneticist after some trait, you've first got to determine that this thing really is genetic. Hamer couldn't just dive for a gene. And he also couldn't dive *just* because the thing runs in families—in that case there'd be an Irish gene for drinking. He needed patterns that look genetic." Traits are *familial* if family members share them, for whatever reason, but *heritable* only if genes are producing the characteristic; Presbyterianism is a familial trait but not a heritable one. Earlier this century, a public health commission studying the disease pellagra in the southern United States found that it ran in families and concluded therefore that it was heritable. In fact, pellagra is caused by a simple vitamin deficiency and is merely familial. It ran in families because of eating habits, not genes. Some common behavioral genetic traits are rolling your tongue into a tube and the less familiar peculiarity of hand-folding. People will, when folding the fingers of both hands together, consistently put one or the other thumb on top, a heritable trait. A simple nonbehavioral genetic trait that can be easily observed is whether the ear lobe is attached or separate.

To establish whether homosexuality was heritable or just familial, Hamer looked at groups of families that had gay members and set about establishing the way the trait bred ("just like you would for pedigree dogs," he adds). The technique is called pedigree analysis, essentially constructing a family tree for sexual orientation. He started

in each family with one gay person and then branched out to brothers and sisters, mother, father, cousins, aunts, uncles, labeling everyone: squares for males, circles for females, clear for heterosexual,
shaded for homosexual.

Hamer and his team members spent months assembling sexual
orientation pedigrees, asking gay men and their families: did the men
have gay brothers, gay sisters? How about their parents and grand-
parents? Their cousins? After collecting 114 families, it became abun-
dantly clear that the trait ran in families. Hamer marked off the first
critical point on his "Seems to Be Genetic" checklist:

> 1. Being gay runs in families. He characterized the finding in
> the extremely careful wording of the scientist: "Having the
> trait run in families is necessary but not sufficient to show a
> genetic influence on a trait."

Using the same data, he then looked at a second crucial question
and was able to check off a second important checklist point:

> 2. "There were increased rates of gay people among family
> members genetically related to each other even when raised
> apart in different households." This finding filtered out "en-
> vironmental influences"—the homosexual subjects had only
> genetics in common—and highlighted the biological corre-
> lation between related genes and homosexuality. "Again,"
> Hamer says cautiously, "this is consistent with the genetic the-
> ory, but doesn't prove it."

The third and last question was the most important. Hamer explains, "The question becomes the scatter of the trait: Do these family members show a random distribution, as if you'd taken the family
tree and randomly sprinkled the trait over it like salt from a shaker?
That's what would happen if there were no genes. Or do you see a
pattern, a pattern that looks genetic?" If a trait is truly genetic, examining family after family in a pedigree analysis will be like repeatedly throwing a pair of loaded dice and watching the patterns emerge:
Everything seems normal at first, but if you're careful, patient, and
tenacious, after a time you'll notice that certain sides turn up more
often than others. The genes guiding the trait, like little hidden
droplets of lead loaded inside the dice, guide outcomes, subtly molding results out of what would otherwise be random. The pattern of

the orientations, the way they turn up over and over in family members as they cascade down through generations, is called, appropriately, genetic loading. To the geneticist's eye, these patterns are as characteristic as a fingerprint.

Hamer had been sifting through the pedigree data for weeks, but no pattern was emerging. Staying late on a Friday evening in an empty lab, he sat at the computer looking through his seventy-six gay subjects and more than a thousand relatives. The data had been compiled and all the subject marked with little tags, red tags for gay, black for straight, and he sat back, staring at it. Nothing much. In the empty lab, the equipment hummed quietly. Outside, night was falling and cars on Old Georgetown Road rushed out of Washington toward the suburbs. He cut out certain categories, switching them around, and the screen was a moving checker board of red and black. Still not much. He rechecked his numbers. They were correct. Some of the families looked like this:

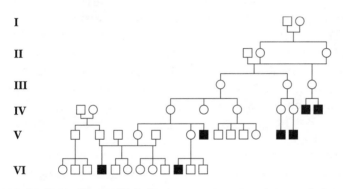

A family pedigree displaying apparent maternal transmission of male homosexuality. The family was selected because many members were homosexual.
(■) Homosexual males. (□) Nonhomosexual males. (○) Nonhomosexual females.

And then he noticed that in some of the families, most of the red marks were aligned at one side of the screen. Hamer looked at these families again. He placed the mothers on one side, and now there were patterns of red.

It was the first telltale to emerge from the mass of pedigree data. Compared to the average rate of homosexual men in the population—Hamer was using the figure of 2 percent—there were, he realized, a much larger number of gay men going down through generations on their maternal side. Not only that, there were extra gay male cousins and uncles, and they too were on the maternal side. And yet—he rechecked the figures to be sure—all the other gay people, all the lesbians, all the gay male cousins and uncles on the paternal sides of the family, were randomly distributed. The fingerprint had leaped out from nowhere, narrowing the search in a single moment from the entire human genome to one-twentieth of it, one particular chromosome. It had exposed the first part of the gene's genetic address.

A genetic address is similar to a mailing address except that it goes not from the most precise location (the street number) to the largest area (the state) but from the largest genetic area, the chromosome, to the smallest, the gene. The address of the gene that, if it comes in a damaged allele, gives you T-cell acute lymphoblastic leukemia, is 11p15. This means that of the twenty-three pairs of chromosomes in the human genome, the gene is on chromosome 11, more specifically on chromosome 11's short (p, petit) arm, and more specifically still in the p arm's fifteenth region. (There are many genes that cause different types of leukemia spread out on a variety of chromosomes, including the seventeenth, twenty-second, and tenth; chronic myeloid leukemia's address, for example, is 9q34, the ninth chromosome on the q (long) arm in the thirty-fourth region.) Hamer realized that the pattern for homosexuality he was seeing in certain families resembled the classic genetic pattern left by blue-monochromatic color blindness and hemophilia B. The genes for these traits are on the X chromosome, the male window of vulnerability, where men have only one copy and no back-up—the hemophilia allele is at Xq27, the twenty-seventh region of the X's q arm, and blue-monochromatic color blindness in its twenty-eighth—and now, it appeared, the gene he was looking for was also on the X.

"If this gay gene for men were on the X," says Hamer, "homosexuality would show up more often on the mother's side of the family, like color blindness, and it would always be passed through Mom, since men always get their X from their mother. But you wouldn't see it in women since they have a second X, a safety backup that compensates. This gets sort of like a complicated family reunion, but try imagining it this way. If you're gay with a gay gene on the X, you're going to notice that more of your mom's brothers are going to be gay while your dad's brothers aren't. Also, you'd realize that your maternal cousins, your mother's sister's sons, are more likely than average to be gay, but all your other three types of cousins—mother's brother's sons, father's sister's sons, and father's brother's sons—are not." These categories, he discovered, were exactly the ones elevated; the others were completely average. And so Hamer crossed off the third and last point on his checklist:

3. More male relatives on the mother's side were gay. *"Only* this pattern, called a maternal effect, was consistent with the theory that there might be a gene on the X chromosome," Hamer says. "It is necessary for the theory—but doesn't prove it."

What would conclusively prove it? Finding the gene.

"At that point," says Hamer, "we decided to escalate the study to the next step: going into the DNA to find the region of the X where the gene was hiding." It was at this point that the team confronted the big problem. "The fundamental difficulty of finding a gene for a trait is completely simple," Hamer says with a sigh. "I'll tell you what it is. The human genome is so damn huge. That's it. That's the problem. And that's *always* the difficulty in genetics. We don't know where or what the hell most of the genes are that we're interested in. The human genome is three times ten to the ninth bases long, about 3 billion Cs, Gs, As, and Ts, and except for little oases of knowledge that geneticists have managed to colonize along this DNA string of informational desert where we have figured out what's going on, it's still mostly 3 billion nucleotides of uncharted wilderness."[2]

"So," says Hamer, "this is my next riddle to solve: How do you find something when you don't know what you're looking for? How do I find 'it,' this gene, when I don't even know what 'it' is? I don't know where on the chromosome to look, I don't know what protein it codes for, I don't know what ACGT sequence to look for, I have no clue. I can't look for it directly. But what I *can* do is look indirectly for it. I can wander along the X chromosome and try to figure out where brothers have inherited the same sequences from Mom. And somewhere inside those sequences should be the gene.

"Presently," adds Hamer, "since the Human Genome Project hasn't yet given us good enough technology to determine the entire exact sequence of ACGT, we have to rely on a somewhat less exact technology, and that is linkage analysis."

In genetics, linkage analysis can be likened to an attack strategy of dropping depth charges semi-blindly from a ship onto a submarine hiding below. It is laborious, inefficient, and primitive, but it works and has been used, for example, at UC Berkeley by Dr. Mary-Claire King to scout out and narrow down the region in which BRCA1, the breast cancer allele, was hiding. The technique is based on the simple premise that genes that are close neighbors on a chromosome are usually inherited together. When genes are passed down from parent to child, the sexual process acts as a blender that mixes them, forming new genetic combinations. The genes that sit far away from each other on chromosomes tend, not surprisingly, to get separated from each other. Hamer had to go on a search through the wilderness of chromosomes to find out which genes were always being passed down from mothers to their gay sons again and again (and *not* to their straight ones). To track these genes, he would use "markers," chunks of DNA he could identify and track, even if he had no idea what they did. Markers are like uninteresting fish that are known always to travel with a certain, valuable shark; if you are hunting the shark but don't know where it is or even what it looks like, find and follow the fish that mark it, and eventually they will lead you to your quarry. Hamer knew that if a genetic marker segregated again and again from mothers to gay sons, he could bet with a high degree of certainty that the gene he was looking for was near it. (The marker that was segregating and the gay gene close by it were "linked," as the fish and the shark are linked, thus the name of the technique: linkage analysis.)

Hamer thought linkage analysis might work for two reasons. First was a fact that seems almost ridiculously obvious: When two people who are related (like brothers) share a genetic trait (like blue eyes), the fact that they share it is not just coincidence. Their parents passed the gene for blue eyes on to both of them. This means that if you look at all the blue-eyed brothers in a family, statistically more of them are going to share this particular blue-eyed gene than would be expected by random chance alone. As Hamer puts it, "If two sons—not twins, but just normal brothers—share a trait, like being left-handed, or blond, or, in this case, gay—and if there's some gene that makes a protein affecting that trait, then we'd expect those brothers to share the same version of that gene more often than not. That's because the allele of the gene that the brother has will influence his phenotype: the color of his eyes, or how tall he is, or the way he throws a baseball or falls in love. Straight men would have one version of this gene, and gay men would have a different one."

The other reason is the equally simple fact that men always get their Y from their fathers and their X from their mothers. A man's father delivers his Y to his son as is; Dad only has one Y to give, after all, so there's no possibility of mixing up those genes, and every man therefore is genetically exactly identical to his father on his Y chromosome (and to his grandfather, and great-grandfather, and so on forever). His mother, however, has two X chromosomes to choose from, and evolution, always seeking to mix up genes wherever possible, has decreed that during reproduction, Mom's two Xs will be mixed, a new X created from the old two, and that will be handed down to her son.

What Hamer had to do was, in principle, simple. He didn't know where the gay gene was, and at that point in the game he didn't care—sending a search party out into the vast genomic wilderness after this one tiny gene was simply too technologically daunting. He simply wanted to track the Xs and see if both gay brothers got any part of the X from Mom a suspiciously high percent of the time. Follow the markers as they swim from mothers to sons, and they will always, in the end, tell which parts of his mother's Xs a son got.

～

So in 1991 Hamer and Pattatucci started advertising in gay newspapers for gay brothers. For those who responded and agreed to come in, Hamer set up appointments at NIH and interviewed them with the questionnaire he and Pattatucci had created.

Eventually they selected forty families that had at least two gay brothers. The brothers were not twins, and because they were looking for a "sex-linked" trait (to the mother), none of the fathers of the gay brothers was himself gay. Similarly, none of the families had more than one lesbian member, because they were looking for a trait that was "sex-limited" to males. Hamer and Pattatucci assembled data, sweated over computer programs and interview questions, booked plane reservations and caught flights to subjects who couldn't come to the NIH. They took blood from the gay brothers and, when possible, from their mothers as well. Then they took the DNA out of the blood and broke it up into letters, dumped it into bowls of gelatin with electric current running through it, separated the letters, and read them up and down, column after column after column.

They ate airport food and conducted dozens of interviews sitting in hotel rooms that looked identical and in living rooms they had never seen. They took phone messages and filled in for each other at last moments when they got behind. They worked nights. They worked weekends. They worked with snow on the frozen ground outside Building 37 and ice on Old Georgetown Road just beyond their windows and then with the air conditioners keeping the swelter of the Washington summer at bay. And then another winter and another summer.

They went after a gene they hoped was there.

⌒

Just after their study is written (and rewritten several times) and run by colleagues and peer-reviewed but before it is to appear in *Science* magazine, Hamer sits down in his tiny cubicle to show what the team found. He cheerfully pushes a pile of notebooks out of the way ("Sorry, it's a mess") and, with the relief of one who has completed a task, proudly places a copy of the six-page article down on the cleared-off bit of desktop. He flips it open.

"The study," he explains, "is actually incredibly simple. It's just one big, fancy coin toss. Seriously, that's the way to think of it. In fact, the entire study can be reduced to a very simple question: Did both gay sons get any region of one of Mom's X chromosomes at rates higher than you'd expect just by random chance? That's the whole study. It's just heads or tails, and if you toss the coin forty times and get thirty tails where by chance you should only get twenty, you know something is going on."

Stroll down the Xs of the forty different pairs of gay brothers, says Hamer, and you'll pass a series of genetic markers on your way, one per region. Stop at each marker and simply compare it in all forty pairs, which Hamer describes as flipping a coin forty times. Did both brothers inherit the marker the usual, uninteresting, average number of times from both of their mother's X chromosomes (like getting roughly 50 percent heads and 50 percent tails)? Or, rather, was one particular marker suddenly inherited from the same X by both brothers over and over, the equivalent of the coin's suddenly showing up tails again and again and again, contradicting the rules of chance?[3] Both brothers have the same marker? Then they got this region from the same X. Different markers? They got their regions from different Xs. "Observe," says Hamer.

He pulls out a pen and writes down "test" and "null." "As with any good experiment, we created two hypotheses: a test hypothesis and a null hypothesis. The test hypothesis is always just the opposite of the null hypothesis. The test hypothesis is: 'There *is* a gay gene on the X chromosome that influences sexual orientation in human males.' The null hypothesis is, of course, 'There is *not* a gay gene on the X that influences sexual orientation in human males.' " The null, Hamer notes, says there's nothing here, and if you group these subjects together based on their homosexuality, you're just going to discover that the Xs segregate according to random chance—no more, no less.

"Now, what if the null is right? What if there's nothing there?"

Every mother has a two Xs, X1 and X2. By random chance, she will give her sons X1 about half the time. The other half the time, she'll give them X2. With one son, it would be like tossing one coin, heads or tails, a 50 percent chance of getting X1 and a 50 percent chance of X2. With two sons, it's like tossing two coins; there are four possible combinations the sons could have, and (divide 50 by 2) you have a 25 percent chance of both being heads or tails. Hamer illustrates this:

SON 1	SON 2	*Random chance of getting a given X*
1. X1 and	X1	25%
2. X1 and	X2	25%
3. X2 and	X1	25%
4. X2 and	X2	25%

"If you come up with this pattern," says Hamer, "the markers are sorting out according to the laws of random chance, and your null hypothesis is right: there's no genetic linkage. So you have to say, 'I guess there's no sexual orientation gene on the X chromosome of the human male,' and you wash your test tubes and turn out the light and go home. End of story.

"Ah, but what if the *test* hypothesis is true? What if there *is* a gay gene?" Hamer is speaking a bit faster now. "We're starting from the beginning again. This is what should happen.

"Say this gay gene comes in two alleles, straight and gay. The H version of the gene makes you gay, and h, the second allele with a slightly different order of ACG and T, makes you straight. Remember, we don't know where this gay gene is among thousands of other genes, we have no idea what protein it makes—we can't possibly find it. But we do know one thing: theoretically, both of our gay brothers had to have gotten the same gay version of this gene from Mom's *same X*. So we can look, quite simply, not for this unknown gene but to see which of the markers that sit along the X the gay brothers got.

"If they consistently got their markers from Mom's same X at a rate higher than random chance, the gay gene should be nearby.

"And that," says Hamer, "is the entire study."

Hamer and his lab began checking markers on the X chromosome. "Fortunately, one of the earliest markers I did turned out to be right in the center of the correct region," he says. "The reason is that I'm a slob. Well, sometimes I'm a slob. In most of our gene assays (measurements), we use radioactive isotopes, which can be somewhat dangerous. But since I'm notoriously sloppy in the lab, all my postdoc fellows and virtually everyone on the fourth floor would get upset when I worked with radioactivity because it would wind up in the wrong places. My technician, Stella Hu, more or less banned me from using radioactive isotopes. Believe me, she runs the lab."

Hu's spot in the lab is on a stool at a desk near the door, a few feet from Hamer's little cubicle. She is usually hard at work and does not seem to take lightly to sloppiness. "So there I am, nothing to do but twiddle my thumbs. But it turns out that the bigger genetic markers, these relatively huge stretches of DNA, are so large that you can assay them without radioactive material. You just ID them with dye on a gel. So, since I wanted to get going on a marker analysis, I started with those, and it just happened that one of the very first markers I tried was the right one. This saved a hell of a lot of time. Once you get one that shows a correlation, you can start looking at others in the same region, see if they're common to both brothers. Say both brothers inherit the same chunk of DNA 70 percent of the time, much more than by chance. You divide that chunk up and do the same thing, and one of those markers is even higher."[4]

As they began sorting through DNA, Hamer faced several major headaches. "When you're a gene sleuth," he says with an ironic sigh, "there are all sorts of plot twists you have to think your way out of." One roadblock thrown at them by the data had to do with figuring out which of the mother's Xs they were actually tracking. "What do you do," Hamer proposes, "if *both* of Mom's Xs have the same marker?"

You become disoriented. If the markers are the only recognizable thing in sight, and they are identical (or homozygous), how can you distinguish which of the mother's Xs they came from? Faced with this murky condition, Hamer would be thrown into the dark. "What do we do with this data?" he asks. "Throw it in the garbage? Because we can't tell if the sons got their markers from Mom's first X or her second." Homozygosity threatened to confuse or even ruin a large amount of data that Hamer's team had painstakingly gathered. "But there is," says Hamer, "a trick." He adds modestly, "It's simple."

The trick is, in fact, extremely simple: Go to another marker. The odds of a woman having two sets of identical markers are small. If she does, go to a third, or fourth, until there are two markers of different versions that confirm which of the two Xs the region is from. (At the same time, Hamer had to be careful that the supplementary landmarks were not too far from the main landmark; if they were, the markers actually may be "unlinked": In the process of meiosis, when the reproductive cells are formed, markers are often separated—it is called genetic crossover.)

Another problem involved awkward gaps in information. "What if you don't have the mother's DNA?" asks Hamer. "Maybe she's dead, or maybe she refused to participate in the study. You can't check her markers because you don't *have* her markers. Again, do you throw out the information? No. You derive her using her sons' DNA and statistics."

Hamer compares the process to reading the thin black bars of the universal product code (UPC). Genes are very much like the UPCs, of which, like snowflakes, no two are alike. The scanner at the checkout reads them easily, but what if it malfunctioned and read only part of the code, maybe only the center bars? The scanner would be acting somewhat like a geneticist reading only a single marker. The bars in the center of the code might be exactly the same, just as the markers could be identical; the scanner couldn't tell the pasta sauce from the cereal from the orange juice because in its narrow view, the bars are identical. Fix the scanner so it looks at the whole code, and it would know exactly which was which.

Hamer adopted this strategy for the markers: He broadened, in essence, the range of bars read. "Say that your marker, R, is a gene. You don't know what it does, and you don't care; R always sits in its little spot on the DNA, you can recognize R when you see it, and that's enough. Now, R comes in six different alleles, six versions of the same gene. But not all of the versions are equally common, which is a familiar concept if you think of a given country where maybe 60 percent of people have brown eyes, 30 percent have blue, and only about 10 percent have green or gray. Some versions of this marker gene are common, some quite rare. Let's say that the following is the frequency of the six alleles in the general population for this hypothetical marker, R.

"Marker R1 is found in 10 percent of the population. Ten percent of the people you see on the subway in the morning are carrying the first version of the R marker gene.

"R2 is found in 20 percent of the population. Twenty percent of everyone has R2.

"R3 is in 5 percent of everyone.

"So is R4.

"R5 is much more common. Thirty percent of everyone has it.

"R6 is the same: 30 percent.

"Add that up to get 100 percent.

"And there's another marker, V, right next door to R. V only comes in two alleles:

"V1 is found in 10 percent of the population.

"V2 is in 90 percent.

"Say we don't have the mother's DNA and her sons both have marker R1, and you have to figure out whether they both got their R1s from only one of Mom's Xs or from both. Use statistics. We know that Mom had R1 on at least *one* of her Xs and we know there's a 10 percent chance that she was homozygous for R1, that she had R1 on both of her Xs. So you're about 90 percent sure that Mom only had R1 on one of the Xs, and there's thus a 90 percent chance her gay sons got their R1 from this same X chromosome.

"But then, because you're a superconservative geneticist, you say: 'Ninety percent's not good enough.' So you look in the sons' DNA for V, the nearby marker. Guess what: they both have V1, homozygous again. Back to your dice. Mom's probability of having R1 is 10 percent, and the probability of her having V1 is also 10 percent, so you multiply 10 percent times 10 percent, and you get 1 percent. Statistically, you're now about 99 percent sure. That's pretty damn sure."

Hamer sits back with satisfaction. "You don't have Mom's DNA, but you do that five times, and you're statistically much more than 99 percent certain that the boys got the same X chromosome."[5]

The intrepid genes, however, soon threw another roadblock in the researchers' path: What if the mother has a gay allele on *both* of her chromosomes? Knowing which X went to which son, the heart of the study's method, becomes meaningless; even if the gay gene is right next door to a perfect marker, Hamer wouldn't see any linkage. Again, the scientists parried with statistics. "Fortunately," he says, "there's a little fact that saves us. The fact is simply that there aren't that many gay people in the world, and if there is a 'gay gene' it must be pretty rare." If 2 percent of X chromosomes carried the gay version of the gene and 98 percent had the straight version, then at most 4 percent of the mothers of gay brothers would have the gay allele on both chromosomes. Because of this factor, Hamer can never expect to see perfect, 100 percent linkage, "but on the other hand, it means you're very unlikely to think there is a linkage when there really isn't one. It's nice; it makes it easy to be hyperconservative, which good geneticists are."

The brothers had been coming in, responding to ads and word of mouth, and as their interviews were completed and their DNA extracted, the researchers analyzed the X chromosomes for each pair, patiently tracking down markers and at each marker asking the simple question: Is this marker shared more than 50 percent of the time? They began to compile the data.

"Lots of people ask, 'When do you publish a study?' The answer is: when the figures get good enough." For his team, the figures got good enough when they reached forty pairs of brothers and traveled through twenty-two markers down the narrow band of genetic wilderness. "And so," Hamer says, his eyes lighting up with evident excitement—he turns the issue of *Science* to page 323,[6] on which is printed Table 2 of the study, "Summary of Linkage Results," and lays it down almost with a flourish. "What did we get?" Table 2 is a list of the twenty-two markers spanning the length of the X chromosome that the Hamer team looked at, twenty-two landmarks labeled Marker A through Marker V.

"This list of markers is the entire story of the study," Hamer says quietly. "You look at the numbers. The story is in the numbers. Start here." He points to the top: Marker A.

Marker A is at the top of the neat column on the left labeled Locus. Under the next column, Location, is the marker's genetic address. A's is "p22," which means it is on the X chromosome's p (petit) arm in the p arm's twenty-second region, or Xp22. This is the remotest tip of the X, in a sense the northernmost outpost on the long thin ribbon of the chromosome.

The next column gives the number of alleles for each marker: A comes in six different versions.

But the two most important columns are toward the right side of the page. "The answer to the question is"—Hamer points to one of the columns—"this number here." It is the column labeled z. Hamer explains that z is the number of gay brothers who shared the same allele (out of Marker A's six possible alleles) from Mom's same X.

SUMMARY OF LINKAGE RESULTS

Locus	Location	AL*	HET#	Sib-Pairs†			z_1•	$2\ln L(z_1)$•	P
				[D]	[S]	[−]			
A. .KAL	p22	6	0.77	5	16	14	0.51	0.01	ns
B. .DXS996	p22	11	0.84	7	14	18	≤.5	≤0	ns
C. .DXS992	p	8	0.87	6	13	19	≤.5	≤0	ns
D. .DMD1	p21	9	0.78	3	10	23	≤.5	≤0	ns
E. .DXS993	p11	6	0.80	3	14	17	≤.5	≤0	ns
F. .DXS991	p	8	0.77	8	14	14	0.57	0.61	ns
G. .DXS986	q	10	0.71	7	20	10	0.65	2.11	ns
H. .DXS990	q	7	0.76	4	19	13	0.55	0.25	ns
I. .DXS1105	q	5	0.48	3	20	9	≤.5	≤0	ns
J. .DXS456	q21	10	0.85	8	20	8	0.75	7.95	0.00241
K. .DXS1001	q26	10	0.82	8	16	13	0.60	1.09	ns
L. .DXS994	q26	5	0.75	7	17	13	0.55	0.26	ns
M. .DXS297	q27	5	0.70	5	21	8	0.71	4.25	0.01963
N. .FMR	q27	17	0.79	6	17	14	0.56	0.45	ns
O. .FRAXA	q27	8	0.72	4	17	13	0.56	0.38	ns
P. .DXS548	q27	6	0.67	7	20	7	0.73	5.21	0.01123
Q. .GABRA3	q28	4	0.35	2	23	3	0.74	2.39	ns
R. .DXS52	**q28**	**12**	**0.79**	**9**	**22**	**6**	**0.81**	**11.83**	**0.00029**
S. .G6PD	**q28**	**2**	**0.36**	**4**	**24**	**2**	**0.85**	**6.38**	**0.00577**
T. .F8C	**q28**	**2**	**0.41**	**5**	**24**	**3**	**0.82**	**6.56**	**0.00522**
U. .DXS1108	**q28**	**6**	**0.71**	**8**	**22**	**4**	**0.85**	**12.87**	**0.00017**
V. .DXYS154◄	**q28**	**10**	**0.71**	**8**	**22**	**5**	**0.83**	**12.84**	**0.00017**
R/S/T/U/V	**q28**	**−**	**0.99**	**12**	**21**	**7**	**0.82**	**18.14**	**0.00001**

Linkage analysis was performed on 40 male homosexual sib-pairs using 22 X-chromosome markers (ref. 30). The 5 marker loci located in distal Xq28 are in **bold.**

*AL is the number of different alleles observed in 62 to 150 independent chromosomes.

#HET is the calculated heterozygosity; $HET = 1 - \Sigma f_i^2$, where f_i = frequency of the ith allele.

†[D] is the observed number of concordant-by-descent pairs; [S] is the observed number of concordant-by-state pairs; [−] is the observed number of discordant pairs; noninformative pairs are not included in this analysis.

•z_1 is the estimated probability that 2 homosexual brothers share the marker locus by-descent (ref. 31).

♣$L(z_1)$ is the ratio of the likelihoods of the observed data at z_1 versus the null hypothesis of z_1 = 1/2 (ref. 31).

♦P (one-sided) was calculated by taking $2\ln L(z_1)$ to be distributed as a chi-squared statistic at one degree of freedom; ns: P > 0.05.

◄ Only the maternal, X-linked contribution was considered for this sex-linked locus (ref. 23).

Since it is Marker A, the number sits at the very top of its column: z equals .51. In other words, Hamer point out, 51 percent of the gay brothers shared the same version of Marker A. "Almost exactly half," he says, "which is almost exactly what you'd expect to get by random chance, by flipping your coin. But as a scientist, you want to know not only what z *is* but what the number *means*, what greater significance does it hold? So you look here." His finger slides right to stop at the farthest-right column. It says p. "P is significance, a statistical measure. P takes a good look at z and tells you if you should get excited about it or shrug your shoulders. P is what you get after you run your z through a chi square analysis, a standard statistical analysis to get the greater significance this marker has for your universe. And this is what p says for Marker A." The significance column says that for Marker A, p is ns.

"Not significant." Hamer smiles. "Nothing. Zero. Zip. Random chance. No meaning. 'Not significant' means your p is too big."

The thing to remember about p is that Hamer wanted a small one. "I mean, you want p to be *tiny,*" he said, "as close to zilch as possible." A p value of .05 means, as Hamer had explained before, that there is only a 5 percent chance that the finding is meaningless, only a 5 percent chance that it turned up through dumb luck. "And it means, conversely," he noted, "that there is a 95 percent chance that you're onto something." Were Hamer to get an even smaller p, like .03, this would be even better: It would mean that now there was only a 3 percent chance that the result was pure coincidence but 97 percent that it was significant. "The smaller your p gets," says Hamer, eyeing the p column, "the smaller your chance that the finding is meaningless. A p of .003, for example, would be truly great."

He looks back at the table. "So let's go to Marker B," says Hamer.

Marker B, next down from Marker A in the Xp22 wilderness of the tip of the X, has eleven alleles and is shared less than 50 percent of the time by gay brothers. Significance? ns. Zip again.

Marker C is ns.

Marker D is ns.

Marker E in Xp11—we're traveling ever so slowly from the tip of the X toward its midpoint—comes in six alleles and is also shared less than 50 percent of the time. P is not significant. Nothing here.

At Marker F, there is a slight change. The z has jumped up a notch; 57 percent of the brothers are sharing the marker. And next down, Marker G is shared by 65 percent. But both are still ns, and G is already past X's midpoint on its long arm, the q arm, so we've already run through half the X chromosome with no result.

Marker H is ns.

So is Marker I, but when we reach Marker J—in region q21—

there's a difference. J is shared by 75 percent of the brothers, and here an interesting p value finally appears: ".00241" says the p column.

At Marker K, the z collapses, falling back to 60 percent. Significance? None, again.

Marker L is the same but worse, only 55 percent sharing.

But Marker M is 71 percent and has a p of .01963.

Quietly, over in its own column, z has begun unsteadily but slowly climbing. We are nearing land's end, the tip of the q arm of the X.

Marker P is 73 percent and Marker Q—we've just entered the twenty-eighth and last region—is 74 percent.

And then we reach the five final markers. They appear in bold type: R, S, T, U, and V, and together they cover the twenty-eighth region. Notice the number of alleles these have, says Hamer. He multiplies 12 by 2 by 2 by 6 by 10 to get, as he puts it, 2,880 possible bars in this universal product code. If those markers work in your favor, he says, the risk of being misled is 2,880 to 1 odds against. So do they?

"When you do linkage analysis," Hamer explains, "you get a bell curve as you go down the list of markers and get nearer and nearer to the gene. The numbers start to go up to a peak. Of the forty pairs of gay brothers, thirty-three shared with each other the Xq28 region—these five markers—and seven did not."

Moving down from Marker R to Marker V, the zs are still climbing to their peak, the five highest z values we've seen: 81 percent, 85 percent, 82 percent, 85 percent, 83 percent. But it is the p that is surprising. Out of nowhere it has plunged, falling like leadweight. The p values of the markers are .00029, .00577, .00522, .00017, and .00017.

"Remember that an overall p value as high as .05 is considered scientifically significant," says Hamer. "But because the genome is big, you'd really like your specific p value to be less than .001 for you to stick your neck out—especially when you're doing the very first genetic analysis of the trait." He looks back to the page. "When you do the stats for Markers R, S, T, U, and V in the Xq28 region, you get"—he glances at the page again, just to make sure—"a 99.5 percent probability of its being a significant finding. Which is terrific. But it's your final p value, the statistical significance, that in the end determines the solidity of your study. The final p we got is—"

Hamer frowns and looks down at the study, as if to reconfirm that the figure is really there. His eyes quickly search the page near the bottom of the p column where Xq28's net worth is dispassionately summarized in small, precise print, locates it, and points. The intended object of the finger is lost for a moment in the pool of numbers reproduced in black photocopier ink, and then, at the very bottom, the statistic jumps into focus.

The p is .00001.

~

In the scientific publication *Mendelian Inheritance in Man,* a huge tome scientists often refer to as the geneticists' Bible, Hamer's locus has been recorded. Its official name is GAY-1. Researchers are now in the process of tracking down the specific gay gene inside the locus.

Hamer's team recently completed two further studies of Xq28 and sexual orientation.[7] The first tested an entirely different set of gay brothers. It confirmed the original finding.

The second tested the heterosexual brothers of the gay brothers. If Hamer's results were valid, there should have been no linkage between the straight brothers and GAY-1. There was no linkage.

Chapter Eight

WHAT DOES "GENETIC" MEAN?

I F THE PUBLICATION of Simon LeVay's neuroanatomical study had been greeted with interest, the announcement of Hamer's genetic study, in the same magazine and almost exactly two years later, caused an international media frenzy. Around the world, newspapers dated July 16, 1993, carried the news. Both Hamer and *Science*'s publicity department had gone into high gear, responding to the questions of the *Washington Post* and the *New York Times*; the *Post*'s front-page headline read, "Study Links Genes to Homosexuality."

Television reporters found Hamer immediately. He, in turn, began to find television, and the rest of it, quite trying. Dan Rutz of *Prime News* on CNN filed an evening report on the study, and Hamer appeared on NBC's *Today* show at 8:00 A.M. after doing CBS *This Morning* at 7:00, where Paula Zahn, with great seriousness, asked him about the outmoded theories of environmental causation. "It's really important," Hamer responded diplomatically, "to realize that there was never any hard evidence, never any experiments done to test that. And in fact when some researchers looked at more than a thousand gay and heterosexual men a few years ago, they found very little difference in environment or parenting style."

There were a few minor mistakes. The following morning on ABC's *Good Morning America*, Charles Gibson announced: "Researchers at the National Cancer Institute have identified a similar genetic pattern in thirty-three pairs of gay brothers, a finding which suggests the tendency to homosexuality may be inherited." Hamer joined

Gibson via satellite and explained he hadn't found a "similar genetic pattern" in the brothers but a high percentage of markers inherited from the same chromosome. (The *MacNeil/Lehrer NewsHour* made the same mistake.) "If," Hamer added to Gibson, "you think of all your genetic information as a single forest of 100,000 trees, we've now narrowed the search down to a thicket of one hundred trees or so, and we're in the process of searching for the single tree in this area."

Gibson asked how sure he was of the results. "Well," Hamer said over his satellite feed, "one good way to test it is to take a coin, flip it forty times, and get thirty-three heads. You'd get that less than 1 percent of the time. That's our confidence level."

And on ABC *World News Tonight* Peter Jennings, anchoring from New York, introduced ABC science correspondent George Strait, who reported, "After studying the genetic makeup of forty pairs of homosexual brothers, scientists at the National Cancer Institute found two-thirds had exactly the same kind of genetic material along the very same section of the so-called X chromosome." Actually, they did not have the same genetic material but had, rather, inherited a locus from the same X.

But these were quibbles, and the reporting of the facts and details of the study was generally excellent. What increasingly struck Hamer was a larger problem, the media's misunderstanding of what, exactly, the word "genetics" meant, its misconception of "a gene." Hamer, the molecular geneticist whose entire career consisted of searching for genes and an explanation of what they would mean, discovered that his first task was explaining what a gene does *not* do and what having a gene does *not*, in fact, mean.

On *NBC Nightly News* Tom Brokaw opened with the Hamer story, going to NBC science correspondent Robert Bazell. Bazell reported: "This latest research . . . shows that homosexuality is a natural genetic variation, like left-handedness or blue eyes." Likewise, on ABC, George Strait reported, "Gay activists intend to use today's discovery to support their argument that homosexuality is not deviant behavior." It wasn't that Strait himself was saying a gene automatically makes a trait "not deviant"—he wasn't, exactly. But if he wasn't, Strait (and virtually all science journalists, many of whom made similar reports) did not understand a vastly more important and fundamental aspect of the genetics, namely that Hamer's findings, and genes in general, do not speak to this question in the first place.

NBC implied that a genetic origin made a trait acceptable, or (ABC's variation on the theme) "not deviant." To Hamer, this reporting was incorrect, or at least fatally imprecise. Having a gene, he noted, doesn't mean anything about a trait's being "natural" as in "good" or even just "benign." By that logic, the discovery of a gene for colon cancer would render colon cancer a "natural" variation of human biological development "like left-handedness or blue eyes." In fact, cancer *is,* in a strictly biological sense, natural. But in that same sense so is having Hodgkins' disease and fingernails. The crucial distinction—one that neither network made—is that Hodgkins' is pathological, that is, destructive to the organism, and in no uncertain terms deviates from the healthy norm, while having fingernails is not and does not. The fact that there is a gene for diabetes doesn't mean that diabetes is "healthy" or "good." It is neither, and having a gene for a trait does not mean the trait is good. A trait can be "genetic," perfectly "natural," and absolutely lethal.

Genes also may play a role in alcoholism, violence, gambling, pedophilia, and other states of mind expressed behaviorally. Misidentifying a color-coded warning sign because one has genes for color blindness does not make that behavior "not deviant."

In the end, genes mean nothing to whether we value or devalue these behaviors. (They might change the way we attempt to deal with them, but that is entirely another matter.) There is increasing evidence that manic depression is genetically influenced; finding those genes will not make manic depression any less a pathological state of mind than it is. It will not make manic depression into NBC's "left-handedness or blue eyes." What determines whether a genetic trait—left-handedness, diabetes, blue eyes, cancer, hair color—is pathological or nonpathological, to be accepted or repudiated, is completely unrelated to any genetic influence. Genes are irrelevant to judging a trait's deviance and pathology or its lack of it. Both pneumonia and diabetes cause measurable physical deterioration and harm and that fact is irrelevant to the fact that one is caused by genes and the other is not. The media is very unwise to argue that the simple fact that there is a gene for homosexuality means it isn't deviant. A gay gene does not mean that homosexuality is not pathological, and it does not mean that it is not deviant. This is not what "genetic" means.

But where Hamer's results truly eluded the media—and where the media perplexed him in return—was on a point so diametrically opposed to basic biology that it was days before Hamer fathomed what he was being asked. The *Today* show announced, "There is new evidence that homosexuality may be inherited in some cases and not a matter of choice," and on *ABC World News* Peter Jennings said: "There is a new medical study which may have a significant impact on the debate about the nature of homosexuality: Is it choice or is it biology?" Tom Brokaw opened with: "There's new medical evidence that homosexuality is genetic, not acquired behavior." On *Good Morning, America,* Hamer was asked, "So if you say that genes are part of the reason, you're really saying it's not a choice." Hamer paused and then responded with an observation about phenotypic variation that bypassed the question. The media could not understand why Hamer would not address "choice"; Hamer could not understand why they were asking him about it.

Ted Koppel conducted the last major interview with Hamer, and the tape of the *Nightline* broadcast illustrates perfectly the crossed signals of politics and science on the issue of genetics and homosexuality. Afterward, sitting in the tiny office attached to his lab, Hamer could analyze it, understand it in retrospect, even look back at it and laugh. At the time, he found it baffling.

Nightline began at 11:30 as usual, announced by the voiceover, "There may be a genetic basis for alcoholism. That doesn't mean we tell everybody to go out and get drunk because you can't help it."

Koppel announces: "Tonight: the genetic link to male homosexuality." He then opens the broadcast with the most popular, and fundamentally incorrect, view of what genes mean: "More authoritatively than ever before, a scientific study is suggesting that a man's homosexual tendencies may not be a matter of choice."

He continues. "Think about it for just a moment, think only about the legal implications. While it is constitutional for example to prohibit certain behavior, it is not constitutional to make status such as race illegal. If the findings of this study are confirmed, it will not quite raise homosexuality to the same legal level as race, but it moves it a lot closer. It would also be true however that if homosexuality is genetically traceable, that some parents-to-be may choose to eliminate the baby."

Nightline cuts to its science correspondent Dave Marash, who has prepared a report. Marash begins with a clip from the play *The Twilight of Golds,* about a woman who aborts her baby because of a genetic test that determines it will probably be gay, introduces the viewer to Hamer, listens to sociologist Charles Moskos suggest bizarrely that homosexuals are "a third sex," and then goes to Robert Knight of the Family Research Council (the Christian organization that mails out Bill Byne's work). Knight, in his remarks, calls homosexuality a "lifestyle." If there is a gene, Marash concludes, conventional wisdom holds that gays will find more acceptance. But he notes that race and sex are involuntary and have never stopped people from hating on that basis.

Back in the studio, Koppel welcomes Hamer, who looks slightly nervous and out of his element, his slate-blue eyes wary under the lights. After the introduction, Koppel turns and asks, "If the findings of the study, Dr. Hamer, are confirmed, will it then be accurate to say that homosexuality is not optional behavior?" It is the very first question, and to a biologist, it makes no sense. Hamer freezes.

He tries to figure out what he is being asked. "Well," he begins, "that portion of homosexuality or heterosexuality that is genetically influenced is, of course, not optional because people don't have an option over what genes they're going to inherit."

Koppel also hesitates. He tries again. "In a sense that's begging the question. Or maybe I don't understand the refinement that you have just made."

Hamer is confused. He has been quite clear that the genetic region that he has isolated—or maybe any number of genes that influence sexual orientation—are not the sole factors involved in the creation of the trait sexual orientation (since in genetics, there are frequently nongenetic biological factors contributing to the trait).

Gamely he attempts another answer. "What we've found is that one specific region of the X chromosomes is linked to homosexuality, at least in some men. And what that demonstrates is that part of being gay, or part of being straight, is determined in the genes. The reason I say that it doesn't mean it's not an option," explains Hamer, attempting to dovetail with Koppel's phrasing, "is that homosexuality is [probably] not simply determined by a single gene, as your eye color is determined by a single gene."

But Koppel is interested not in genetics but in the political debate over whether homosexuality is chosen. (Hamer thinks the interview is about genetics.) Koppel attempts, "Will it be possible, at least, to say that it is not a purely—behavioral thing . . . ?" Hamer, trying to be helpful, jumps in. "There are definitely inherited characteristics which are very important," he says supportively. "That's correct."

This is not the answer Koppel wants.

"And," Koppel tries, "how important—" Surely Hamer must realize that politics are central here. "I'm just trying to get you to put it in as commonplace language as you can so that we all understand it."

"Sure," says Hamer, clearing his throat. (Afterward, he will recount, with a thin smile, that he was furiously wondering why Koppel would not accept his scientific answers.) Hamer is trying to make it simple, and after all, if he is on the program, then the subject must be science. Right? He tries again. He talks about the Bailey and Pillard study, he cites precise estimates of the percentage of which genes have "a role in whether a person is gay or heterosexual." He talks about his study, gives a genetic estimate, and confirms—yet again—that the evidence says a substantial component of homosexuality is inherited. "We can measure it," Hamer says clearly, "we can assess it in the laboratory. The *important* finding," he says, reaching his point, "is really the *proof* that there is such a component." He means proof that there is a genotypic component influencing this trait, not proof about "choice" or "volition" or "will" or some phenotypic intangible. For good balance, Hamer refers to some of the other possible factors that might be involved, such as other genes, perhaps hormones. "What's important today," he concludes, "is that we've clearly demonstrated that genes are involved."

After a commercial break, Koppel introduces the Reverend Peter Gomes, professor of Christian morals at Harvard University, and Art Caplan, the director of the Center for Medical Ethics at the University of Minnesota. The mood seems slightly altered, perhaps a shade more tense. There have clearly been some hurried negotiations during the break because Koppel then adds, "And joining us again—and I should point out that Dr. Hamer wants only to refer and to comment on the scientific aspects of the story—is Dean Hamer."

Koppel tries Gomes for an answer to the political question. "Professor Gomes, my sense is that those who want to be accepting of homosexuality are going to be able to use this information to help their case and, as Professor Caplan was just suggesting, those who do not want to use it will be just as able to use it for theirs. Your thoughts?"

Gomes, unlike Hamer, identifies the question as political, not scientific, and answers concisely in political terms, stating in his distinguished, Harvard tones his belief that yes, in the end, the research will be more helpful politically in resolving the debate than harmful. "I think part of the whole conversation about homosexuality has been to confuse it with some deliberate choice of lifestyle, confuse it with a 'lifestyle,' and suggest that it is somehow an option that other people who are normal, as it were, do not have," he says. "In the sense that homosexuality is now to be seen as part of the equipment with which some people are born into the world, in some respects I think normalizes the debate, and I think that's helpful."

Having finally received an answer to the "choice" question, Koppel visibly relaxes. He circles back one more time. "Let me come back to the science of this, Dr. Hamer. To what degree is that kind of reassurance for gays warranted? In other words is—is it appropriate, based on the findings that you have reached, that gays can say look, it's not a matter of choice, it's predetermined?" asks Koppel. "In a sense," he adds quickly, just to make things clear. "Genetically," he adds.

And Hamer pauses, very briefly: He's finally gotten it. He takes a reluctant breath and, gingerly, gives the answer Koppel has been searching for. "I think," Hamer says, tightly and conservatively, "all scientists that have studied sexual orientation already agree that there's very little element of choice in being gay or heterosexual." And with that, he turns and heads back into science, pointing out tersely, and correctly: "The *question* is whether there's a defined genetic component to homosexuality and if we can ultimately understand how that works."

And now Koppel understands this biologist on the screen in front of him. "Why's that so important?" he asks encouragingly.

Hamer lays out his bottom line, the view of the geneticist: "I think it's important because without concrete evidence, it's really all just hand waving." He cites the old psychoanalytic theory of homosexuality's "cause" as the distant father, "an idea that was widely accepted but was *never* based on any fact. That's why I think it's really important for scientists like myself to try to get good, concrete data using standard experimental techniques."

Weeks later, Hamer will reflect that among credible researchers and clinicians, this "choice" debate is no longer seriously discussed. No one with an accurate basic understanding of what genes mean would bring it up, Hamer says with a sigh. Discovering the gene for sexual orientation is no more essential to knowing the nature of the trait than is the discovery of the gene for left-handedness essential to knowing whether or not people choose to be left-handed. But the public is confused about what the gene means and about the politicized scientific debate about "choice" that flows from it. Hamer realizes reluctantly that in this political debate, the gene is a symbol. That is all. It is a symbol of certainty, of a reality that is too politically distasteful, unpalatable, even terrifying to many people, a reality they cannot accept without help.

And this is what "genetic" means politically.

The irony is that, in the end, even politically this much-discussed gene may be irrelevant: As Dave Marash noted at the beginning of the show, those who hate will hate.

Nightline is almost over. They've survived. As a wrapup Koppel, now slyly playing devil's advocate, allows himself a final question by returning to sociologist Moskos's weird interpretation of Hamer's genetics. "Are we here," he asks, "and this may seem like a ludicrous question to each of you, but are we talking about a third gender, a third sex?"

Hamer looks as if he is going to raise an eyebrow, but he doesn't. It is, indeed, a ludicrous question, but then, for anyone familiar with biology, Moskos's was a ludicrous point. And again, not what genes really mean at all. As always, Hamer chooses to be diplomatic and biologically precise: "There are, of course, some people who have been called a 'third sex,' essentially hermaphrodites. The locus we've discovered is not related to that in any way, shape, or form, and as far as we can tell from our research results, the locus we've identified is just a normal variant similar to the sort of variants that cause differences in eye color or handedness or perhaps in other behavioral traits."

It is his last comment, typically scientific. Gomes's terse response, once again, is more to the point. "It hasn't created a new sex," he snaps. "It has broadened our capacity to cope with the two sexes we have."

∼

Hamer views his experience with the media with good-humored amazement. He recalls the CNN reporter who intoned solemnly before the camera that Hamer's possible finding of a gene "indicates homosexuality might not be a choice," Hamer sighs and then laughs. "It's the exact reverse. The phenotypic fact that homosexuality is *not* chosen was one of the mandatory biological preconditions to concluding there was a gene for this trait." More simply, you have to make sure first the trait isn't chosen before you look for its gene. He pauses, incredulous. "Can you imagine any sane, reputable biologist spending years of their life and their resources looking through chromosomes for a gene for something that's *chosen*? I suppose you could do it, but you'd have to be a complete idiot, because it would be the genetic equivalent of staking your entire scientific career and reputation on finding the gene for"—he searches for an example—"being a Methodist." He sits in his office, trying to imagine it.

∼

What genes *really* mean, one exhausted geneticist growled, is constant headaches.

Every day we do more research on genes, and every day we know less about them. Yesterday we were confident they were a construction blueprint, instructions for wiring here, drywall there, plumbing over there. Today we believe that, in fact, they're really not a simple, straightforward blueprint at all, but if they're not, we're not actually sure *what* they are.

We used to know that a gene was a gene was a gene, whether inherited from a man or a woman. This was straightforward, unquestioned Mendelian law. Now we know (or at least we think) that differences between oogenesis—generation of the egg—and spermatogenesis—creation of the sperm—make it matter very much whether certain genes come from Mom or Dad. With Huntington's disease, the sex of the person that passes on the gene means a huge difference in the age of onset. We don't know why. Yesterday we were at least sure that one gene controlled one trait. Today we are at least unsure that this is true and, at most, utterly confused about the matter. Tomorrow we may be sure of something else.

An example. Yesterday (in this case "yesterday" is almost literally accurate) we were absolutely certain, at the very least, that identical twins had identical genes. And we were thus certain that left-handedness and homosexuality couldn't be 100 percent genetic, because in identical twins the traits were concordant only 12 percent of the time for left-handedness and 50 percent of the time for homosexuality. (The exact figures Bailey and Pillard arrived at for homosexuality in men is 48 percent discordant, 52 percent concordant; for women, they got the reverse: 48 percent were concordant, 52 percent discordant.) The other factors are reported by the media to be "environment," a word the media still mistakenly believe means "society." This assumption is still common in most current textbooks, because it was thought impossible for things to be otherwise. Monozygotic twins are just that, from "one zygote," one new cell from an egg and a sperm. The cell divides, two cells, four, eight, and then—and like so many other things, we don't know why—the almost unimaginably tiny human embryo of eight cells splits apart to become two embryos, which will become two people with (we thought) exactly the same DNA. We were thus certain it was impossible for identical twins to have different genes.

Today we are only certain that this impossible thing happens all the time.

"There are at least two ways," explains Lisa Geller, "that identical twins, with basically identical genes, can differ genetically."

Geller is a molecular geneticist who researches the genetics of Alzheimer's disease at Harvard Medical School in Boston. She has short, dark blond hair and is at once intense, engaging, serious, and enthusiastic. She talks fast. In a small meeting room one floor above her lab and high above the Medical School's long, imposing, marble-surrounded rectangular campus just off Longwood Avenue (the campus looks like a grouping of Roman mausoleums), Geller hunted down the last nub of chalk and wrote on the blackboard, "developmental noise." "That's the first one," she says. "It's basically a fancy name biologists use for 'blind luck.' "

How it works, Geller explains, is no more or less surprising than the way anything works. "There's a randomness in all biological development," she explains. "Every single cell in your body is not predetermined as to where it's going to go or what it's going to do. There's a fair amount of plasticity, so, for example, if there's a gene for handedness, maybe it directs you to go in a right-handed direction, but if you're anchored as an embryo to the left side of the uterus, maybe it doesn't." The process is called stochastic, a mathematical term meaning resulting from a random process and one that geneticists use similarly in biology. One geneticist described it as "When your body is getting ready to make a heart, genes direct the heart cells to fuse together and then move on to the next step. If they don't fuse in time, they've missed the train. Ninety-nine times out of a hundred a gene will direct cells correctly, but one time it will just happen not to. That's stochastic. Random processes." There are nongenetic biological stochastic processes as well, such as the growth vitamin biotin. Biotin is used in a variety of ways, but its concentration can vary randomly, and thus the processes that use it also vary randomly.

"We'll never be able to control it all," Geller says, adding, "which, personally, makes me kind of relieved." But the question here is whether there is developmental noise in identical twins. She turns back to the blackboard and, under the words "developmental noise," writes "genetic differences." "Here is the second possible way," she says. "This is speculation, but we're just starting to ask something fascinating, whether it is possible that you could have a trait, like handedness or sexual orientation, that is genetically determined *and yet* have discordance for it in what we always thought were genetic clones: identical twins. And there almost certainly is." Considering that question, Geller says, has forced scientists to alter their concept of alleles. "We used to think they were all distinct, specific versions of genes. Cystic fibrosis alleles, for example, are all standardized, always spelled the exact same way. Now we're asking if there are animate, chimerical pieces of DNA that actually move and change before our eyes and defy our easy classifications." Geneticists also are now asking if the concept of dominant and recessive is quite so fixed as they have always thought.

The best example of this rethinking process comes from a new discovery that certain triplets, such as ACC or GTG, repeat themselves in certain genes over and over again, actually multiplying themselves for some reason not yet understood and changing the effect of the gene. "We already know of five or six single gene traits where this happens," Geller says, taking a pen and a piece of notebook paper. "Okay. Let's use a gene named HD. It has several allele versions, and one of these does something well known. You've heard of it. I'm not going to tell you what it is for the moment. I'm going to leave you hanging.

"Here is what I'll give you. HD makes a protein, like all genes. Its address is 4p16: fourth chromosome, p arm, sixteenth region. And it's a dominant, not a recessive, so all you need is *one* copy of the allele that causes this trait, and you've got the trait. At least, that's the way we've all been taught dominants are supposed to work, right?" She raises an eyebrow, clearly enjoying this, and, as she talks, begins to draw.

"Like all genes, HD is made up of a long string of A, C, G, and T nucleotides, the four letters of DNA. At its beginning you've got your standard START HERE punctuation—the triplet ATG, that signals the gene's beginning—and you've got your STOP HERE period-triplet, TAG, at the end, and all the other triplets inside that code for their various amino acids which you snap together to build the protein." Geller turns the paper around. She has drawn the gene TAG at the front and a string of bases trailing behind it.

"Now, here's what's interesting." She returns to sketching. "What you've got in this particular gene is something called triplet repeats, which are exactly what they sound like, triplets of three nucleotides that repeat over and over in the gene. In HD, CAG is the triplet that's repeated, and you have CAG CAG CAG CAG CAG CAG again and again." She draws these. The gene on the page grows longer. "What we've discovered is that the number of these triplets is very unstable, and it actually changes. Whether you get the trait or not depends, in turn, on the *number* of triplet repeats in your own HD gene. If you have over a certain number of those repeats, you have the trait. Under a certain number, you don't. Most people have thirty-one or thirty-two CAG repeats in their HD. We now know that if you've got somewhere over about forty-five, you get the trait. But if you have under around forty-five, you don't. Why is this? We don't know. What's the trait? Huntington's disease. Your HD will give you Huntington's or not depending on this, and this is, for us, a new way genes operate."

What has amazed geneticists is that the number of repeats appears to be quite variable among people. Geller points out that neither she nor I have Huntington's, but the number of repeats I have may be quite different from the number she is carrying, and one or both of us could be carrying the allele that creates Huntington's—or, for that matter, the allele for Fragile X, which creates mental retardation, or for myotonic dystrophy, or for homosexuality.*

As nearly as we can make out at the moment, if one parent has only a slightly high number of repeats, then chances increase that the child will have an expanded number. What this means in "real life" is that out of nowhere, the child will get Huntington's, which is a so-called genetic disease, even if there is no family history of it at all and it has never, ever shown up in the family's pedigree. The trait is, surprisingly enough, heritable without being familial. This has naturally forced a rethinking of the meaning of "genetic." Add to that the fact that while Huntington's is a dominant, and a child needs to inherit only one copy with more than 45 CAG repeats to get the trait, it is a "single-gene trait" caused by a "dominant" gene that may or may not (again, for reasons we don't understand) create the trait. That a dominant "may or may not" create a trait, Geller notes, breaks with the simple yes-or-no idea of a dominant gene that we've always had.

Another thing we don't yet understand is that a child has a much greater chance of repeats jumping into the danger zone if the father has a high but still normal number of repeats rather than if their mother does. So not only is this "dominant" a sort of "recessive dominant," if one can imagine such a thing, its action also depends in part on which parent passes it on.

Geller sighs and sits back, reflectively tapping the pen on the tabletop. "Very interesting stuff, but it makes it much harder for us geneticists. The big question now is when do those repeats jump, when does it happen?" She literally throws up her hands. "The numbers change from one generation to the next, and nobody knows how or why. You can have a difference in the genes of identical twins, something we'd always assumed was impossible, but understanding this

---

*Geller notes that it appears to be rare for the number of triplets ever to decrease. They seem to ratchet in only one direction: up.

phenomenon now hinges on the question of when this expansion occurs." The point is, Geller notes, that homosexuality could work by this newly discovered genetic mechanism, a trait created by an allele of a gene with a number of repeats hovering *barely* under the trip-wire that triggers it. "Say your mother had sixty-two repeats in the gene that Dean is tracking down in Xq28," Geller says, "but in you, something happened—maybe some stochastic event, random chance—that bumped it up to sixty-three. And that was the line. And that's why you're gay. But your brother has only sixty-two, so he's straight. So heterosexuals are carrying this allele around, passing it on, and not expressing it because they're hovering right under the line without knowing it."

Geller pauses for a moment, musing on something. Then she says, almost to herself, "If we apply this mechanism to something like homosexuality, seeing that around 50 percent of identical twins are discordant for sexual orientation, then the jump has to happen after there's been one division. One cell would jump, making that twin gay, and one cell wouldn't, and that twin would be straight. That's a lot of stuff to happen. But it's possible. The time when genes are most likely to be rearranging themselves is at the sperm or egg stage, or when they are first fertilized. After that, it might be less likely. The cells are sort of doing other things, mostly they're concerned with dividing rapidly, so it isn't a good time to be fooling around with your genes. But maybe that's when a mistake is more likely." She looks out the windows onto the Harvard Medical School quad. "I'd say later is more likely. Just a guess. But you know," she says to the quad, "you can guess forever with biology, but until you run that one experiment and get the answer, you don't know. You just don't know until you know. One really interesting question—"

Geller stops and sits still. Then she blinks and smiles. "It just occurred to me how you'd go about finding out. The thing to do here is to go look at identical twins and see if there is concordance for Huntington's or Fragile X or myotonic dystrophy, all of which are caused by genes that have these variable repeats, and that will give you the answer right away. That would tell us something about when it happens." She laughs. "I think we've just invented a new experiment."

She scratches it down in the margin of the already crowded piece of paper. "I'll call someone up and ask them about the concordance. I know some people who do this stuff. Either they'll know the answer or they'll find out." She is scribbling: "Trinucleotide repeats: 100 percent concordance MZ twins for diseases re: repeats appearing newly?"

"We'd be looking for people who didn't inherit it from someone," she muses rapidly. "People who are the first ones to get it. My guess would be that twins who are monozygotic with new cases of disorders caused by triplet repeats would be very rare. It'd be rare because not that many people get these disorders, and because twins aren't all that common, and identical twins are even less common—so several things have to happen, and someone would have had to keep track of them."

She finishes writing, looks up, and laughs. "This is why I went into science. Not to save humanity. That came later. I went into it because science is the only thing I know of where you can sit around a room and talk about this and that and suddenly someone says, 'Wow, I wonder how that works, that thing you said two sentences ago.' You realize nobody knows the answer, so you go off and try to find the answer. You're a little limited by techniques sometimes, and always by money, of course, but it's really incredible to ask 'How does this all fit together?' That's on a good day, of course." She becomes serious. "Without sounding too sappy, I hope, the scientific search, this asking questions and having some persistence in finding out what the answer might be, is one of the great endeavors that humans are capable of that other animals don't seem to be. It's a thing that distinguishes our humanness."

At Brandeis, Jeff Hall cited another now well-known way in which identical twins are not actually genetically identical, one that flows from a surprising discovery in immunology and antibodies, the weapons the body manufactures to combat bacteria and viruses. "For a long time," Hall says, "one of the biggest mysteries of biology was the question of how our immune systems manage to make infinite numbers of antibody weapons, weapons against invaders the body had never seen before and which had never even *existed* before on the planet, yet weapons which were precisely and specially engineered to fight each new invader."

When, in October, the winter's flu begins filtering in from Beijing or Delhi (flus are named after the places they're first identified), each year's version is a unique viral innovation. The new virus invades the body, and the alerted immune system sends out special cells as scouts to locate and identify the intruders. Then something incredible happens.

"Your immune cells," Hall explains, "B-lymphocytes produced mostly by the thymus and in bone marrow, start turning out antibodies. But not just any antibodies. Each virus is met by your body with a specialized antibody designed exclusively to attack it. When we discovered this, it seemed, for decades, a miracle, a mystery that the body could do this. Do we have a gene for every single antibody we might need? No. That's completely impossible. Do we have some sort of genetic record of every virus that might come along? Ridiculous. It was as if the body were looking into the future."

Finally the mystery was solved. The machine responsible for engineering an infinite number of new antibodies was genetic rearrangement. "We've found that it's genes that control the immune system," Hall says. "The virus stimulates your immune system to actually rearrange your DNA. This is what allows your cells to make this brand new antibody that it's never made before for a virus it's never seen before. Tonagawa at MIT got a Nobel Prize for this." In immune cells, for example, a gene called RAG 1 produces a protein that acts in the brain, rendering the brains of adult identical twins not in fact identical and providing clues to the discordance between left- and right-handed, and gay and straight, identical twins.

Jane Gitschier, a geneticist at the University of California, San Francisco, confirms this. "Although it's the standard, textbook thing to say, it isn't really true that every cell in our body has exactly the same DNA," she says. "For example, your immunoglobulin cells undergo actual genetic rearrangement in the genes that create our antibodies, so your DNA is actually constantly being altered. You also need to remember that, for the most part, when they take your blood to get your DNA, what they're getting is the same genes that are in the other cells in your body *unless* you happen to have had a mutation somewhere during your own embryogenesis, your creation from the meeting of a sperm and an egg. There are mutations in your liver that are not in your blood cells. Mutations happen all the time. They are absolutely necessary in terms of evolution, and if we didn't have them, we would never evolve. The bad side of mutations is of course that they can kill you, but in fact you never see the effects of *most* mutations."

Dean Hamer makes a similar observation. "The fingerprints of identical twins are not identical even though both have the same genes that formed their fingerprints. That's because there's tons of random stuff in development. Even though the genes are the same, perhaps just because one twin was tilted a bit differently in the womb, they wound up with different fingerprints. But the difference doesn't affect the way their hands function; they still work exactly the same and are interchangeable—*except*—except when a cop is trying to fingerprint one of them.

"In the brain, where sexual orientation resides, you might have the same thing happening. A few neurons make a connection earlier in one twin than the other, some axon extends a millimeter farther, and suddenly you've got differences in brain structure that can create differences in behavior, emotion, thinking, anything having to do with the brain. Identical twins are not identical in the ways they think and feel. They're different people. I *would* be surprised to see differences in identical twins in things like height, or eye color, or length of arms, and we've observed that they're usually, although not always, extremely similar in these ways. But I don't find it at all surprising that for things controlled by the brain, monozygotic twins can be completely different."

There are about 10 billion neurons in the cerebral cortex and more possible combinations of synaptic connections between them than there are subatomic particles in the known universe. Before birth and the final fixing of the brain's structure, certain neurons roam freely from place to place like nomads inside the developing brain. Some divide into new neurons, some die, and some glue themselves with cell-adhesion molecules to other neurons they encounter in their wanderings. These adhesion molecules make up synapses, bridges that span brain cells. But the bridges, like the dense streets of a medieval city, develop randomly this way and that into a network between cells that happen to meet during their own independent drifting in a sea of neurons, haphazardly growing into the stable circuitry of a mature brain. Genes supply general guidelines of where neurons should go, but the final setup of the brain, including the brains of identical twins, is determined by nothing more than chance.

So we are again faced with the question "What does it mean when you say that someone 'has a gene that creates a trait'?"

How do we know, in fact, that a trait is *a* trait? Jan Witkowski of Cold Spring Harbor Laboratory asks, "Could what is commonly considered one trait, like Alzheimer's for instance, be under its surface many?"

Medicine has always described Alzheimer's as a disease, but it turns out that what we have been labeling Alzheimer's is actually, at least at the genetic level, three diseases. They just look alike.

"There are, we now know, at least four completely different genes that can create what we've always called by one name, Alzheimer's," says Witkowski. One gene, the APP mutation at 14q24, produces early-onset Alzheimer's. A second gene sits on chromosome 21, and a third, called APO-E4, is in the centromere of chromosome 19 and causes a late onset of the disease. "This work on Alzheimer's is relevant to work on sexual orientation," Witkowski explains, "because as with Alzheimer's, we may find that what seems to be one clearly defined phenotype—homosexuality—is, in fact, due to three completely independent genetic origins, forcing us to perceive it differently, divide it up and understand it in ways we don't now. We looked into the genetics of Alzheimer's, and it shattered, literally, the way we regarded the trait."

Will geneticists find that there is *a* trait called sexual orientation, with two expressions, or is it, Witkowski asks, that what we call homosexuality is, in fact, several distinct genotypic traits with distinctly different genetic etiologies, and the same for heterosexuality? On the phenotypic level, these traits could be identical in every way that matters to us—how we feel sexually, how we perceive our desires and emotions and needs and instincts, what we experience romantically and erotically, and so on—but at the same time be caused by completely different genes, have completely different genetic etiologies. In one person's case homosexuality might be created by a gene in the middle of the long arm of the seventeenth chromosome, for someone else by *two* genes on the X and the eleventh chromosomes working epigenetically (together), and for a third, a gene at the tip of the second chromosome. The genes could act at different times and in different biochemical ways to produce what to our eye looks like a single trait that we call homosexuality. It is entirely possible.

In her office just two floors below Jeff Hall's in the Brandeis biology department, geneticist Fran Lewitter added on to Witkowski's point. In a discussion about the Xq28 study, she began by commenting, "Hamer's was a very good study. I have no doubt that he found something here. But remember that Dean's been quite careful and conservative in saying this is a gene which may act only in a very, very specific group of men, not 'the gene' for all homosexuals.'

"In other words," she added simply, "it's just one gene."

Enough geneticists have repeated this phrase, "it's just one gene," that it has taken on the proportions of a mantra. Hamer himself stresses it repeatedly. But why is it important? After all, all that might be needed is "one gene," correct?

Yes, agreed Lewitter. But at the same time, it is entirely possible that one would not be enough. It is the question of how genes work. "Hamer is saying," she explained, " 'if you have this gene, you *might* be gay.' But the operative word is 'might.' That's the responsible, and probably accurate, way for him to present his results. Why? Because, to use genetic diseases as an example, there are hundreds of diseases you can develop by carrying one and/*or* another gene.

"I'll give you four brief examples," said Lewitter, "all different, but all illustrating the same basic point." To start with, she notes that there are more than ten different kinds of a disease called retinitis pigmentosa, and they are caused by genes on several chromosomes. She ticks them off rapidly: One kind, called autosomal dominant retinitis pigmentosa, is caused by a genes on any of chromosomes 7, 8, 17, or 19. The gene for another kind, autosomal recessive retinitis pigmentosa, can be on the first or sixth. Peripherin-related retinitis pigmentosa is on the sixth, and there are at least two different RP-causing genes on the X, one at Xp11, one at Xp21. "My point is simply that you only need one of these genes to get the phenotype, but if Hamer had been looking for RP genes, and if he had found only the two on the X chromosome, that might have been a great discovery, but he'd only have accounted for 6 percent of all retinitis pigmentosa. The origin of 94 percent of RP would still be unknown. Maybe Hamer's gene at Xq28 accounts for, say, only 15 percent of all male homosexuality. That's fine, that's 15 percent, but he has, then, not yet located other gay genes on other chromosomes." Alzheimer's, she notes, has loci on 21, 14, and 1. Colon cancer, 5 and 2. "And remember," Lewitter points out, "you can get colon cancer from *either* 5 or 2. If you've only found one, you probably don't have the full genetic story."

After RP-type entanglements, geneticists have to worry about another possible trap: ethnicity. Hamer's subjects were mostly white males, and certain mutations are more frequent in certain ethnic populations. As an example, Lewitter mentions that the highest rate of the mutation in hexosaminidase A (HEXA), the gene that, when mutated, gives you Tay-Sachs disease, is found for some reason in Ashkenazi Jews. They have a four base pair insertion in this gene. French Canadians also have a high rate of mutation in this exact same gene and thus a high rate of Tay-Sachs disease, but they have a 7.6 kilobase (kb) *deletion,* which is a gigantic chunk missing from the gene. "Again," said Lewitter, "Dean is saying, wisely, that he doesn't know if he's found the gene that gives you the trait one way or the gene that gives it to you another." The Tay-Sachs disease that a doctor would see in the hospital from the deletion in a French Canadian kid cannot be told from the Tay-Sachs disease due to the insertion in an Ashkenazi kid. There are more than fifty completely different mutations known in HEXA, but they all produce the same outcome: Tay-Sachs disease.

And, Lewitter noted, if Tay-Sachs is a single disease produced by different versions of a single gene, there are also what appear to be completely different diseases produced by a single gene. "Duchenne muscular dystrophy, for example, is caused by a gene on the X chromosome," she said. "It's in the Xp21 region. But there's a different mutation that causes a milder form with a later onset called Becker's muscular dystrophy. Where's the gene? In Xp21. It's just different versions of the exact same gene."

And, last, genes sometimes have to work together. Hemoglobin, an important molecule present in the blood, is the key to oxygen transport in the body. The genes that make hemoglobin are on chromosome 11 (beta locus) and 16 (alpha locus), and both must be fully functional to produce hemoglobin. "If you're missing one," said Lewitter, "you're in big trouble."

And so here we have four different traits and four different ways genes operate to create them: For retinitis pigmentosa, completely unrelated genes cause the same trait. For Tay-Sachs, the same gene but with different alleles causes the same trait. In the Duchenne and Becker muscular dystrophy cases, the same gene causes two different traits. And hemoglobin requires two different genes operating together. And *that,* Lewitter concludes, is why scientists—including Hamer—keep underlining the point. Hamer's got *a* gay gene (maybe). But it may be just the tip of the iceberg. There may be many more gay genes, scattered around the population, varying in their rate of incidence according to race, gender, and other factors. Perhaps, without a second gene, many people could carry one gay gene and still be heterosexual. Hamer's gene might even have several different mutations that produce a homosexual sexual orientation.

Indeed, Hamer himself has been careful to point out that in the original 1993 study, 18 percent of the gay brothers did *not* inherit the same portion of the Xq28 region from their mothers. It is possible the discordance arises from the mother's having a copy of the gay allele on both of her X chromosomes, but most probably, as with Lewitter's example of retinitis pigmentosa, the source of these brothers' homosexual orientation is a completely different gene or an endocrinological etiology. Hamer and other researchers are now looking for loci that influence sexual orientation on chromosomes other than the X.

Genes, paradoxically, simultaneously split apart our perceptions of life and thrust them together. Philip Reilly of the Shriver Center for Mental Retardation in Waltham, Massachusetts, outlined this splitting force. "Defining disease is something we've been working on for a long time," Reilly notes, "and genetics only makes the job harder— not easier, as you'd think—because genetics makes us splitters instead of lumpers. Ten years ago there was a disease of the lungs called, sim- ply, cystic fibrosis. Some kids had it severely, some mildly, but it was a disease. Now that we've found the CF gene and identified 350 com- pletely different mutations inside it, we've discovered there are peo- ple with cystic fibrosis mutations who have no perceptible lung dis- ease at all in this 'disease of the lungs.' We've found people with CF mutations where their only manifestation of the disease is congenital absence of the vas deferens so that they're sterile. And you have to say, *wait* a minute, this is *CF?* We have new entities splitting apart under the classification 'cystic fibrosis,' and genetics is what's causing us to splinter."

Molecular geneticist Maxine Singer pointed out the flip side of this, the unifying force of genes. "People who have thought about evo- lution from Darwin on," she said, "were always puzzled by the fact that eyes in different animals operate in very different ways. They asked, 'How can it be that such very complex organs—and eyes are very com- plex—could have arisen completely independently?' " The answer sci- entists supplied, a bit reluctantly, was the term "convergent evolution," the assumption that these organs were originally from separate genes and arrived at the same function—seeing—by completely different developmental paths driven by natural selection. "No one," said Singer, "was ever really satisfied with that, frankly."

Now "convergent evolution" has been unceremoniously dumped. "This is the latest wrinkle on this amazing universality of biology," said Singer. "We've found that the genes that are used to construct a fly eye and the genes used for a human eye are actually the same genes, but they are called on in completely different ways to do the same job. This is incredibly exciting from the point of view of evolutionary the- ory—the idea that one gene could go about a job in completely dif- ferent ways is rather amazing—but it's also exciting because it tells us again that we can learn an awful lot both about our own cells and about the organisms with which we share the planet by studying any one of them. We can get that knowledge and apply it to the others, and we don't have to study every organism. We can study an insect. Or a bird."

Singer added, "For me, really, the insight into the whole structure of living things on the planet is tremendously exciting because it shows us the extent to which organisms are interdependent. Organisms live in symbiotic relationships with other organisms, and we evolve this way, which means we've actually taken advantage of the natural selection that has operated on *other* organisms. The notion of the interdependence of all life is the contribution of genetics to the understanding of our ecology and environment. We think today of genetics as applying primarily to medicine, but I believe as a geneticist that in time the science of genetics will emerge as just as important to our knowledge of the broadest aspects of life in the biosphere and to our environment."

The splitting and unifying power of genes inevitably goes beyond knowledge of bits of molecules and forces us to perceive elemental aspects of ourselves differently. The process began with Darwin and has only been accelerated by molecular genetics. David Cox, a geneticist at Stanford, asks what genetics does to our ability to define a "trait," particularly one with such strong political aspects as sexual orientation. "Since Darwin," says Cox, "we've been really struggling with this matter of how things should be grouped. The thing about genetics is that genes group things according to etiology, causes, where we humans have always grouped things based on visible differences we can see and identify, differences in which we often put great political or theological stock. The genetic grouping and—for want of a better word—the human grouping can be completely different. This isn't bad or good. It just is. Take medicine."

Medicine, Cox points out, has always grouped diseases according to our ability to identify and cure them. But say there are a group of people with pneumonia, and they all die. You look at this situation, says Cox, and you figure out that they're dying because of bacteria in their lungs that prevents them from breathing. Logically, you classify people as those who have pneumonia and those who don't. Since you're a great doctor, you find an antibiotic like penicillin that cures half of those people. And you feel pretty proud of yourself and your drug.

But why do half the people who get penicillin still die? Dig a little deeper, Cox explains, and you find out that your classification system doesn't really work. "People With Pneumonia" is actually two groups of people with different types of bacteria, one type that is killed by penicillin and one that isn't. So you have to classify people not just by the trait but by the cause of the trait. And now you can come up with an antibiotic that helps the other half.

"The names of these two types of bacteria," Cox concludes, "are pneumococcus, a gram-positive bacteria killed by penicillin, and a type of *E. coli,* a gram-negative bacteria penicillin can't touch and one that must be killed with other antibiotics. Making this distinction, grouping things according to etiology and not the trait we see, keeps 100 percent of people alive. This is a medical example. Genes provide us with thousands of genetic examples. This seems a straight-forward distinction. But what you're *really,* inevitably saying is that there's a discrepancy—a contradiction, actually—between classifications of things based on what we see of them and how we hold them to be socially or politically important and classifications based on etiology. Most nonscientists see them as the same. They aren't. At all. And this discrepancy throws into question many of these gravely important distinctions politicians and religious leaders draw between groups. Like between, say, heterosexuals and homosexuals. Etiological classification, which genetics is based on, forces you to ask: Is there really any difference?"

～

Among the odder features of genes that has emerged is a facet of their behavior called penetrance. It is an effect that refers to the action of *other* genes or the environment, and it provides a way for millions of people to carry a gene expressed only in a tiny minority of them. "There's actually a pretty easy way to think about penetrance without a lot of genetic jargon," says Eric Lander, of the Whitehead Institute for Biomedical Research in Cambridge, Massachusetts. Lander is big and practical, and strides powerfully in and out of rooms with little time for a lot of genetic jargon. "The penetrance of an allele is basically its seniority in the corporate hierarchy of the genome. The question is simply how many genes can it override and how often can it get to do what it wants—and how many other genes can override *it.* Penetrance is about power."

An allele that's 100 percent penetrant, says Lander, is 100 percent powerful. "The allele that gives you Huntington's is 100 percent penetrant. If you've got that allele, it gives you the trait, period, end of story, and no other genes can overrule it. Now, let's take a hypothetical. Say this gay allele Dean is looking for in Xq28 has, I don't know, let's give it 5 percent penetrance. It has only a 5 percent chance of getting its way and making you gay. Not a hell of a lot of leverage, this allele. Even if two brothers inherit this gay allele from Mom, its low penetrance ensures that 90 percent of the time neither brother will be homosexual, about 10 percent of the time one brother will be, and only about one-quarter of 1 percent of the time will both be gay. The reason is that homosexuality is created by a weak allele with very little political influence in the genetic decision-making process."

Lander points out that in real life, discordance for traits in identical twins is not really that rare, so sexual orientation's 50 percent discordance figure is genetically not unusual. Left-handed concordance is only 12 percent—much less than homosexuality—and another well-known example is diabetes, specifically Type 1 diabetes (autoimmune), which is caused in part by a gene allele on the p arm of Chromosome 6. "The penetrance of this little allele is only 30 percent," Lander notes. "If an identical twin has diabetes, his *identical* brother has only a 30 percent chance of having the same trait." He lifts an eyebrow. "I know they have the same genes, but I can't predict who's going to get diabetes and who's not. And remember, *most* genetic traits are like this."

Sexual orientation, heterosexual and/or homosexual, may turn out to operate this way. If we hypothesize that homosexuality is governed by a single gene with two alleles—one gay, one straight—this creates an interesting statistical oddity: You'd think, Lander points out, that there would be, on average, a 50/50 chance of a man's getting the gay allele from his mother, thus a 50/50 chance of being gay. In fact, if the gay allele is only partly penetrant, his chances are much less than 50/50.

This means two things. First, says Lander, it accounts for Hamer's use of affected-trait linkage analysis—only studying those who are "affected," that is, in Hamer's case, gay men—for which he was noisily criticized by those who didn't like his results. "He should have studied heterosexuals as a control," went the criticism, but Lander responds, "It's because of penetrance that geneticists only study affected people since people with the trait are the only people you can be certain are carrying the allele you're looking for."[1] And this is tied to the second point: "There could be hundreds of millions of straight men walking around with this gay allele but who are straight simply because it didn't penetrate." Penetrance, in a sense, presents a genetic "accuracy in labeling" problem. Doing an amniocentesis test on an embryo at week 10 or 11 might reveal clearly that he is carrying the gay gene allele. But this knowledge, like the gene, might be only 5 percent or 10 percent powerful. If the test is done with an eye to aborting a child who will be homosexual, those doing the test need to be aware that an abortion would carry a 90 to 95 percent chance of aborting a heterosexual male fetus.[2]

The penetrance problem makes prenatal amniocentesis testing

for the gay allele much more difficult. Find the gay allele—and it may mean nothing. But penetrance presents only one difficulty in unlocking genes. In so many ways our concept of genetics may be faulty. Lisa Geller, for one, thinks the way we perceive genes is fundamentally incorrect. "Everyone, geneticists included, talks about genes being a 'blueprint,'" she says with clear disdain for the word. "I don't think genes are at *all* a 'blueprint.' They are more like a palette of paints. If you're going to do a painting, you need your palette of colors, but you need a lot of other things too, a canvas—and there are all shapes and sizes and kinds—one brush or ten, ranging from fine camel's hair to thick bristles, an image or twenty you want to paint, some kind of schooling—neoclassical Ingres or modernist Pollock— and, oh yes, talent. You may be limited by your palette. If you only have black and white, your painting is going to be limited to black and white, but that's just limiting your *colors.* Picasso or Rembrandt could do more with black and white than I could ever do with a hundred colors."

Geller maintains that we've got to come up with a way to think about genes that will help us deal better with genetic testing and diagnosis. She uses Gaucher's disease as an example.

Gaucher's can have an onset at any time during life, though on average it is middle age, and it usually is not fatal if treated in time. The gene was found—named GBA, it produces a faulty enzyme called glucocerebrosidase that causes a deterioration of the liver and/or other parts of the body—and a test developed. Geller notes that the test was met with great excitement. "Everyone was thrilled that we could test people," she recalls rather grimly. "So—and this is the punch line—would you abort a fetus for a disease that it might not get till middle age and which isn't absolutely fatal? Just because it has the gene? Because guess what? They've started testing people for this gene now, and they've discovered that so far about two-thirds of those who have *two* copies of this defective gene and who, according to the rules of genetics as we understand them, are absolutely, positively supposed to get the disease never get the disease."

Geller pauses for a dramatic moment, then leans forward across the table. "So *now* what are you going to do? You're dealing with a disease that has a wildly variable onset which is impossible to predict, and now you find that even having two copies doesn't necessarily guarantee you're going to get it. The same is true of Huntington's. Even though it is always fatal, it varies widely in both its onset and the order in which you see symptoms, a variation which the genetics can't entirely explain. We just don't understand what modulates these things. All we know at the moment is that having the gene doesn't tell you everything. There is a gene—and so what? Having the gene doesn't matter. What matters is the phenotype, the person. Gaucher's is telling us we have a lot to learn about what it means to say that something is 'genetic.' The gene can be the *cause* and yet at the same time not be *determinative.*"

When BRCA-1 was discovered, an apprehensive Frances Visco, president of the National Breast Cancer Coalition in Washington, warned, "You're talking about giving [women] a test telling them they have an 85 percent chance of getting a disease that we don't know how to prevent, and for which there is no known cure." BRCA-1, says Geller, is a tough call. "But," she says with a grim jauntiness, "I can go that one worse with GBA, because it turns out there *is* a drug that treats Gaucher's. The drug is fantastic, it works wonderfully. And it's the simplest thing in the world: it's just the enzyme that this broken gene doesn't produce. You give the enzyme to people with the disease, and they're fine. The catch is that the average cost for this drug is about $100,000 a year.

"So. You're a doctor in a clinic. You have a test for GBA. A family comes in, gets their fetus tested, when the test comes back, it tells you the fetus has two bad alleles. You're sitting in your office with the lab results, the family is waiting outside. Here's what you have." Geller, sitting forward with her elbows on the table, calmly and ruthlessly ticks off the points on her fingers. "The fetus may get this disease. But she may not. You don't know. So first you're faced with a discussion of abortion. You also have information which the family's insurance company badly wants which will affect the ability of all the people involved to get and maintain insurance for the rest of their lives. If the parents opt not to abort the fetus, and if the child who is born gets the disease, she may get it in childhood and die. Or she may live a crippled existence her whole life. *Or* she may get it in very old age after a full and happy life. You don't know. You can tell the mother her fetus has a bomb ticking inside her, but you aren't even sure the bomb is *armed*. If the baby is born and doesn't get the disease as a child, she may show no signs of the disease as an adult. If she shows no signs of the disease, if you give her the drug, *if* she can afford the drug, then it'll give her doctors a better prognosis if she *does* develop the disease. But if she shows no signs of the disease, she may never get the disease at all. You have no way of knowing. If she's never going to get the disease at all, she will have mortgaged her home for hundreds of thousands of dollars for a medicine she doesn't need."

Geller sits back. "This is what genetics means: an incredibly complicated dilemma." If BRCA-1 is a tough call for genetic counselors and GBA is even tougher, who knows what GAY-1 will be?

～

One of the more popular misconceptions of genes is that they are "for" traits, "for" a certain eye color or neuron or blood type. This misconception is tied to our most common popular (and popularized) misperception of evolution. Evan Balaban explained the problem using handedness and a child's toy.

"Ninety-two percent of the population is right-handed and 8 percent is left-handed," says Balaban. "People know this simply from observation of phenotypes—people. But then, having observed this thing, they ask *why* are 8 percent of us left-handed, these weirdos. And they just assume that the question itself is valid, that we can always ask 'why,' that this biological fact has some biological or evolutionary Meaning, capital M. Well, we know that sometimes this question itself, this asking Why, is meaningless."

People generally assume about evolution, explains Balaban, that if a trait is common, it's both "good" and there's some important reason for its existence, that nature must have wanted it that way. People mistakenly call this assumption evolution. There is actually another possibility, which is often true, which is that this thing is purely an accident of development. "People assume nature has a 'reason' for you to be right-handed and that because most of us are right-handed it's an 'adaptation' that over millions of years has been selected 'for.' 'For' means capital M Meaning." In fact, says Balaban, the trait we observe may not be there "for" any purpose at all, and the importance we ascribe to it may be fabricated; this is Darwinism exploited by those who misunderstand evolution to give political Meaning to something that, in a real evolutionary sense, might have absolutely none. Like left-handedness. Or homosexuality. It may be that, purely by accident, other things constrained it to be so.

"Philosopher Elliot Sober talks about the difference between the selection 'of' something and selection 'for' something. There is a child's toy that consists of a number of disks with round holes in them stacked on top of each other. The holes start out large and become smaller as the disks descend. The child drops different-sized marbles of various colors through the holes, and of course the mar-

bles separate out as they descend through the levels, each disk selecting certain marbles according to size. What if it just happened that
the marbles were colored by size? At the bottom, the marbles are all
small, but maybe they are all red too. So there was selection 'for' size
but, as it happened, selection 'of' color as well. Evolution didn't care
about color. It just happened by chance that the smallest marbles had
a given color. The genes have been selected for because they are
doing something we can't see: selecting for the size of the marbles.
The genes see size, size, size. All we see is red, red, red. So if you look
at the marbles on the bottom and they're all red and you say, 'Ah
hah!'—well, 'ah hah' nothing. Handedness could be size or it could
be color, and *you* have to figure out which it is. And homosexuality—
is it color, 'of,' or size, 'for'? How do you figure it out?"

So far geneticists have identified two situs genes in rodents, genes
that determine whether the heart or other organs are reversed, called
situs inversus. What does it Mean? Probably nothing. "All the stuff that
people read into asymmetry is garbage," says Balaban. The human
heart is usually on the left side of the body, the liver on the right, and
the question when they are reversed is: Is it usually one way or another
because of a biological Purpose? Or is it just because?

"Every discussion of biology that invokes evolution," Balaban says,
"makes certain assumptions about, one, the original relation between
color and size among the marbles, and two, which one of them the
principle of selection is really operating on. Color might actually have
been *very* important in some era long ago and then ceased to be, or
it may be that it's been very important all along but only on a certain
level, like developmental genes that are turned off and then show up
later by chance."

Developmental genes are genes that operate only during very specific, limited periods of embryonic and fetal life, performing their al-
lotted tasks in overseeing cell division and growth and then shutting
themselves off completely, forever. The problem is, scientists have discovered that many of these developmental genes can be triggered accidentally decades later, when some factor turns them on again—viral
infection, radiation, fats, or chemicals. They start dividing uncontrollably and are then called by another name, oncogenes: cancer genes.
So far, we know of sixteen developmental genes, and fourteen of them
can become oncogenes. And here is yet another contradiction of our
old way of perceiving genes. We often speak, without giving it much
thought, of diseases that are "genetic"—caused by genes—as diseases
that can be "inherited." Cancer is an example of a genetic disease that

is not, the great majority of the time, inherited. We think of diseases one gets from genes as different from, say, sexually transmitted viral diseases, which one contracts from behavior. Yet cancer is genetic, unlike AIDS. And cancer is behavioral, *like* AIDS.

It is cancer that once again makes us reevaluate what we see as genetic. The reason, says Jasper Rine of the University of California, Berkeley, is that *"all* cancer is probably genetic."

Much cancer is caused by smoking and eating high-fat foods, which is "behavioral," but Rine raises a hand, slicing the air into two important distinctions. "There's no evidence for any cancer that is not genetic," he says, "but that's *not* to say that all cancer is heritable, that is, can be transmitted from one generation to another. Many mutations—and cancer is a mutation—are caused by genetic changes. But only a fraction of genetic changes are heritable, and that fraction is only the cancers that change the germ line, the tissue that makes sperm and eggs, which allows parents to pass them to children."

Retinoblastoma is one of the clearest examples. A cancer of the retina, retinoblastoma progresses to affect a number of other tissues and, if not operated on, is invariably fatal. (The operation, unfortunately, usually requires removal of the eye to save the life of the person.) This cancer runs through generations due to a mutation in a gene on chromosome 11. But a person also can get retinoblastoma even if it doesn't run in the family, essentially by bad luck: a spontaneous mutation in this gene. "So what do you have?" Rine asks with a smile. "A heritable cancer that you didn't inherit but which is completely genetic."

Cancer itself fits into some unknown evolutionary place in the child's game of colored marbles. "The funny thing is," Lisa Geller notes, "by curing so many other diseases, we've essentially created cancer. Cancer is largely a disease of old age, and by making people well enough to live long, we've now got this disease caused by these genes that people virtually never saw before created by genes that were never supposed to turn on again. Is cancer 'normal' evolutionarily?" She shrugs. "This idea that you live with one person for fifty years is hardly normal evolutionarily. Alzheimer's we get at sixty-five, but given that people used to live to twenty or thirty, this may be, evolutionarily, an abnormal life span."

Balaban agrees that "the concept of 'a normal life span' is very difficult to deal with in biology. If things go haywire only at fifty or sixty—like cancer or Alzheimer's—then maybe it's because it's beyond the 'normal life span' of people and in former times it didn't *matter* if these things happened because nature never 'intended' people to live that long. So as far as oncogenes go—for hundreds of thousands of years nature said, 'Who cares! It's irrelevant!' " It would seem that developmental genes selected "for" the functions we need them to perform in development, but there was, probably from time immemorial, although we didn't know it, a selection "of" cancer.

⁓

If we profoundly misperceive what "gene" means, we do greater violence to the concept of "environment." We divide the indivisible. We create concepts that do not exist. We say "genes or choice," we say "biology or environment." Evan Balaban sighs. "This 'environment versus biology' division is ridiculous. Everything is both." Bill Byne: "People have this vague Freudian idea that it's the way you're raised or something, which in psychoanalysis is fine but is completely incorrect in biology and the legitimate sciences. Environment to a biologist *is* biology, by definition." Lisa Geller: "This thing called environment, at least the way most of the national media uses it, doesn't exist." Richard Lewontin: " 'Nurture/nature' is bullshit."

If there are a million incorrect ways to express the concepts of what are popularly called biology and environment, there are at the same time two correct ones.

One approach is to say that environment and biology are opposite sides of the same coin, but the important thing to remember is that it is *always* one coin, one human being (or animal, or plant), axiomatically indivisible. The embryo and its genes exist in an environment as do flour, eggs, and butter that must be measured and mixed and poured and baked at a certain temperature for a certain time. The stirring action, oven temperature, and time are the environment, measurable physical inputs, inextricably part of the cake, which is more than simply the sum of the ingredients transformed. For humans, biologists consider food and light and air and medicine to be the environment, without which genes are nothing but tiny molecules floating in a cell. The two, put together, suddenly disappear, and a new, indivisible thing takes their place.

The other approach, which seems to contradict the example above but actually is just a different way of expressing the same thing, is to say that the popularized idea of "environment" simply doesn't exist. And never has. There are only genes and the results of genes.

In the end, the two mean the same thing. "Environment *is* biology." Or "Environment doesn't exist."

"We might be trying to do the impossible," says Jan Witkowski, "looking at sexual orientation in terms of genes *or* environment, because then you're not looking at sexual orientation, which is both, of course. When I say 'environment,' however, I'm not saying only social training but inputs that have an effect on you such as your hormones or the enzymes made by your genes."

The idea that hormones constitute "environment" conflicts with the popular notion, but that is Witkowski's point. "People talk glibly about 'the environment,' " he says. "I know the popular concept is"— he opened his eyes wide in the direction of the wall and expelled a breath, sifting through examples—"oh, I don't know, your mother giving you a strange look or your teacher scolding you or something; journalists use the term 'environment' in both a very outmoded and a very narrow way. Scientists, too, would once have said that for cystic fibrosis, where a simple mutation in a gene produces the disease, environment is irrelevant, but few biologists would say that today because in science we now understand environment quite differently. Changing medical treatments, which we now understand to constitute the environment, have helped patients with cystic fibrosis live into their early thirties whereas two decades ago they died in their late teens. You've still got the cystic fibrosis mutation, but your expression of it, your phenotype, is modified by your medical environment."

Witkowski cited phenylketonuria (PKU), a disease caused by a gene that can't make an enzyme that breaks down phenylalanine. The phenylketones used to poison children born with the defective allele. Today they simply avoid phenylalanine and live almost completely normal lives. "So diet is 'environment,' " says Witkowski. "Give diabetics insulin, and they don't go into diabetic coma. The disorder is there, and so is the gene, but the biochemical environment ameliorates it. So you can either say 'insulin is environment' or you can say that 'environment' is our ability to treat people. Either way of thinking of it would be quite correct."

Geller agrees. "By 'environment' I don't mean at all the old Freudian stuff about your parents or people you see on television or whatever. Environment to a biologist means sensory inputs that have a biochemical or neurological impact. It can take any of a zillion different forms, none of which we really understand and absolutely none of which we can either control for in a study or measure accurately. I'll put it in slightly cartoon terms for a moment. It's like, maybe on the fourth week of your gestation your mom ate broccoli, which has certain nutritional properties, and these nutrients—which are 'environmental' because they come from your environment—interacted with a gene you have, and you wound up left-handed." She made a mild face. "It's outrageous stuff, but first, that is what biologists mean by 'environment,' and, second, it's absolutely possible that things are just that obscure. We don't know. The more you look, the more complicated it is."

During the age of Freud, "environment" was equated with another problematic word, "psychology," and a line was drawn between "biology" and "psychology." Neuroanatomist Simon LeVay rejects the division. They are, he says, one and the same. "This distinction between biology and psychology," says LeVay, "is, in the final analysis, total garbage because the mind is nothing but the brain doing its job. So everything mental has a physical basis in the brain, although when they hear this very simple observation, people think you are saying that thought, desire, preference for a certain work by Bruckner over a similar piece by Mahler, and other mental phenomena are physically 'determined' when it's simply an observation that, again, everything mental is physically represented in the brain. What is a biological question? *That's* a biological question. Our biology both reflects and creates what we like. It's not reflects *or* creates. It's both.

"Biology and psychology are merely different ways of looking at the exact same thing," he continues. "People are trying to get at the difference between what is innately determined and what's culturally determined, but they screw that up and say that's the difference between biology *versus* psychology. It isn't. Biologists just look at the mind from the bottom up, at the level of molecules and synapses, and psychologists look at it from the top down at the level of behavior, trying to get to the bottom of it. When I say it's biological, I don't mean to say that it's immutable. It's still maybe totally culturally determined. But even if something is totally culturally determined, like which music you like, if it has a representation in the brain—and of course it does—it has to have some biological substrate in terms of synapses, synaptic function, chemistry, something like that, and therefore is open to neurobiologists to get in and unravel it."

The neuroanatomist pauses and then says tersely, "To be frank, psychology as an isolated discipline is threatening to reach a dead end. It ignores a whole way of looking at the mind that is much more ahead of it. Psychology is not sufficiently constrained, and it's too easy to build theories that are impossible to validate, either for or against. Freudian psychoanalysis is by far the worst example of this, how you can build up an elaborate house of cards and not be constrained by reality in any way. Psychology sees the brain as an untouchable, mysterious black box, like a television set that is always beyond our direct reach, and it thinks the best it can do to figure out how the television works is to feed a signal into the machine and watch how it turns that signal into a picture. Psychology says 'We can never really get inside there and look at the nuts and bolts to see how the machine actually works. We can only look at it from the outside, slip things in and see what comes out.' But biology says 'Look, however carefully you compare what comes out of the brain with what goes in, there's always a range of possibilities about what really lies inside. The only way to answer that question is to go into the brain and *look*, find out how the hell it works, how many impulses are being fired, how many neurons, how the chemistry works.'

"The great thing," LeVay says, "is that now psychology and biology are fusing, which is really exciting. People are learning to figure out how learning works. People have studied learning at a psychological level for God knows how long. They've learned some useful things." He laughs. "I know that sounds incredibly patronizing—but now we're starting to understand in molecular terms and neuroanatomical terms how that happens. So there's a coming together that's really interesting. Ideally, psychology would provide a description on a mental level of what needs to be explained by the biologists. The word 'psychology' is coming to be used as what it really means, 'biology,' but just talked about from the perspective of describing what biology does."

The Sphinx-like problem of defining "environment" directly confronts the hyena team as well. "All it would take," Laurence Frank says, "to create this effect in the female hyena is one gene changing the concentration of androgens in the womb—a change in the hyena's fetal environment, but that's environment in quotes, the scientific sense of environment, not the popular sense. So environment is biology."

"Environment," adds endocrinologist Paul Licht, "has to be both things that are acting on you and their method of getting to you." Many animals, rodents in particular, regulate their reproductive cycles by the amount of estrogens they get from plants eaten in their normal diets. Plants make estrogenic compounds called cytoestrogens—*cyto* means plant—that affect the animals that eat them. Ecological changes that lead to changes in the plants alter intake of all-important, powerful steroids. "But then," Licht notes, "man-made pollution is also a source of steroids." He motions outside at a Berkeley bus jetting away from the curb on a blackened plume of filth that bumps expectantly against the large window and then sinks away reluctantly to melt into the air. "You breathe some of these synthetic steroids into your lungs everyday. They can get to a fetus, human or hyena, by going through its mother." DDT acts as a steroid, says Licht, as does polychlorinated biphenyl, which is used to insulate the electrical transformers that are seen everywhere. PCB is released into the air by burning plastic, and it acts both as a carcinogen and as an estrogen. In Florida, huge populations of alligators are no longer breeding and their eggs no longer hatching due to trace amounts of pollutants in the environment. "This is stuff generated in such minute amounts we thought for sure they were at safe levels," Licht says. "Now they're showing that there is no such thing as a safe level. They can't find a low enough amount." He shrugs. "The sperm count today in human beings in the industrialized nations is only 50 percent what it was in the 1930s and '40s. Sperm count has gone down by half. This could be pollution. Could be stress. Both are environment. DDT wiped out whole populations of birds by changing their reproductive systems."

Environment is so powerful that it can make two exactly identical sets of genes produce two utterly different animals. If certain insects in African and Arab countries grow up in low population density, in their adult stage they are green in color, relatively inactive, unsocial, solitary, and they don't fly much. These insects are called by a familiar name: grasshoppers. If they grow up in crowded conditons, however, they are dark black in color, have longer wings, are intensely social, live in flocks, and fly in huge swarms to greener pastures. In this case, they are known by a very different name: plague locusts. The same genes. Two social environments. Two completely different animals.

The human species produces similar examples of culture having a biological result. Dairy owners add the enzyme lactase to their milk products, not generally for the benefit of European Americans but rather for their African American and Asian American customers, many of whom as adults don't have the allele of a certain gene that produces the enzyme needed to digest lactose, a disaccharide sugar contained in milk. (The undigested lactose will simply sit in the guts of those lacking this version of the gene, breeding bacteria and leading to explosive diarrhea and other medical problems.) This phenomenon is an excellent example of culture's biological influence on genes and, by extension, all biology. Americans from historically nondairy cultures, such as African Americans and Japanese Americans, produce little or no lactase as adults, while Norwegian Americans and Hispanic Americans with European genetic ancestry produce abundant amounts. There are, however, a small number of African herder tribes that have historically had dairy in their diet, and they produce the lactase enzyme. Their African American descendants have no more trouble digesting milk than do Norwegian Americans.

The observation that there are biological differences between the races usually elicits a hysterical reaction, often—as everyone should at this point know—for the good reason that many patently false claims of such differences have been made with devastating results. But hysteria aside, there are numerous biological differences between the races. The presence of the lactase allele is only one. Another is proposed by a Northwestern University Medical School professor, who points out that it is entirely possible that in terms of mental ability, many American blacks are biologically inferior on average to whites, but that this is not the same as saying they are genetically inferior. Furthermore, and perhaps most interestingly, this observation cuts in favor of the political arguments made by liberals. The reason for the disparity could well be nutritional deprivation. Consider, this professor notes, that for centuries, blacks in America have on average had lower incomes, worse medical care, and an inferior diet. This nutritional deprivation could lead to a general biological deficit that falls along racial lines.

But if so, what is the cause? Is it nutrition—which would popularly be classified, again without much thought, as "nature"—or poverty—which would be labeled, equally thoughtlessly, "nurture"? Neural architecture is formed by nutrition, whose intake is determined by educated eating habits and money, access to which is determined by the state of the economy and the social climate, the availability of jobs, and laws and culture. Is "biology" or "environment" responsible for the biological/neural condition of African- versus European- versus Hispanic- versus Native- versus Asian-Americans? The answer is: an inseparable blend of the two.

This same professor notes that Northwestern's Medical School had launched an affirmative action program for black students but had had to suspend it; the black students were committing suicide at a significantly higher rate than the Europeans, Asians, and Hispanics. "And why?" he asks. "Because they hadn't been educationally prepared to function intellectually the way you need to function—the logic, the rigor—in medical school. That's a matter of training. Could their biology have a role? It could, and they could be less neurologically able, but that could be reversed in a single generation of eating right, resting right, and keeping healthy. Then, if they are raised on a rigorous intellectual regimen, I think it's clear they perform identically to the other students." Politically, this professor describes himself as a "realistic liberal." He sighs. "Obviously environment has a biological impact. I'm sure that each time these kids see an act of violence on Chicago's South Side it changes their brain."[3]

Generally conservatives arm themselves with biological arguments while liberals argue for the power of social forces and thus in favor of social programs. Yet politically, the above observation works more for liberals. That social forces have an impact not only on opportunity but on biological ability is an argument for programs that regulate those social forces. Even more important, the observation is that biological mental differences do not run, necessarily, along lines of race at all but rather along lines of culture and economic class. Change those, and the biological differences disappear. But on a more fundamental level, the observation is evidence for the conservative side of what is an almost inconceivably divisive issue. It constitutes a reminder that biological differences in intellectual capacity *could* be not only biological (that is, "environmental") but genetic—*heritably* genetic—as well. At one point LeVay observed: "People don't think it's strange that Swedes have genes that make their hair blond, physically different from African hair, and yet they get very upset when you suggest that there are genes that make some brains physically different from other brains. We know that the building of the brain is governed by our genes. And thus the mind is under the influence of genes too. And it is therefore logical to assume there are differences in the mind." When he made this statement, LeVay was not specifically referring to the IQ debate, but that is inexorably where it leads.

Finally, after all the momentous issues posed by genes, there is one that baffles geneticists personally. People perceive genes as freeing them from responsibility. Sitting in his tiny cubicle, Hamer's phone rang one afternoon. He answered, listened, got a somewhat pained expression. He took a deep breath. "Well," he said to the caller, "I think the best thing I can tell any parent of a gay person is that homosexuality is a natural, if minority, variant of human sexuality." He listened some more, tried to say reassuring things, and hung up somewhat exasperated. "That was another mother of a gay son," he said. "She wanted me to tell her 'It's okay, it's genetic,' and that it's not her fault." He looked perplexed. "Of course it's her fault! It's her genes!"

Lisa Geller has remarked, "I often wonder why people are looking for the gay gene. What happens the day we find it? Nothing. Is anyone any different the next morning? No. Straight people are still straight. Gay people are still gay." She paused. "Unless they try to change it." She shook her head. "I think what I'm trying to say is that I don't really see the point. For a disease trait, there are obvious, if complicated, medical reasons for finding genes. But for *homosexuality?* . . . Why? Sometimes I think it's just that the word 'gene' is this mesmerizing incantation. Say 'synapse,' and no one reacts. Say 'gene,' and people read all sorts of meanings into it. The word makes people bonkers. I don't know what this weird fascination is with genes that makes everyone think they're so much more important than the rest of the biological pieces and processes that make us up. Let's face it, you are the way you are. Regardless of how you got to be that way."

∽

In his book *Lonely Hearts of the Cosmos,* Dennis Overbye discussed the greatest mystery of quantum mechanics, the difficulty of knowing the absolute truth about anything we look at.

> Quantum theory said that there was an ineluctable fuzziness, a kind of chaos at the microheart of reality. . . . Observed on the smallest scale, nature exhibited a curious duality between wavelike and particlelike behavior. An electron, for example, could act like a wave or a particle depending on the circumstances. One of the most mystifying consequences of this duality was an epistemological nightmare called the uncertainty principle, which seemed to draw a fine line in the dust against mankind's centuries-old quest to know the world in finer and finer detail. The uncertainty principle stated that knowledge came with a price. The more precisely you knew one trait of a particle like an electron—say, its position—the less precisely you could know something else, like its momentum.

Talking one day in her lab, Angela Pattatucci unintentionally and almost uncannily echoed Overbye. She was, however, discussing not particle physics but biology: "We define this thing called sexual orientation," she said thoughtfully. "We have to in order to do the study. But there's a paradox in doing research, and that is that in biology, when you try to learn about a general phenomenon like sexual orientation, the more you attempt to find something meaningful about that phenomenon, the more you have to narrowly define it, because you realize it's too complex to study all its aspects. But that narrowing, while it increases your ability truly to understand the trait, at the same time reduces your ability to say anything meaningful about the trait that you set out to study in the first place. The more you know, the less you know. You have a lot of information about specific components, but maybe the information says little about what it is you were originally after. It's really almost impossible for us to understand the biological phenomenon of sexual orientation and what creates and comprises the biology of sexual orientation at the same time. We extrapolate our knowledge, and we actually may have no right to do so.

"Of course," she added, "this is true for about all scientific investigation." In physics, the blend of particle motion and particle position is called the wave, a fiction we construct that enables us to ascribe both qualities to a given particle simultaneously. Waves have the advantage of allowing us to conceive of the particle itself. They have the decided disadvantage of conceptually disintegrating whenever you ask a more pinpoint question, either "Position now?" or "Momentum now?" We cannot know both. As Overbye says, "At the moment of actual scrutiny of an electron or a baseball, [Danish physicist Niels] Bohr concluded, the wave function magically 'collapsed' to a specific answer to whatever question was being asked. But the scientist had to ask—otherwise nature didn't answer."

Pattatucci was saying that in biology, the question likewise perversely defines the answer, yet without the question, the cells and neurons and nucleotides give no answers. A botanist talks about a deadly bug infestation on the roses. An entomologist discusses the colony of aphids living on their food supply. The concept of disease collapses. The answer depends on the person asking the question. The observer assigns the meaning.

David Cox of Stanford also holds a moral relativist view of genes. "You can't assign a unique property or task to a gene," says Cox, "because depending on the context it's in, it can become, instantaneously, something else." He thinks for a moment. Genes are like ethics, he says slowly. An ethical principle by itself, standing alone, seems clear, even absolute. Put the principle in the real world, with different people and cultures and assumptions, and suddenly—he holds up his hands and, with a smile, shoots the ethics into the air, opening his fingers as if releasing doves—this clear principle isn't so clear. Situational ethics. That's genetics. Genes aren't clear and they aren't absolutes. They depend on the context they're in.

Of course, just because one is assigning meaning to an observed phenomenon does not mean the meaning is incorrect. It means merely that it is necessary to *remember* that the subjective assignment of meaning is part and parcel of the scientific search, and the questions must be both exquisitely precise and constantly suspect. You must, Pattatucci instructs with cool irony, distrust your definitions and believe them wholeheartedly at each moment. "Of everything I've experienced and learned doing this research," she says seriously, sitting on her lab stool beneath a cliff of Pyrex jars, "the most illuminating experience for me has been discovering that the way we ask the questions reveal what sexual orientation truly is. From the beginning, when I started doing the interviews, I would often preface the central question by saying to the homosexual men and women I was talking to 'Okay, I'm going to ask you a question, but I want you to completely divorce your behavior from your answer. I want to know what's on your interior, and not your exterior. For example, sometimes we're boiling mad inside and yet outside we act calm, an example of how our behavior and interior feelings can be incongruous.'

"And then I asked the question: 'Looking back over your life, seeing it in retrospect, do you feel that who you are now, your homosexual orientation, has always been part of you, part of who you are, even though you might not have always recognized yourself as "being gay" and even though your sexual behavior might have been with members of the opposite sex?' When I ask the question without the preface, I usually get a recital, sometimes sort of rambling, of people's sexual activity, what they've done, and then a lot of confusion when they try to reach conclusions based on this behavior and not on their interior feelings. People talk about 'becoming' gay at age thirty-eight or whatever, but what they're giving you is what they've *done*, not what they've felt. When I preface it, when I ask it that way, almost always and in every case, people answer simply: yes."

Pattatucci has interviewed innumerable lesbians who have told her that through their adolescent years of dating boys, they were, as they often put it, "performing for society." They appeared on the outside to be happy while inside, they say, they weren't happy and they weren't satisfied and they didn't feel "right." "Sometimes they didn't know exactly why at the time, and sometimes they did. It actually is fairly rare, even when I talk to people who identify as bisexuals, that they say their interior, true sexual orientations have changed. Their *behavior* may have changed, but their homosexual core has always been there. That's the important thing. The behavior is irrelevant compared to the core.

"I want to back up for a moment," Pattatucci says calmly, "and emphasize that people who are opposed to science and to this research for any number of ideological reasons, both right and left, should not be comforted by this observation that asking the question helps determine the answer. Must we, as scientists, constantly mistrust our definitions? Of course. But I am not saying there is no truth, that none of the definitions are meaningful and that none of what we're seeing is solid. People hear you say 'How the question is asked will determine the answer,' and they think this means they can discount the result. What it really means is that one needs to ask the right questions, and when you've done your homework and you ask the right questions and the data starts coming back, you're thrilled because you realize you've tapped the vein. You're on the right track, and you're seeing the real picture. And you're simply thrilled."

Currently there are very few physical tests for those human traits that reside in our minds as feeling and imagination and desire and intellect. There is one, however, though it has yet to be recognized as such. At a sidewalk dinner table on a warm spring night in Washington, D.C., Rick Weiss, a science reporter for the *Washington Post*, was talking about psychopharmaceuticals. Weiss had been consumed with a piece he was writing on the drugs Prozac and Zoloft, and he had talked during the meal about the profound, almost unnerving changes in personality that were being created with these seratonin-reuptake inhibitors. "We've come a long way from the earliest drugs," he said, shaking his head, "the old MAOIs with all their contraindications and side effects.[4] But what these new drugs are doing, honestly, makes you wonder what parts of ourselves are really all that deeply anchored."

Psychopharmaceuticals are a hot topic—one of the best-selling books this decade is Peter Kramer's *Listening to Prozac*—and we marvel at how they change personality. People discuss the remarkable transformations the drugs have brought about in them. Lithium erases the depression that has been a lifelong character mark and creates (it almost seems) someone else; Effexor lightens mood; Prozac turns a dour person sunny; Haldol returns sanity and logic and perspective to someone always characterized, until the moment he or she began taking Haldol, by a very different temperament; Clozaril and Risperdal not only cure apathy but alter reality itself.

But all these drugs have revealed something about ourselves that we watch warily and with a certain unease: Our individual personality, something we feel is so deeply a part of "us," something that indeed comprises the essence of our "us-ness" and of who are are, is increasingly disclosing itself to be plastic and malleable, a surface reflection of the ebb and flow of neurotransmitters in the brain. The drugs, Weiss had had to conclude, force us to ask what is truly profoundly rooted in "us." So what is personality? I asked, rather uninventively. Well, he replied, frowning slightly, might it not be better to approach the question from a somewhat different angle? After all, what a drug *doesn't* do can often be as revealing as what it does. It seemed that the more interesting question raised by these drugs was, What is sexual orientation?

Personality has been uprooted, modified, ironed, tamed, altered. And through it all, no Prozac capsule, no psychopharmaceutical, no hormone, no drug, no surgical procedure has ever changed anyone's sexual orientation. It seems, Weiss noted with a raised eyebrow, that our sexual orientation is an even more deeply rooted part of us than our personality.

∼

There is, amidst all the fevered talk of "looking for the gay gene" and "finding the gay gene" and "cloning the gay gene," one crucial question that is entirely ignored. It is also entirely obvious.

The same fate met the discovery of BRCA-1, the gene with the breast cancer allele. Amid the confetti and excitement, a scientist involved in that research quietly noted that, after all, the discovery was "just the beginning." "It's going to take a lot of work to figure out how this gene works," she reminded her colleagues.

A gene's function is to make a protein. That is all. Finding a gene simply means having to figure out what protein it constructs. Figuring out the protein means then figuring out how it works to create the effect we see.

And all of a sudden we say wait a minute. What would a gay gene actually *do*?

Chapter Nine

~

HOW THE GAY GENE MIGHT WORK

Hᴏᴡ ᴡᴏᴜʟᴅ a gay gene work? What exactly would this string of molecules do?

~

Flying back from a long scientific conference at Harvard, Hamer was staring out of the cool plastic oval at a thunderhead just beyond the wing. The US Air shuttle from Logan was comfortably empty. Hamer's computer was open on his lap, but the screen was dark and hibernating. The jet's engines appeared to be coaxing him into sleep.

Suddenly, he was completely alert. "Did you hear," he said, "they just found the *fruitless* gene."

Hamer was intensely but composedly excited. When his mind begins to race, his eyes become slightly wider. A new study was just out that had finally tracked down the "gay gene" in the fruit flies—it was, he said, "beautiful"—and scientists had figured out how this gene operates. "It's right smack in the middle of the genetic sexual differentiation pathway for the brain." Being male or female in *Drosophila,* he explained, is determined by how many sex versus non-sex chromosomes the fly has. This important ratio controls a protein that makes the fly's body either male or female. "Now it turns out," Hamer says, "that there's an entirely different, parallel pathway for wiring the central nervous system—i.e., the brain—into either male or female. In other words, the sex of the fly's body on the one hand, and the sex of its brain on the other, are determined by different pathways from a common starting point. And the *fruitless* gene is controlling the brain-pathway. *Fruitless* is apparently creating this so-called gay male fly by giving a fly with a male body a partly female brain."

Hamer rapidly drummed his fingers against his tray table for an instant, tapped the side of the computer, scowled as he pursued some theoretical genetic lead down corridors in his head, and then appeared to be mesmerized once again by the thunderhead.

∿

The field of astrophysics presents probably the best example of a two-prong approach to cracking scientific problems. Practitioners of cosmology are divided into two species: theorists and experimentalists. They work symbiotically, and sometimes resentfully, in an enforced alliance where the theorist speculates about the workings of particles and matter and the experimentalist puts these speculations to physical test, disproving some theories, which are discarded, and proving others, upon which the theorists then heap further hypotheses. The theorist guides the experimentalist. The experimentalist confirms the theorist.

In biology, by contrast, researchers frequently will wear both hats. When a biologist, in the role of experimenter, finds a locus for a gene, or even the gene itself, he or she must then slip on the theorist's hat and speculate as to how this string of chemicals just discovered might, in fact, operate. The theories then guide the experimental search deeper into the chromosomes.

At Harvard, where Dean Hamer had just taken time from a conference to give a guest lecture on Xq28 to his undergrad genetics class, Richard Lewontin, a highly respected population geneticist, suggested that the gay gene may be a gene that controls the way a human embryo perceives heat.

He tossed this unusual idea out casually as a barely recollected thought. "Look," Lewontin explains simply, "this is speculative, but once you've found a locus, your next step is to conceptualize what this gene's modus operandi might be. It might be a developmental gene that turns on in utero, that determines your response to heat, stimuli that come into the womb and mold one person's neural architecture differently from another's. That creates a different sexual orientation." He turns his palms up in a why-not? gesture. Geneticists call this effect epigenetic, a genetic influence that interacts with an environmental input to create an effect. "Let me give you a case that we talk a lot about in class."

There are what scientists call "maze-dull" and "maze-bright" rats that do, respectively, poorly and well in mazes, and rats can be selected and bred for these traits, because there is a gene for running a maze. What is this gene? It turns out that the maze-bright rats are partly blind. The reason they do so well in the maze is that they don't get distracted by other cues due to having genes that interfere with their vision. Lewontin imagines that human IQ is a similar story, that certain people are less distracted because they are less genetically sensitive to sights and smells when they're taking IQ tests. "We think of genes as 'making you smart,' " he says impatiently, "as if genes were little units inside us labeled High IQ. In reality, these so-called IQ genes could be labeled Bad Eyesight, Needs Glasses. Saying, as Dean is, that there's a gene *for* sexual orientation is like saying there's a gene for running a maze well. What I'm saying is that there is a difference between saying that a gene causes sexual orientation or raises IQ or the ability to run a maze well, and a gene that has an effect on the ability to hear and therefore affects perception and, thus, behavior."

There are, in fact, a multitude of genes that work like this. Virtually every animal has receptors on cell surfaces that detect physical changes around it, gathering information and passing it in to the cell.

David Botstein of Stanford, speculating about the way the gay gene might operate, points out that it could conceivably be a different gene in every single person. The example he uses is race. "There are no single alleles that correspond to any one aspect of race you can pin down and point to 'Oh, here's the eye shape gene, here's the skin color gene!' You can see that in the offspring of marriages where one parent is, for example, Asian and one is white, because the kids have a combination of features across a spectrum. That tells you right away that skin color, eye color, hair texture, the shape of the eye, et cetera, which together make up race, are not traits controlled by single genes, since if race were just one gene, the kids would be either all Asian or all Caucasian, each of their features distinctly one or the other." Botstein adds that there could even be, for instance, an allele that influences skin color in a South American Indian but that has a completely different physiological consequence, say, on height or the texture of one's hair—"remember, this is the same allele," says Botstein—in a Zulu woman or a Japanese man.

It seems amazing, considering the ironclad rule that one-gene-makes-one-protein, that each protein would perform multiple jobs and that a single gene could create completely different traits in different people. Given the complexity of biology, however, it is *not* really amazing. Each gene might make one protein, but that protein interacts with perhaps thousands of other proteins—as would the protein made by the gay gene—and will have different biochemical effects.

Botstein's example is the gene allele that causes lactose intolerance. "By Caucasian standards," he says, "East Asians, due to a gene version many of them have, are intolerant to lactose. And to alcohol, but I'm talking about lactose intolerance here, which is a major health issue in Asia because of children having problems drinking cow's milk, and people having problems ingesting alcohol is not exactly a major health problem." The intolerance is due to one gene allele, but it is clear now that this allele alone is not responsible, that there are other genes and other *things* at work because of the simple fact that *both* whites and Asians can carry this "lactose intolerance allele," but it has a different consequence in Asians than in Caucasians. Same allele, same gene, different results segregating by race.

The enzyme process works like petroleum refining where the raw petroleum is altered step by step into different substances, from crudely refined petroleum to diesel and finally clear, high-octane gasoline. Each step in the metabolizing of raw lactose is controlled by one gene: one gene, one enzyme, one metabolic transfiguration. The process goes from lactose, to glucose, to glucose-6 phosphate, and on to the end. Say that in most Western Europeans the gene that makes the first enzyme works perfectly, providing more than enough enzyme to refine the lactose into glucose. The gene that makes the second enzyme also works well, and the gene that makes the third, and the fourth. "These people take in lactose," says Botstein, "and it gushes right through the pipeline and is metabolized very nicely."

In many Asians, however, say the gene that makes the third enzyme doesn't work well. The process functions smoothly until the third gene, which cannot process the intermediate product it is given. This only partly processed substance begins building up in the body, unable to move forward, powerless to go back. "Your body can't have these intermediates hanging around," Botstein says. "They're supposed to be transformed and moved out of there. If you start out with lactose and end up with metabolized lactose, neither of those will make you sick, but everything in between, if it hangs around in large quantities, does. The other genes are working away, this stuff is gushing through the pipes, a different product after every process, lactose, glucose, glucose-6 phosphate—and *boom*. The third gene can't refine glucose-6 phosphate, this weird intermediate crashes into the wall, and it starts flooding the machinery. And when the doctor sees them, the diagnosis is: lactose intolerance illness."

The point, Botstein says, is simply this. The allele of the third gene in an Asian causes this buildup. But the exact same allele in a European may cause no problem at all. The reason is not the allele but the other genes, upstream and downstream, with which this allele must interact. Although it's the same allele and the same gene, in one case there is a problem, and in the other there is not; and it has nothing to do with the gene itself.

There are thousands of examples. Several of the steroid enzymes involved in making secondary sex characteristics have imbalances known to cause problems, and there are other more familiar and more tragic examples of gene interaction. A deletion—an entirely missing gene—means entirely missing an essential biochemical bridge from one place to another. Muscular dystrophy, a deterioration of the muscles (dys-trophy means "faulty growth"), is one of these missing genetic bridges. Those missing a gene that makes factor 8 cannot cross a different biochemical bridge and may get hemophilia A; those missing factor 9 may have hemophilia B. True albinos, people with no pigment in their skin or hair, are missing a genetic walkway that supplies a protein forming one of the bridges to melanin.

It also matters if a person has zero, one, or two functional copies of the allele. In familial hypercholesterolemia, having no good copies means the body can't make the LDL receptor that processes cholesterol, which builds up in the blood to cause heart attacks at two years of age. Two good alleles—the normal state—make the receptor normally. But having one working allele and one bad one means the good copy will make a little of the receptor, and cholesterol builds up more slowly. Either the person consumes a low-cholesterol diet, or he or she will suffer a heart attack at forty. And, as mentioned, genes are on at different times, some only during development (certain developmental genes in the sexual differentiation pathways control menopause and make hair turn white and fall out), some all the time. Some turn on and off throughout life, and some are on only in certain tissues. Virtually any combination you can think of, says Botstein, has a real-life counterpart.

In the next room, someone sneezes. Botstein looks down and says with immense satisfaction, "Bingo. Allergies. There you go. IGE, an antibody made by one gene. Some humans have a version of this gene that makes a lot of IGE, and they tend to have allergies. Some—about half the population—make virtually no IGE and they don't get allergies. People, if they think about it, think allergies are some sort of something caused by dust in their eyes. But allergies are genetic."

The gay gene could operate through any of these molecular mechanisms. It could determine our sight or sense of smell or the way we perceive heat. It could form some change of direction in any of millions of biochemical cascades pouring through our bodies or refine a protein slightly differently in some people than in others. The answer may "fall out of the blue," as Botstein put it, from research that has nothing to do specifically with sexual orientation. "People are really beginning to understand a lot about protein structures," he says, "and serious numbers of structures of really important molecules are coming out in all sorts of areas. They're beginning to give us an idea of genetic mechanisms that were unthinkable ten years ago."

~

Complete candor about the molecular operation of a politically loaded gene can lead to trouble, and some of the public statements made by scientists to the press about the gay gene are purposefully misleading. Take the word "complex." A "complex" trait is more difficult to decipher because it is almost certainly polygenic—due to many genes— and almost certainly *not* monogenic, and is therefore more safely out of reach of scientists trying to decipher it. Most biologists state for the record that homosexuality is a "complex" trait, decades and decades, perhaps centuries, away from being understood.

Ask a reputable Harvard biologist when he thinks Hamer will have the whole thing doped out, sexual orientation, the genes, the proteins, everything?

"Never. The *whole* thing? He himself says never! There are a lot of places all over the genome that are going to affect sexual orientation. Why? Because human sexuality is complex. Nobody knows how many genes are involved in hearing, for Chrissake. Or how many genes are involved in height. Nobody has the slightest idea. And this makes height look like child's play."

This has, up till now, been dogma. Hamer himself says it. It is the standard sound bite offered reporters.

Across the river in Boston, Cassandra Smith sees things differently. A geneticist at Boston University, Smith is a slim, quiet woman with dark hair and a precise, direct way of speaking. She very casually says, "I expect sexual orientation will come down to just one or two genes," which is utterly at odds with the standard line, as she acknowledges. Her reasoning, however, makes a crucial distinction. "Sexual orientation is a simple trait. Everyone says it's complex, but it's not complex at all. People say 'Sexuality is complex.' And this is correct. But don't confuse sexuality, which is quite complex, with sexual orientation, which is quite simple."

"Sexuality" is not, Smith explains, "a trait" but rather "a common inheritance." Botstein made the same observation of race, and Smith's distinction between "sexual orientation" and "sexuality" is expressed by David Botstein as "the difference between a baseball and a baseball game." "Sexuality" is a huge rubric under which fall a vast number of traits: sex drive, the age of puberty, physical build, the preferred age or race of sexual partners, the size of one's genitals, one's moral, political, and religious attitudes toward sex and the ways they create and govern ideas of romance and desire, sources of erotic stimulation and response, the number of desired partners, and so on.

Sexual orientation, by contrast, is the *gender* one finds sexually and romantically attractive: men or women. And that's it.

Within the gender one finds attractive, however, there is enormous variety of tastes and preferences and perceptions, and these are the other aspects of sexuality. A heterosexual woman is attracted to men. But what kinds and shapes and colors of men? What male personalities and male social statures and male attitudes? What approaches, among these men, to sex, sex acts, frequency of sex does she like or dislike? All of these vary widely and all are aspects of sexuality, but the sexual orientation component of sexuality is one small, specific piece: This woman is attracted to men. Smith observes that sexual orientation is, thus, simply a directional marker, male or female. (Richard Pillard's informal definition, both simple and practical, underlines the "directional marker" definition. "Sexual orientation is, when you pass people on the sidewalk at lunch hour," says Pillard, "which sex do your eyes flick on involuntarily: men or women?")

"Such a trait may easily be just a few genes," Smith says. "Maybe just one."[1]

Like Smith, Philip Reilly of the Shriver Center for Mental Retardation believes sexual orientation will be discovered to be under the control of a small number of genes. "I imagine it will come down closer to one," he says. The reason is that sexuality, in all its aspects, is one of the most fundamental phenomena subject to evolution, and in Reilly's view there has to be basic biological wiring that controls all the various traits that make it up, established at an early point. "I'd speculate that sexual orientation is linked to a very early event in embryogenesis and thus possibly could involve just a few fundamentally important genes that start that process unfolding. I really would be surprised, for example, if we were to learn that the gene turned on late in fetal development."

Prudence in discussing this is not necessarily misplaced. The suggestion that sexual orientation is genetically uncomplicated raises hackles. In a relatively candid moment, Hamer admitted, "Originally, we used 'sexuality' and 'sexual orientation' almost synonymously, but the more work we do, *really* looking at exactly what you're trying to find, the more it becomes clear the original 'complexity' of the trait was due to that confusion." How complicated does he expect it to be genetically? He thinks about it. "I'd say sexual orientation will probably be the same order of genetic complexity as handedness." He adds that this analogy must be understood clearly. Handedness, the neurological ability to move the arm and fingers in a way that allows writing, swinging a racquet, or opening a jar, almost certainly involves hundreds or thousands of genes (this includes, for example, the genes that construct the muscles in the hands themselves). The *direction* of that handedness, however—left-handed or right-handed—may be determined by very simple biological factors, maybe only a few genes. Or one. "The overall machine," he notes, "which you could call sexuality, may be, genetically, incredibly complicated, but the directional switch for that machine, straight or gay, may be incredibly simple."

One of Hamer's colleagues, who will not allow himself to be identified, was much more frank. "Look, you'll never get me to say it publicly, but I think it's clear that this is really a pretty simple trait. It's just not politically wise for us working in this field to say it yet. It makes people nervous and upset; the simpler a trait is, the easier it is to research, and that's threatening to a number of different types of people on all political sides for a number of very different reasons. We'll say it when we get more evidence, which is as it should be, but if you look at where the data are going, there's not much question."

In fact, in certain circumstances some of the so-called complex traits, such as height, turn out to be not only not complex but, we now know, controlled by a single gene. One case in the literature where a single gene has a major effect on a "complex" trait is found in Pygmies. If a Pygmy has a child with an African of average height, all the children are either of average height or of Pygmy height. The trait is one or the other, clear cut, and the reason is that the trait is controlled by one gene. What does the gene do? The most logical thing in the world: It helps manufacture the receptor molecule for growth hormone. Without the receptor, the hormone cannot dock, and it floats helplessly outside its destination, ineffectual, until it disappears. People often think of sexual orientation as a truly "singular" trait, yet there is another trait that, in its essentially universally bimodal division, in the deep, certain way in which it is felt from the youngest of ages, in its stability and persistence and its marked, consistent emotional and behavioral expressions across cultures, in the way a small minority obstinately and in the face of intense social pressure and the logic of their own bodies insist that there is a mismatch, resembles sexual orientation closely: gender.

Perhaps the most intriguing similarity between sexual orientation and gender is their mutual use of a firm, clear, unshakable mental image of maleness and femaleness. For the trait sexual orientation, the image in the brain is externally directed: "This is the gender to which you are attracted." For the trait gender, it is internal: "This is the gender you yourself are." Gender, a discrete, well-defined trait, might be controlled by very few genes, and the key to its operation may unlock not only the biological mystery of sexual orientation but another trait whose phenotypic parallels—and thus possibly its biological origins—to homosexuality are astounding: transsexualism. Not only the details of the trait itself but the personal histories of transsexuals recall with astonishing similarity those of homosexuals.

In an article titled "The Body Lies" published in *The New Yorker,* clinical social worker Amy Bloom examined the phenomenon of F.T.M.s, "Female to Male" transsexuals, XX women born with an unalterable male gender identification, a feeling deep inside themselves that tells them they are men. (M.T.F.s are the obverse, Male to Female.) "Male is not gay or straight," wrote Bloom, "it's male. We may not know what it is, but we know it's not about whether male or female stimuli inspire your erection."

Through sex reassignment surgery—Dr. Don Laub, a surgeon, calls it sex confirmation surgery—the women Bloom talked with had made their bodies male. She confronted the crucial distinction between gender orientation and sexual orientation in the person of Louis Sullivan, an F.T.M. who knew irrevocably and profoundly from toddler girlhood, as transsexuals do, that he was male, and who also knew, irrevocably and profoundly, as homosexuals do, that he was attracted sexually to men. Because of cases like Sullivan, doctors and researchers recognize two clearly discrete traits: the gender one feels one is, on the one hand, and the gender to which one is sexually attracted, on the other. Dr. Ira Pauly, chairman of the Department of Psychiatry at the University of Nevada, notes that the prevalence of the two traits is strikingly different: For males, homosexuality is a trait carried by five in a hundred men, where transsexualism appears in one in fifty thousand. The appearance of both in one person is thus, statistically, an extremely rare event.

Two different traits, but anyone familiar with the experience of gay men and women will instantly recognize the astonishingly similar story of Michael (the pseudonym he chose), a serious, black transsexual man who concluded Bloom's piece. "I grew up in a nice, materially comfortable, middle-class life," he said to Bloom. "But I carried a deep, dark secret around with me. . . . I hate to sound like Marlo Thomas, but I just wanted to be free to be me, whatever that was. And I didn't know, although I kept going to the library, trying to find out . . . [G]oing off to school, horrified that I had to go in what felt like drag, sure that everyone would laugh at me, I knew that I'd better get used to it, because this body was not becoming male and it clearly made a difference to the world. . . . But Joan of Arc did it for me, explained me to me, when I encountered her in school, at the age of nine. I thought, well, here we go, and when I was twelve, finally, I found a book on transsexuals."

Michael thought maybe he was a lesbian. "Could be—I know I'm attracted to women. I went to [lesbian] consciousness-raising meetings, and I'd listen and feel like a fraud. One girl said, 'What makes each of us feel like a real woman?' And while they went around the room, answering, I thought, nothing—absolutely nothing on earth makes me feel like a woman . . . I'm just a plain old heterosexual man . . . I don't think of myself as a transsexual anymore. I was one, I made that transition, now I'm just a man. . . . I'm a decent person, I'm not ashamed. I don't know why this condition chose me. We, people who have been through this transition—we are among the few people in the world who have overcome obstacles and fulfilled their lifelong dreams. All these obstacles, and I am who I dreamed I'd be, who I wanted to be."[2]

∼

How would the biochemical mechanics of a gay gene work, which proteins hitting which other proteins and refining what chemicals, to create a desire for men or women? How would a chemical create a feeling of being male or female? To answer this question, theorists look for clues at how other genes operate. A few examples:

"My son, Nathaniel," says Dr. Mike Baum of Boston University, "has ocular albinism." Baum's office and lab are in a new, bright building across the river from Cambridge, and Baum, a friendly, balding middle-aged professor, works with his students on ferrets and their sexual and behavioral makeup. "He has ocular albinism," says Baum, "because he has a defective gene at Xp22.3, the 22.3rd section of the p arm of the X chromosome. It's an X-linked recessive, so it's passed down the maternal side. His mother is a carrier, and my son's sons won't have it, but all his daughters will be carriers, and on average half of his grandsons will have ocular albinism. Unless he does some kind of genetic screening, of course."

What Nathaniel has is the same syndrome that exists in Siamese cats, where it is universal. And since they all have ocular albinism and are all cross-eyed, "Siamese cats can't see diddley," Baum says bluntly.

The genetically fascinating aspect of ocular albinism is that it is a profound neural malfunction caused by a gene that has absolutely nothing to do with the neural system. The job of this gene is a completely unrelated task: coloring the eye's retina.

In its damaged allele version, the gene hinders the development of natural eye coloring, producing an albino eye. This is, however, only the gene's direct effect. The result of not having color in the eye is what leads to a dramatic result: the entire visual nervous system is damaged. For reasons not understood, the lack of melanin, a biochemical that colors tissue, results in 30 percent more ganglia nerves projecting themselves from the retina to the opposite side of the brain, with terrible consequences for visual acuity. Since there are an excessive number of nerves crossing, much more information than normal is projected onto the brain's opposite side, and the information the brain is receiving is severely mismatched. And yet, despite this devastating neural result, the genes controlling the nervous system are all fully intact and functioning.

"The first evidence we had that Nathaniel had this," Baum recalls, "was when he was a baby, and we were living in Holland. Our kitchen wall was tiled, and it had a large black stripe in the tile. We noticed that when we held him up near the wall, he'd laugh and giggle and have a big emotional response to the stripe. Then you'd hold him away and he'd calm down. The reason is probably that it was the first thing he ever really saw."

The ocular motor of ocular albinos is also distorted, and their eyes move back and forth jerkily. "My son uses binoculars in class to look at the blackboard, although he doesn't use a cane. He sees certain things in familiar environments, and he told me I was on the wrong side of the road once." Baum raises and drops his eyebrows thoughtfully. "Normally I wouldn't have thought he could see that. He can't drive." He concludes simply, "He basically gets through life without seeing a lot."

At the University of California, San Francisco, Jane Gitschier is following clues to track down another gene that operates in a very different way. UCSF is a postgrad medical and research campus set in the hills that cut down the backbone of the city. The hills—cliffs, really—into which the university is tucked are covered with tall, willowy eucalyptus trees that, when the wind blows in off the ocean, crunch faintly like new crinoline and wash the city's sea air blowing through them with a clean, astringent smell. On the opposite side of the street from Gitschier's small but spotless and well-lit office, one looks north down a steep hill toward the Golden Gate Bridge, or west across the Richmond to the Pacific, when the view isn't a sheet of fog.

Gitschier was one of the leaders of the hunt for the Menkes gene. There had been linkage studies of Menkes in families, and the general location of the gene had been mapped to Xq13, near the centromere of the q arm of the X chromosome and not far from Dean Hamer's GAY-1. The gene itself, however, had not been identified, said Gitschier, and the protein it produced was unknown. But Menkes has extremely characteristic effects, and it was one of these that gave researchers the necessary clue to figuring out how the gene operated. (The statistical prevalence of Menkes in males is equal to that of transsexualism in males.)

She counts off the clues on her fingers. "Menkes boys are born very floppy. You notice this almost immediately. They have almost no muscle tone at all because they're neurologically impaired. They are hypopigmented, which means they have very light skin and hair color. But you'll notice they're not albino, they just have lighter skin and hair than their siblings do. They have connective tissue defects, and something called arteriol tortuosity, where if you do a scan of their brains the arteries sort of move around in patterns you don't usually see."

Gitschier takes out a series of plastic slide holders and runs a finger down the rows. She selects one and holds it up to the desk light. It is a brain filling the inside of a skull. It looks odd, not like a photograph but like the slightly hazy electronic outlines of shapes that appear on radar screens. "This is an MRI scan of the brain of a little boy with Menkes disease," she says. "I don't have the scan of a normal brain to compare it with, but the neurologists will tell you it's already showing some signs of being abnormal."

The finger runs down the plastic packet. "This is the same brain eight months later."

She smoothly extracts a different slide. When she holds it up to the light from the desk lamp, the image—when one finally realizes what it is—is repulsive. The brain looks like it is shrinking, which is,

Gitschier says, exactly what it's doing. The brain is now a shrunken mass of tissue floating in the middle of a cavernous pool of darkness inside the skull.

"These little boys have an inability to thermoregulate," continues Gitschier, "to keep their body temperature at 98 degrees. And they have a very characteristic type of hair. It's extremely brittle, like a scouring pad, and they call it steely hair. When they lie on their beds, their hair just rubs off their scalps onto the pillow. This is a photograph of a boy with Menkes disease."

Another slide: A pale blond child, looking solemnly past the photographer. A strip of spiky hair runs down the center of his skull like a Mohawk, since all the brittle hair has been rubbed off the sides of his head.

Scientists knew these aspects of the phenotypic profile of the trait created by this gene. But what in the world was the gene actually *doing*? David Danks, a medical geneticist in Adelaide, Australia, made the connection. "He'd seen a lot of sheep down there," Gitschier says, "and he realized that this steely hair one sees in Menkes boys was reminiscent of the wool of sheep who grazed on grass that grew in copper-deficient soil. Copper-deficient sheep produced unmarketable wool, unmarketable because it was so brittle. So he sort of made a leap. He thought, 'Maybe what's wrong with these boys with Menkes disease is that for some reason they don't have enough copper.' And it turned out that was exactly the case."

On Danks's speculation, scientists isolated cells from Menkes children and tested their copper levels, and they found that they were severely copper deficient. Knowing the general region of the Menkes gene, several groups, including Gitschier's, isolated the DNA that encompassed it, narrowing its location to a small segment of genetic material. Researcher Christopher Vulpe and coworker Barbara Levinson together then found the gene itself and sequenced it. "And it turns out," says Gitschier, "to be a copper pump."

The gene makes a protein, she explains, as do all genes, and this particular protein's job is to span the membrane of a cell, form a protein bridge, and pump copper across that membrane. "Actually," Gitschier explains, "we're not sure yet whether it spans the membranes of cells or the membranes of organelles, like the endoplasmic reticulum and the mitochondria, and vesicles within cells. That's what we're working on now. But we *are* pretty certain just from the sequence of amino acids in the protein this gene makes that it's a copper pump." Its sequence is homologous to other proteins that pump cations (positively charged ions, like calcium). "You look at the protein, and it's got a certain combination of amino acids, a motif you recognize. Motifs are similar patterns. They're like a signature tune you might hear repeated in an opera by Wagner and you recognize it easily." She smiles and hums a few bars of something, and the theme is instantly familiar: "The Ride of the Valkyries."

Gitschier reaches for something on her desk. "Here is another Menkes brain." She holds up a slide. "See? The arteries are all twisted and gnarled. The failure to thermoregulate, this extremely brittle hair, it all comes down to the inability to pump copper. A lot of enzymes in our bodies require copper, a very important cofactor, and if they don't have it, they don't work as well, and sometimes not at all. For example, there's an enzyme called tyrosinase that is important for making melanin, the pigment that colors your skin. And if that enzyme doesn't have enough—guess what—copper in it, it doesn't work. That's why these little boys are so pale."

Jane Gitschier sits back. "So now you know. All of this, the color, the thermoregulation, the hair, all the various aspects of the trait, is an example of what one gene that makes one protein does. A gay gene might operate in a similar way: a gene that produces an iron pump, as yet unknown, supplying some important cofactor we are not yet aware of that's needed in some chemical process we can't even imagine, a process that may have absolutely nothing whatsoever to do directly with sex or gender or desire or any of the other obvious things."

If there is any popular canard, any near-universal misunderstanding of biology regularly applied to the biology of homosexuality, it is the notion that evolution could not select for a gay gene. A superficial reading of Darwin leads to the assumption that genes that directly decrease reproduction are selected against and die out. The key word is "directly." "Many people," Eric Lander of the Whitehead Institute points out, "who have a gut reaction against homosexuality will express their feeling by saying that it is 'against nature'—that's the way they generally put it—because nature only selects alleles that enhance reproduction, and thus homosexuality, which is nonreproductive, must be a perversion of the natural order." He pauses, then warns, "This is based on a fundamental misunderstanding of evolution. People assume that a gay gene would lower reproduction. You can't assume that at all." In fact, E. O. Wilson of Harvard, the father of sociobiology and a powerful biological theorist, hypothesizes that nature has selected for the gay gene because it brings with it an indirect advantage. Suppose, says Wilson, that it is a million years ago, and there is a gene that stops some members of *Homo erectus* from reproducing because it directs their sexual desire to members of the same sex. And say this group makes up maybe around 5 percent of the population of each tribe. Wilson observes that since they do not produce their own offspring, these tribe members are available, if their brothers or sisters are killed on a hunt or in an accident, to help raise their siblings' children and assure their survival. Even without the death of the parents, where some or all of the siblings' offspring might starve, with a childless homosexual aunt or uncle around to help out, they might all survive. Wilson calls this kin selection.

But if those searching for the gay gene have this theory to work with, they also have a very common example of a gene allele creating a trait with which everyone is familiar, a gene that has been evolutionarily selected for over millennia while directly decreasing reproduction. And it does not merely discourage reproduction in the somewhat roundabout way of redirecting its carrier's sexual desire toward the same gender. It does it by killing those who carry it. This gene is lethal.

"There's this apparent conundrum," says UC Berkeley's Jasper Rine, "which is how come gene alleles that kill you don't die out of the population? Isn't natural selection supposed to select for genes that help you? The answer is 'Well, yes—but there's a catch.' The catch is that nature works on averages, not on individuals. Natural selection picks the genes that help more people than they hurt." There is, Rine notes, this rather amazing allele of a gene that sits at 11p15.5, at the end of the p arm of the eleventh chromosome. "This particular allele, which has been selected for generation after generation, has the power to protect you from death or to destroy you, two real extremes. It all depends on whether you've got one or two copies of it."

If, Rine explains, you're one of the unfortunate people who have two copies of this allele—both your main copy and your backup copy—you develop a well-known disease of the blood that cripples and eventually kills you. But if you're one of the lucky ones and only have *one* copy of the allele, it endows you with a powerful genetic protection against a deadly parasite. "What is the parasite?" Rine asks. *"Plasmodium,* the organism that causes malaria. It grows in human red blood cells, and one copy of this allele protects its carriers from the malaria parasite. What is the disease? The terrible, horribly painful blood disorder sickle cell anemia." (The gene controls the manufacture of Beta globin, an enzyme that helps carry oxygen in the blood; two bad copies of it destroys the crucial oxygen transfer process.)

In Africa and the Mediterranean, where the malarial parasite has existed for thousands of years, nature has selected for this paradoxical allele. While it kills thousands of people with sickle cell anemia, on average it saves hundreds of thousands more from malaria. Rine smiles and cocks his head. "It depends whether you'd call it a deadly mutation or a God-given life saver. But the fact remains that nature selects *for* this deadly gene. If someone said to you, 'Listen, there is a genetic allele that has existed for millennia killing enormous numbers of people, and nature *intentionally* maintains it in the population,' you'd say, 'Get out of here! You're out of your mind.' And yet it's true. It's absolutely true."[3]

The gay gene, Rine notes, could operate in a similar way: selected for because of, say, the cumulative increase in reproduction in the tribe suggested by Wilson and perpetuated by heterosexuals who carry it invisibly.

Possible genetic models for the operation of a gay gene include epigenesis as illustrated by Lewontin's gene for heat, Wilson's kin selection, penetrance, and Rine's point about net versus individual reproduction. The strangest one by far, however, is pleiotropy.

Pleiotropy occurs in genetics the way side effects occur with drugs. Homosexuality, one researcher theorizes, could be a genetic side effect.

The classic example of pleiotropy is found in an article by Richard Lewontin and Harvard paleontologist Stephen Jay Gould titled *The Spandrels of San Marco*. It's an architectural example of the biological question, Why is this thing—this effect, this trait—here? Spandrels are triangular panels created when arches are placed next to each other, such as the arches holding up the dome of San Marco Church in Venice. Painters took advantage of these triangular spaces to paint beautiful frescoes. The confusion of a gene's primary with its secondary effects would be like a researcher of sexual orientation going to San Marco and saying "Look! Evolution has produced these spandrels as a place to put these beautiful paintings." And she would be completely wrong because the spandrels were created as by-products of the architectural structures called arches whose primary purpose is to support the building. Miss this fact, and the spandrel is misinterpreted. Pleiotropy thus poses the tricky question: Is homosexuality the arch, selected according to kin selection to create a small number of nonbreeders who will help breeders increase their reproduction, or is it the fresco-adorned spandrel, merely a secondary by-product?

It is ridiculously easy for genes to create secondary effects. Genes can make only a limited number of enzymes, and a single gene can, and often does, do two or more jobs because the same chemical products must be used many places in the human body. Redundancy is standard in the chemical pathways to any biological result, and the same genes that man assembly lines to make a certain hormone receptor might man different assembly lines at different points to make collagen. Some scientists believe genes work not linearly, each exclusively on its own job, but in parallel, double- triple- and quadruple-teaming their workloads by joining up with different groups of genes to do different jobs. These scientists argue that without such cooperation, we would not have enough genes to create ourselves. With millions of overlapped jobs, it would be unusual if unintended pleiotropic effects were not generated.

Pleiotropic effects can take any number of forms. Evolution selected Northern Europe an alleles for fair skin to combat rickets; rickets is a Vitamin D deficiency, and sunlight helps produce Vitamin D in the skin, more in light skin than in dark. (The reason Vitamin D is added to milk is to protect against rickets. In Australia there is no Vitamin D supplement because there's plenty of sun.) These same alleles, however, have the unexpected pleiotropic effect of increasing skin cancer in regions with more sun.

Then there is what biologist William Rice calls sex-antagonistic pleiotropy, side effects that benefit one sex and hurt the other. Hemochromatosis, a recessive genetic disease in which excessive amounts of iron are deposited in the liver and skin, may be an example of sex-antagonistic pleiotropy because it is more damaging to men than women.

The allele responsible is of a gene at 6p21. Basically half the Chromosome 6s in the world (and the genes they carry, like this one) exist and have always existed through thousands of human generations in women, and half have been in men. The primary job of this allele (and the reason evolution has lovingly conserved it) is to enhance the ability to pick up iron from the intestine. Biologist Jared Diamond argues that since women lose iron when they menstruate, this allele might, as a secondary effect—a pleiotropic effect—help women who would otherwise be anemic and lacking iron to maintain good iron levels. Since men don't lose iron to the same degree, unfortunately the secondary effect in men—the male pleiotropic effect—is to leave them with an excess of iron, which starts settling into their tissue and making them sick. This is the "sex antagonistic" part. Is the gene good or bad? The fact that it has been conserved shows that its benefit to women on average outweighs its harm to men. Again, we're talking about a gene that is found 50 percent of the time in men and 50 percent in women. All genes on all the chromosomes exist in an even male/female 50 percent/50 percent split—*except* (and this is the one glaring exception) the sex chromosomes, X and Y. *They* are different.

Since women are XX and men are XY, genes on the X chromosome are found 66 percent of the time in women, while only 33 percent of them exist in men. (The Y is even more skewed—the ratios are 100 percent of the time in men and 0 percent in women—but the X, a much larger chromosome with many, many more genes, is more interesting to biologists.) And the X just happens to include Hamer's gene somewhere inside Xq28. "The cool, and somewhat bizarre, thing about a gene on the X," Hamer has remarked, "is the math." The math, in this part of the genome, strongly skews the established rules by which natural selection plays, as if the laws of physics were annulled on one corner of the planet, and gives X genes a decidedly different perspective, sort of a "female point of view." This means the good things X genes do for women can be much less than the harm (decreased reproduction) they do to men and still—from the gene's way of looking at things—come out ahead. Why? The beneficial effects for women are, essentially, being multiplied by two (66 percent versus 33 percent). X's attitude toward men is, What have you done for me lately?

In fact, the break-even point isn't even close. Hamer points out that gay men and lesbians have always had children, but it's not unreasonable to suspect that they have fewer than their heterosexual counterparts. If an X-linked gene reduced the number of children men have by, say, 50 percent, it would only have to increase the number of children women have by 25.1 percent—half that number plus a little bit—in order to tilt the balance scales and convince nature to select for it. It could increase reproduction in women through any number of possible beneficial effects. Perhaps the gene changes sex hormone metabolism in a way that makes women go through puberty a little early or menopause a little late—and as a side effect just happens to make men gay. This means the gene would be an "increased childbearing gene" in women that, on balance, caused increased reproduction and was only secondarily a "gay gene." Nature would concentrate on the first trait while we would notice—and fight about—only the secondary trait.

If the gene's "primary effect," the reason nature is selecting for it, is to increase significantly reproduction in women, it could do just about anything to men, including killing them. Nature wouldn't care. Dean Hamer knows that most people assume he is tracking a gene whose primary effect is homosexuality but which may actually be a gene that creates homosexuality as a secondary effect. The problem with the effects is deciphering which is which.

Aside from genes on the X, there is another exception to the genetic rule of random segregation. There are, in fact, genes so outside the normal genetic arena that they don't even sit on our chromosomes. They live in mitochondria, and they behave completely outside normal parameters, because they can be passed on only by women.

Mitochondria are tiny, mysterious organisms that live inside cells as independent units. They have their own genes, their own ribosomes, and they produce their own RNA and proteins. Mitochondria are, we believe, the descendants of an ancient bacterium that existed when the world's original atmosphere had very little oxygen but a great deal of carbon dioxide. The first plants burned the carbon dioxide in the air and released oxygen, the levels of which increased steadily. To our ancestors, animals that breathed carbon dioxide, these plants were generating a deadly pollution, and eventually animals and all organisms that couldn't burn oxygen began dying. Then a billion years ago, oxygen-eating bacteria came to live in animal cells, allowing them to adapt to the oxygen-poisoned air. These bacteria evolved into the mitochondria that live in the cytoplasm of human cells, taking oxygen and carbohydrates and burning them, and giving off carbon dioxide and energy to power its cell-host. They are the true energy generators of our bodies.

Had the mitochondria not infiltrated cells and made their lives inside them, planetary life would be utterly different from what it is today. Except for certain oozes in swamps and some organisms that live in the gut called anaerobes, which are among the most ancient life-forms on earth, almost everything that cannot burn oxygen is dead now. Because they lack mitochondria, they can survive only in places where they won't be poisoned by oxygen. Giardia, the intestinal parasite known well to campers, is an anaerobe without mitochondria. The smell of sulfur in a swamp where there is still water, mud, silt, and rotting leaves on the bottom results from anaerobes taking sulfur compounds instead of oxygen, burning them, and releasing hydrogen sulfide, the rotten-egg-smelling gas. The bacteria at the top of the swamp water burn the oxygen, sealing off everything below and creating an atmosphere hospitable to mitochondrialess life-forms.

A number of diseases are caused by mutations in mitochondrial genes. Most of them are respiratory illnesses due to the mitochondrial link to oxygen metabolism, but their causes are little understood. Some scientists believe that since mitochondria have few genes, they are unlikely to have significant influence over behavior. Others disagree, pointing out that they govern much steroid refining. In fact, the initial processing of cholesterol into a sex hormone—cholesterol is the most important sex hormone raw material—takes place in mitochondria.[4]

Humans cannot live without mitochondria, and mitochondria can't live without us. But they have their own genes, which differ from ours, and this means, strangely enough, that some of the genes in our bodies have their own evolutionary interests that diverge from ours. To a great extent, these interests are dictated by the peculiar fact that mitochondria are transmitted only through egg cells, never through sperm. Every egg a woman produces has around 250,000 mitochondria; every sperm has 12, near the tail, which are used to power the sperm to its target and are destroyed on penetration. This means, from the mitochondrial genes' point of view, that they spend not half or two-thirds of their time in women, but 100 percent of it. And none in men.

These facts have led one biologist, who declines to be named, to a theory of how homosexuality might work genetically. The theory is based on two things. The first is the standard evolutionary hypothesis that genes desire to replicate themselves, not us, their vehicles. The second is the fact that mitochondria exist only in women. When a gene spends 100 percent of its time in women, sex-antagonistic pleiotropy—any effect on men—is now limitless. Whatever the gene does to males is irrelevant to it. Even the slightest, most ephemeral beneficial effect in women would outweigh virtually any negative effect in men. Suppose the mitochondrial gene made men sterile, preventing them from having any children at all. If at the same time it gave women even the slightest advantage, then from the gene's point of view it has come out ahead since its survival depends only on women.

The theory of the modus operandi of the gay gene, though it flows from these two facts, occurred to this scientist after contemplating an experiment carried out in another species: parasitic wasps.

The most striking feature of the Trichogramma wasp was that it was entirely female. There were no males at all, and none had ever been found. Females produce daughters through parthenogenesis, from the Greek for "virgin" and "birth," asexual reproduction.

Ordinarily it is to a species' advantage to produce offspring in equal ratios of male to female, as humans do, but there are insect genes, thought to have been derived from viruses millions of years ago, that find it to their benefit to distort the sex ratio so severely that the species produces only females. Thus Trichogramma was not considered surprising when it was discovered. After all, ants and bees, it was known, also were produced by virgin birth without fathers.

But something different was operating here. It turned out that these Trichogramma wasps were infected by a bacterium, one that behaved exactly like the sex-ratio distorter genes and the mitochondrial genes. Wasp eggs contain a great deal of cytoplasm, which—as with the mitochondria—is where the bacteria live, but wasp sperm has no cytoplasm and thus no bacteria. Only female wasps can pass on the bacteria, so as far as the bacteria is concerned, only females are useful. Males are useless and of no consequence. And so the bacteria in the eggs somehow converted every wasp male into a female.[5]

"Given this discovery," says the biologist, "a researcher thought to give this all-female wasp species antibiotics to kill the bacteria. And suddenly they produced males. Simply with antibiotics, he 'cured' them of their parthenogenesis, the asexual method of reproducing that had been a universal trait in the species and through which they had bred for perhaps millions of years. And they became sexual, of two sexes.

"You see," he concludes, "mitochondrial genes, sex-ratio distorting genes, and these bacteria don't care about male reproduction at all, so they can get up to all kinds of tricks. Homosexuality could very well be the sex-antagonistic pleiotropic effect of a mitochondrial gene or even a bacterium transmitted exclusively through females. Or rather through women, since we're talking about humans."

He sits back and pauses for a moment. "To phrase it more simply, what I'm saying is that homosexuality could be the side effect of some gene or virus or bacterium that has hijacked part of the human sexual organism. Obviously I'm not going to say this on the record. I have nothing professionally invested in the idea; it's simply a theory, but we're at the point now in the research where we are starting to develop these theories to try and figure out how a gay gene might operate. *Could* this be the case? Absolutely. Is it likely? I have absolutely no idea. I hope not. I very much hope not. There is more than science here. This is positing that homosexuality may be a type of bacterial infection we've just never encountered before, one that we may eventually be able to eradicate with an antibiotic."

⁓

The most completely thought-through theory of how the gay gene might function in creating a homosexual sexual orientation has been proposed by Boston University's Richard Pillard. "In the middle of this century," begins Pillard, "a French doctor named Alfred Jost observed something very odd about cows."

Like humans, cows usually have a single offspring but occasionally they produce twins. Jost noticed that male-male twins and female-female twins were almost always anatomically normal, but when the twins were one male and one female, the female twin often had masculinized genitalia—the clitoris was enlarged, approaching the size of a penis—and she often looked like a little bull. "It's an old and well-known phenomenon in animal husbandry," said Pillard, "and the name for these female calves is, for some reason, 'freemartins.' "

Jost did some investigating and found that the degree to which the female's genitals were masculinized was directly related to the degree of anastomosis in the womb, that is, the number of blood vessels the twins shared during gestation. The calves are attached to the placenta, and sometimes many blood vessels go from one calf to the other. If the twins shared a large amount of blood circulation, the freemartin phenomenon would be much more likely to occur.

"Monsieur Jost then conducted an experiment," says Pillard. "He surgically removed fetal rabbits from the womb of Momma rabbit, the doe, castrated the male babies ("Jost deserves credit," Pillard notes with admiration, "for the microsurgery involved, taking these tiny fetuses out and snipping off their nearly microscopic testes"), put them back inside Momma and let them go through gestation and be born. What he found was something rather amazing: these genetically male rabbits had become anatomically perfect females. From these observations he constructed a mighty theory.

"He theorized that there were two different hormones that are secreted by the testes of the male fetus even while it's still in the womb, and these two hormones have two different actions. One of them masculinizes the animal, giving it male genitals, and this is the hormone that creates freemartins out of female calves. This hormone, which the brother calf's testicles are pumping out, gets into his sister through the blood she shares with him and causes her clitoris to enlarge into a penis next to her vagina. If you cut open Momma cow early in her gestation, you can see that the female calves are completely female. It's only later when their brother's maleness chemicals hits them that they start masculinizing."

It was the second hormone, however, that was more surprising.

While examining his masculinized female cow, the freemartin, and his feminized male rabbit, Jost had noticed something: the male-female reversal wasn't balanced. The cow, who had female genes, had both male and female features and was thus what could be loosely called a hermaphrodite, although she had neither prostate or testicles and so could not manufacture sperm. She was a very incomplete male, but partly male nevertheless; the rabbit, on the other hand, had male genes—and yet was *completely* female. It was not male on any level, not even slightly hermaphroditic, except for a Y among its chromosomes. There was a strange biological asymmetry here, and its implication seemed to be that *femaleness* was, somehow, much stronger.

"Jost realized," Pillard says, "that this second, mystery hormone must be a kind of organic poison, biochemically and inexorably stripping the male of his femaleness, killing off all female features." The theory made sense, and Jost, logically, called this process *de*feminizing the male. This must be why, he reasoned, cutting off the testicles, the fetal source of male hormones, could turn what genetically should be a male rabbit into an anatomically flawless female. These were rabbits that had vaginas and ovaries and could give birth, and yet they are XY. The mystery hormone must have stripped away everything feminine.

"Jost realized that he'd stumbled across something important, and he took it to Lawson Wilkins at Johns Hopkins University. It was the 1960s, and Wilkins had written the preeminent textbook on pediatric endocrinology, the study of hormones in fetuses and children.[6] So Jost presents his findings and his theory and Wilkins says, 'Hm, well, I don't know. Can you explain this baffling phenomenon I've been seeing of women with no body hair and testicles inside their bodies?"

"This shows the power of Jost's discovery. Jost says, *'Certainement,* my theory accounts for this exactly.' " Jost knew, thanks to the technique of karyotyping, which in the 1940s first allowed scientists to lay out a person's chromosomes and look at them, that these women were actually XY genetic males, just as were the rabbits. These "men" (because they were genetically male) had Y chromosomes, and a gene on the Y is needed to make testicles. Their testicles were producing testosterone just as they should have been, a fact Jost and Wilkins knew because testosterone levels are relatively simple to measure. In fact, oddly enough, these levels were higher on average in these "women" (because they are physically female) than in men. "What was going on," says Pillard, "we now know is actually very common in biology, and familiar to anyone who knows a diabetic."

Due to a malfunctioning gene, a diabetic has lost the cell clusters in the pancreas that make a molecule called insulin. The problem this creates is simple: Insulin is a chemical pass-key that allows sugars to enter the cells of the body wherever they are needed. Without insulin to receive these sugars and usher them into the cell, cells cannot absorb the glucose they need to make the chemical products they're supposed to be making, the diabetic urinates the glucose out of his system, and ultimately dies. But injections of insulin, doses of pure hormone, can now be used to treat the disease. Cow or hog insulin was used, but now human gene cloning has made possible the use of humulin—human insulin—as the pass-key.

What was occurring in the case of the women with testicles was quite similar, except the gene that was malfunctioning controlled not the production of insulin but of Androgen Receptor Molecule, which functions, much like insulin, as a pass-key. When androgens are released into the system to masculinize the body, the Androgen Receptor stands at the door and admits them into cells so they can do their masculinizing work. The gene that makes Androgen Receptor is one of many necessary for male sexual development, allowing androgens to make genitals grow, build penis tissue, develop the prostate, and create hair growth. (When we completely understand Androgen Receptor Molecule, we will have found a cure for male-pattern baldness. The trick is making it *fail* to convey androgens into the cells of our scalps.)

The women with no body hair and testicles had a condition called Androgen Insensitivity. The gene that was supposed to allow them to process androgen didn't work. The gene on their Y chromosome had given them male testicles, which were pumping out testosterone, but as in diabetes, the testosterone was helpless to get inside the cells to masculinize them. Since the fundamental sex of human being is female, these genetic male humans became females.

"Jost," Pillard says, "had discovered that the book of Genesis got it backward. The default gender is female. Adam was really made out of Eve's rib, not the reverse. Without male hormones, we'd all be women, regardless of whether we are XX or XY. This is called the Eve principle. To become male, you've got to add something that will create the change."

But Jost pointed out something else as well. These genetically male women had breasts and hips and external genitalia. All of their secondary sex characteristics, such as distribution of body weight and body hair, were fully female. They secreted a *lot* of testosterone, so their testosterone levels measured normal to high, but it was being converted into an estrogen, estradiol, which was making them even more feminine. They were frequently quite tall and buxom, and all the initial cases in the 1950s and 1960s were discovered in airline stewardesses and runway models. But Jost noted there was one huge difference between the women and the rabbits. The women weren't female inside. They had no internal reproductive equipment, no uteri, no ovaries, and they had what are known as "blind vaginas," vaginas that looked normal on the outside but were shallow and incomplete. Like the rabbits, they were genetic males, but unlike the rabbits, they were imperfect females.

"The obvious difference, Jost pointed out, was that he had cut off the rabbit's testes *in the womb,* depriving them as developing embryos of all their testosterone, but the humans had their testes, which were functioning perfectly. Okay, so they couldn't absorb testosterone, sort of a sexual diabetes, but that wasn't the question. Since the rabbits had proven the Eve principle that the female equipment should just spring into being, the question was why they didn't have ovaries and the rest. It seemed that the testes they had were pumping out this second mystery hormone. What was it doing? Killing off their female equipment, *de*feminizing them."

The hormone has now been identified and is called Mullerian Inhibiting Hormone, or MIH. The process by which MIH and testosterone turn fetuses either into males or females is called sexual differentiation, and Pillard's theory of homosexuality is built on it.

Every man reading this will no doubt be surprised to learn that he has, inside of him, a vagina, as well as ovaries, a uterus, and fallopian tubes. Women will be equally interested to hear that each of them is carrying a seminal vesical for transporting sperm and a male vas deferens, but none of this equipment is functional.

At conception, every human embryo is basically female. In fact, biologists consider the default gender for all mammals to be female. (In point of fact, the embryonic body is, except for one microscopic bit of genetic material—a Y or an X—completely identical.)

Through the first two trimesters each fetus has two separate, complete packages of male and female interior sexual equipment, the Wolffian duct (for the male equipment) and the Mullerian duct (which contains the female equipment). These tubes of tissue are situated in the fetus's lower abdomen behind the bladder. Next to them is an undifferentiated sex gland and an indistinguishable genital tubercle. Into the third trimester, the tubes wait, either to blossom into being or to die. Every egg carries an X chromosome, and male and female first diverged when the egg was fertilized; sperm carry either another X chromosome or a Y, and from that point two very different scenarios start to play out.

For those whose sperm carried an X, the sex chromosome became XX, and things happened as follows. The Mullerian duct simply flowered, automatically making them internally female. (Not even estrogen was really required here. While it does help develop secondary sex characteristics such as breasts, estrogen is not thought essential for developing the internal reproductive organs.) The Mullerian duct grew into a uterus and fallopian tubes connected to the sex gland, which swelled and metamorphosed into ovaries. The genital tubercle became a clitoris nestled at the top of the labia. The Wolffian ducts, with their neatly packaged male equipment, bravely awaiting the testosterone that would burgeon them into being, waited in vain. Not receiving their testosterone punch, they simply shriveled up—the remnants of the vesicle for carrying semen came to rest on top of the ovaries, where they remain through adulthood, and the person was thus a fully developed female. But it is, biologically speaking, easy to become female. The process of becoming male is much more complex and thus more vulnerable to technical snags along the way.

For those fertilized with a sperm containing a Y, that Y brings a gene with it, up on the tip of its short arm, called TDF, Testis Determining Factor, and the following chain of events must happen to arrive at successful masculinization. First, the TDF gene has to function correctly. If it does, within weeks after conception it begins secreting its protein which, by the second trimester, will turn the sex gland into testes, which descend to their location between the legs. The genital tubercle becomes a penis. But at this point, where women need no hormones for the Mullerian duct to develop and the Wolffian to die, men must have *two:* testosterone from the testes to develop the Wolffian duct into the seminal vesicle and vas deferens, and Jost's second substance, Mullerian Inhibiting Hormone, to kill off the durable Mullerian duct. (Men carry the remnants of their Mullerian duct— uterus, inner vagina, and fallopian tubes—under their prostate and next to their testes.) For men, the process consists of two steps and is more complex: first, masculinization must occur; second, *de*feminization. Without MIH, the fetus will develop both male and female organs.[7]

The intriguing question for Pillard was where to find the control switches governing the overall process. Certain genes design the Mullerian Duct with the means of its own destruction: MIH receptor molecules studding its surface, waiting to receive the poison hormone. The production and deployment of this receptor molecule resides with a gene we know is in the sex chromosome that directs the testes that produce MIH controls.

MIH was, Pillard says, difficult to track down, but it leaves a distinctive footprint. "Its effects were first inferred," he says, "because if you have a boy with an undescended testicle on one side that withers, which is fairly common, it won't make Mullerian Inhibiting Hormone. The Mullerian duct on that side is not inhibited by anything, and, on that side, you get an ovary, a fallopian tube, and no testicle, where on the other side you've got your testicle and you get no ovary and fallopian tube. It is very localized action, this hormone. MIH from the right testicle does the right side and MIH from the left testicle takes care of the left. MIH turns out to be a polysaccharide, a big clunky molecule that doesn't go very far. Neurotransmitters have a molecular weight of around 100. This stuff has a weight of something like 6,000, so it's really a big old tank of a hormone, and we've always figured it doesn't travel very far around the body. But then it doesn't need to. Its target, the Mullerian duct, is right in its neighborhood."

But Pillard began to wonder if this was in fact the case. Transsexuals started to puzzle him. Why were they so different from homosexuals? "It would be logical to assume," he points out, "that pre-transsexuals, who are sexually and romantically attracted to people of their same physical sex, would be in most ways on the same end of the spectrum as homosexuals. But if the spectrum is gender self-identification, which seems central to both traits, transsexuals are on one end with a deep, unalterable conviction that their gender is mismatched with their body, and heterosexuals and homosexuals are together on the other, perfectly at home with the gender of their bodies." Pillard was also aware of an observation UCLA's Roger Gorski had made about altered rats that allow themselves to be mounted by members of their same sex. "What they're saying," Gorski stated, "or would be saying, is 'Help! All of a sudden I'm a girl trapped in a boy's body.' What we've created with the altered rat is not a model for homosexuality at all. It's a model for transsexualism."

There were other pieces of the puzzle to fit together. Pillard also knew, of course, that gay men are physically as completely masculinized as straight men. They cannot be distinguished in any part of their bodies—except, if LeVay is right, by differences in the brain, and this would accord with yet another puzzle piece: the only difference between homosexuals and heterosexuals seems to be in orientation of sexual and romantic feelings, which, of course, originate in the brain.

So Pillard began to theorize that homosexuality, at least in men, originated from the process of sexual differentiation. What if, he wondered, there were some additional function of MIH of which we knew nothing? Rat brains are differentiated by hormones—testosterone makes them male. We think MIH is just some hormone, mused Pillard, that, a few months into gestation, lumbers a couple of inches from the testes to the Mullerian duct and poisons it. What if it were doing more than that to us? What if in the womb it were acting not only on the duct but on the embryonic brain to defeminize that organ as well? And in most men, this is exactly what happens.

But just maybe, in gay men, while the testosterone works perfectly both in brain and body, the MIH does its job *only* in the body and fails somehow to work in the brain. What would this mean? It would mean that gay men had brains that were masculinized but never defeminized. In other words, their brains were male *and* female.

Pillard was theorizing, in effect, that the sexual centers of gay men's brains, such as the hypothalamus, were partially feminine. "There turn out to be some receptors for this Mullerian Inhibiting Hormone in odd parts of the body that you wouldn't expect, like the corpora lutea, where the eggs develop," he says. "If MIH comes lumbering along, it should latch onto the corpora lutea and kill it. After all, it's a killer hormone, and that's its job. So what I'm wondering is, maybe it *does*, in some form or another, somehow get around the body, in contradiction to what we've always assumed. What if it does get into most men's brains somehow and defeminizes them. And suppose that in gay men, the brain is not defeminized. Obviously MIH is working in gay men's bodies, but maybe not in their brains, so gay men have masculinized bodies and defeminized bodies and masculinized brains but they do not have defeminized brains.

"So maybe the gene that Hamer found in Xq28 is involved on the assembly line making the MIH receptor molecule that should be in the brain. And in gay men, this gene is defective."

Like a good scientist, Pillard has considered the reasons his theory could be wrong. "It seems absurd to hypothesize that MIH gets into your brain because, first, it's supposed to have a very local action, and if it does get around, why wouldn't it get over and nail the Mullerian duct on the opposite side of the body? Second, this big, awkward hormone can't cross the blood-brain barrier that prevents noxious chemicals from getting into the brain; this is a barrier that will even stop HIV, although we now know HIV can hitch a ride inside other organisms. And third, if it's true that gay men are not defeminized in their brains, how come they *are* defeminized in their genitals? To respond to these problems, you would have to theorize that MIH goes through some chemical change at different points. Of course, that's not at all unlikely, but this is all pure speculation at this point. Still, we may very well wind up finding that genes that have effects on sexual orientation function along these lines."

There is evidence to indicate that some of Pillard's problems with his own theory are not insurmountable. Patricia K. Donahoe, chief of Pediatric Surgery at Massachusetts General, has been looking for the MIH receptor in other areas of the body, and she has been able to raise antibodies to it in the spinal cord and brain of the rat. "If the receptors that receive MIH are in the nervous system," notes endocrinologist Mike Baum of Boston University, "it makes sense that MIH would be there too." And there is new evidence that MIH divides into smaller pieces, which are more mobile. This may be the way it gets around.

Other intriguing evidence also supports a theory resembling Pillard's. It accords with what Dr. Richard Green calls male vulnerability. Green points out that for every 100 human females conceived, there are approximately 130 male conceptions, a huge imbalance in favor of boys. At birth, however, the sex ratio is virtually identical at 1.02 females to 1.04 males. Clearly there is a fairly high death rate in utero for boys, usually, it is believed, through spontaneous abortion; girls conceived as girls usually end in a live birth. Why? It's tough to become a male; the more elaborate masculinization process is filled with things that can go wrong, and the MIH/testosterone case is just one example. The question is, is male heterosexuality equally vulnerable to disruption? Long-known evidence from the phenotypic profile of sexual orientation indicates that it is: There appear to be approximately two gay men for every gay woman, which is consistent with Green's observation.

Hormonal evidence also supports the idea of the stability of being female and the instability of being male. Genetic males can lack male hormones in many ways, says Mike Baum, giving the example of five alpha reductase. Men need five alpha reductase to convert testosterone to dihydrotestosterone, or DhT, the supertestosterone critical for adult hair growth and prostate function. (Merck's popular new drug Proscar counters prostate growth by inhibiting five alpha reductase and, thus, DhT.) But DhT has another specialized purpose in men: it develops the external genitals. A well-known genetic defect in men called 5 alpha reductase deficiency acts like a natural Proscar, but it happens at a time when males *need* DhT to develop their penis and scrotum. Men who don't have the enzyme don't develop external male genitals. Naked, they look just like women, although internally they are fully male because only ordinary testosterone is needed for the internal Wolffian duct. Lacking male hormones creates a variety of effects but is far from fatal.

But there are, on the other hand, no clinical cases of a lack of estrogen in any animal we know of, Baum points out. "The reason is probably that you need estrogen for survival and if you don't have it, you die—at least we infer this from the fact that there are no cases of any animals not having it. It's not the same for testosterone. You can survive just fine without it, it's just that men who don't have it look like women. We have many examples of people who lack androgen receptors and are viable human beings, but not a single gene mutation that makes anyone lack aromatase (the enzyme that changes testosterone to estradiol) or estrogen receptors. To the extent to which androgens are precursors of estrogens, every healthy, living person's sex hormone synthesis pathways function. There is not a single known person in whom the chemical chain from cholesterol, the basic precursor, all the way up through the female hormones to progesterone does not operate.[8] But then you reach the male hormones—and these can be a mess. It seems nature just doesn't care that much after that. This is, again, a case where biology makes it harder to become male. This fits with Pillard's theory and with the observation that because homosexual orientation is present at age two or earlier, it would have to happen in embryonic development, when MIH is active.

Laura Allen notes another curious imbalance that has long been discussed: Men and women, for no reason yet discovered, have different rates of certain disorders. Women are twice as likely to suffer from major depression and anorexia as men. (Of the men who suffer from anorexia, a large percentage are gay.) Men are more likely to have neurological disorders such as dyslexia—three times more men than women are dyslexic—and autism—seven times more men. More than twice as many women as men get migraines, a striking difference. As for schizophrenia, it occurs in equal rates, but men and women have a noticeably different course and prognosis.

Allen thinks the higher rates of disorders in men may come from the higher possibility of error in the difficult process of male differentiation. It may be, she thus suggests, that on our way to finding the key to human sexual orientation, we will uncover answers to innumerable questions about the origins of a host of syndromes that have plagued humanity, unlocking the isolation of autistics and providing a cure for hundreds of human ills.

~

At an evening talk at the Carnegie Institution in Washington, D.C., Dr. Francis Collins, the head of the Human Genome Project, commented to a packed auditorium that the project means the end of the insurance and pharmaceutical industries as we know them.

What's happening now, said Collins, what's going on, is speed. Simply speed. Consider, for example, the rate at which we are learning about our genes. On April 25, 1953, James Watson and Francis Crick published their one-page paper, "Molecular Structure of Nucleic Acids," introducing the world to DNA. By 1967, fourteen years later—he clicks on a slide and twists around for a moment to verify that it's the right one—our map of the human X chromosome looked like the graph on the next page. We knew that ocular albinism was somewhere in there and, near the bottom of the q arm someplace, hemophilia A.

Then, Collins said, two papers were written. In 1980 Stanford geneticist David Botstein and others, in the *American Journal of Human Genetics,* published "Construction of a Genetic Linkage Map in Man Using Restriction Fragment Length Polymorphisms," or RFLPs. RFLPs were tools that allowed us to develop linkage analysis. In 1985 Kerry Mullis published a paper called "The Polymerase Chain Reaction." If Botstein pointed out a genetic bloodhound capable of tracking down elusive genes, Mullis built a photocopier capable of producing unlimited quantities of the genetic material needed by scientists for further research. In 1994 our map of the X chromosome was laden with genes scientists had discovered. (See page 156.) Collins drew another chart. If you start with what we know now here, he said—he made a mark at zero on the left—and consider the work we're scheduled to do in sequencing the human genome, by the year 2005 what's going to happen is this. He drew a curve that skyrocketed to an almost vertical climb.

This is what we're aiming at, said Collins. In order to meet these goals, in each year our knowledge must grow logarithmically, and we will have to sequence more DNA than in all previous years put together.

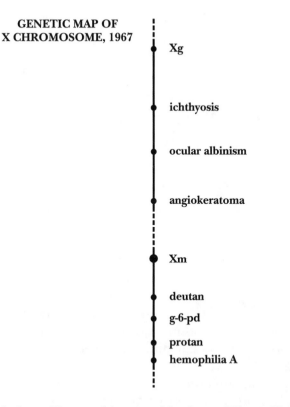

GENETIC MAP OF
X CHROMOSOME, 1967

Xg

ichthyosis

ocular albinism

angiokeratoma

Xm

deutan

g-6-pd

protan

hemophilia A

This knowledge will turn things upside down. What will it mean to an insurance industry that functions entirely on actuarial tables, calculating thousands of comparative risks and averaging everyone together, when by looking at a person's DNA we know what the risk are for each individual? What will we charge them? What will it do to an industry based entirely on disease unknowns when the diseases are known and predicted? What will it mean to an industry that protects itself by weeding out those with preexisting conditions when we all have preexisting conditions? Collins mentioned a man he had met whose daughter has muscular dystrophy. She couldn't get insurance, and the man said, "You know, I'm actually glad these genetic tests are increasing, because eventually *everyone's* going to have a 'preexisting condition.' And then they'll have to change their policy. Or change the industry."

Every single person will soon be known to have some preexisting condition. We just don't know what they are today. Soon we will.

Chapter Ten

HOW GENETIC SURGERY CAN CHANGE
HOMOSEXUALITY TO HETEROSEXUALITY

IN HARVARD's red brick Museum of Comparative Zoology on Oxford Street, an imposing, vaguely gloomy structure recalling the
Victorian world of late-nineteenth-century science—one expects
to see Charles Darwin wandering its halls—Hamer delivered an early-
morning lecture on his search for the gay gene to Dr. Richard Lewontin's undergraduate genetics class.

Lewontin is a highly respected, highly politicized, controversial,
and quite well-known Harvard population geneticist, tall and hand-
some. His politics, which his colleague David Botstein of Stanford cat-
egorizes bluntly as "pretty hard-core traditional leftist," engender his
staunch dislike for the search for the biological origins of sexual orientation and an unconcealed antipathy toward Hamer's work. This
dislike is well known around Harvard and even at other universities,
and Lewontin's scientific reputation and power are such that geneticists measure carefully what they say about this research. Lewontin
does not take the point of view toward it of, say, Bill Byne. He is not
merely a critic of the research. Lewontin is an opponent.

That morning, before the undergraduates, Hamer had been quite
nervous, which he revealed by talking even more rapidly than his usual
lightning pace. Lewontin and Evan Balaban (whose office was two
floors below Lewontin's) had installed themselves in the back row, two
menacing intellectual presences, watchful as hawks. The students
nervously occupied the seats in the contested middle ground of the
classroom as the rat-tat-tat of methodological explanation flew over-

head; when Balaban put a question not to but *at* Hamer, they all but ducked. The search for Xq28 was narrated across the blackboard under swipes of Hamer's hand, the clues racing in lines of chalky white and yellow, rising to a crescendo. The students looked increasingly convinced, which further raised the tension. At one point, as a sort of palliative, Hamer cleared his throat, backtracked a bit, and stressed that, on the other hand, sexual orientation is so "complex" and "no doubt" due to so many genes, they probably would never be able to figure it out. The audience found this reassuring.

Lewontin had been remarkably quiet throughout, his eyes on the chalky data, taking it in. Afterwards, on the way back to his office in a new annex, up the museum's dark, old wood staircases and through a labyrinth of large rooms filled with obscure animals in glass displays, he remained silent.

Sitting at his desk, he looks grimly meditative. "The question that I asked Hamer downstairs that he didn't answer because he didn't have time is why do you want to do what you're doing? What happens when we find a difference on a chromosome whose variation between individuals has some effect on a behavioral—or for that matter any—trait? What do I do with that information?" He sat back in his chair. "Now, Dean gave a partial answer to that. Politics. 'What do I do about it' is *always* political. People think that if they find the gene for a trait, it'll affect people's lives. And he gave an example, he said here's this guy, this right-wing nut"—Hamer had referred to Ken Adelman, a conservative columnist for the *Washington Times*— "who is antigay, and the moment this guy hears homosexuality is biological he stops being antigay. Now that is the political point of view Dr. Hamer has, but I disagree. I'd ask *why* is knowing how sexual orientation is created biologically a good thing to know?"

And how does he, Lewontin, answer that question? He scowls.

"It's irrelevant! I don't care! What difference does it make to me which genes affect sexual orientation? None whatsoever. That's what I say to my gay friends, that's what they say to me. You get this right-wing guy who thinks a particular sexual orientation is bad, but now that he knows it's genetic, he thinks it's okay. So he's reached the right conclusion. Good for him. But it's stupid! He must be one of the very few people in the world who's become convinced that something is not a defect for the reason that it's biological. The response to that is so simple it's mind-boggling: cancer is biological. Does that make cancer good? There are a million biological defects. It's not even logical. And—moreover—you have to realize what the research implies.

First, that the sexual orientation is bad and should be changed. If you don't have that view—and I include Hamer in this—why would you care? That's my question, and that's my problem with everything that's been done on this."

What about Hamer's methodology? Lewontin has in the past maintained publicly that the genetics of certain traits will be too complex to dissect, and his colleagues have become used to his challenges to studies on all sorts of methodological and technological grounds. This is why it is a surprise to hear what he says next:

"Oh," he says mildly, "I thought what he said this morning was quite good. This is essentially a better, cleaner pedigree method. The trouble with the old pedigree method is that it's extremely sensitive to small changes in the data, that's why you get all these retractions— manic depression, alcoholism, all those studies that looked good at first and then were disowned—whereas the sibling pair method is *not* fragile. If you make one erroneous determination it doesn't screw up the whole thing. Linkage analysis *is* fragile. But in principle the method he's using should do very well. I don't doubt for a moment that he's already found a region of the chromosome that has some influence on sexual orientation."

The problem lies elsewhere. Perhaps it is motive that bothers Lewontin. He brings up the possibility that Hamer's interest in the origin of sexual orientation is powered by the fact that others are interested, but he dismisses that as illegitimate. "In that case, he's only reinforcing that view that being gay is bad, he's saying okay, I'll show you it's biological so you can't blame me, whereas the right answer is that the issue of blame is not there in the first place. You're going along with the game, 'Yeah, I'm bad but I can't help myself.' The proper struggle for gay people is to say "Why the hell are you blaming me?" Don't blame me at all. Why do you care about the gene? I mean, if God appeared to me in a dream and told me which genes they were, what would I *do* with it?" He laughs.

"But the second thing that's implied is that just because you have the gene, you can't change the trait. Excuse me, everyone's looking for the genes for diabetes. Are they doing this for *fun*? The equation of genetic with unchangeable is absolute garbage! You find a gene that makes some difference in your physiology, but nobody ever said you couldn't change people's physiology, they've been doing it with diabetics for seventy years now by giving people insulin. Is this not obvious? People like Hamer reason as if they were unaware of all these men and women in other labs saying 'If I find the genes for all these other traits like diabetes and cystic fibrosis and Huntington's and Tay-Sachs maybe I'll be able to cure them, and I'll sure as hell try.' And maybe they will and they certainly should try. You can do genetic engineering to get rid of the gene altogether, which is nice and fancy and everyone's talking about it, but it's really unnecessary. You don't have to get rid of genes to change their effect. My hair color is a consequence of my genes, and on a genetic level I could interfere with pigment production, but why even *bother* with that if I can dump something on it from a bottle? So people who think that if they find the gene everyone will accept that they can't change it—that's bullshit! My genes gave me my nose shape but I can get an operation any day of the week.

"So I come back to my original question: Why does Dr. Hamer want to know whether there are genes segregating in humans that have an effect on sexual orientation? What is he going to do next?"

So it's a philosophical objection that he has.

Lewontin scowls and looks extremely irritated. "It's not philosophical," he says loudly, "it's *political*! All human genetics is politics! That's not to say it's on the left or the right or whatever. What I'm trying to say is that over and over again there is one central question in human genetics: *Why do you want to know the answer?* Why are you asking *that* question? All questions are not equal in biology. That's my problem with the research of homosexuality. I'll say it differently: The asking of a scientific question has the property of saying that that's an important question. Otherwise you wouldn't spend your time and someone else's money on it. And asking the question 'Are there any genetic differences between homosexuals and heterosexuals?' is stating that the answer to that question is an important answer. That's my problem: You've got to tell me why you believe it's an important question."

Across the continent in Palo Alto, David Botstein considers Lewontin's point of view. Botstein is the discoverer of RFLPs, restriction fragment length polymorphisms, a seminal discovery in modern biology. Jan Witkowski dates the beginning of human genetics from 1980, when David Botstein published a paper on using RFLPs for linkage analysis. Botstein's persona—imposing, a large bearlike physical presence—matches his professional reputation. He is a man who speaks congenitally ex cathedra. He works on yeast, trying to extract the general principles of how cells are organized, and he is involved in the genetic mapping of manic depression. "We are working on manic depression here," he says, swinging a large arm. "Our biggest problem is figuring out who is depressed. It's very *difficult*, figuring out who is depressed. Sometimes we get very *depressed* about trying to figure out who's depressed." He has a large, sunny office overlooking a courtyard at Stanford and, as a harried administrator, two secretaries just outside the office door.

Botstein reads Lewontin's words: "It's not philosophical, it's *political . . .* there is one central question in human genetics: *why do you want to know the answer? . . .*"

Botstein puts the paper down, sits back, squints intensely for a moment at nothing, and then begins to talk.

"I think Richard Lewontin is being extremely up front by saying it's political. I have been saying for years, in private of course, that the problem with Dick and people like him is that they are politically motivated. And he's entitled to that political motivation. He believes that as a matter of *policy,* not science, it's a bad thing to know these things. He no longer denies that they can be known, which *used* to be his position. But science marches on, da da *da,* da da *da*"—he waves a hand back and forth at the ceiling—"and our technology and knowledge get better and better. It's no longer the case that you can't find these things out, and so he has to live with that. And now he's completely aboveboard and says flatly that he doesn't want to know, and he thinks none of us should want to know because of the way the knowledge can be misused. Actually I should restate that. He believes that this is a kind of knowledge which can *only* be misused.

"You have gotten a wonderful quote from Lewontin, which I love. And if that gets into print, I will be extremely happy and you will have done something good, because Lewontin has, for the first time, admitted that all this stuff is about politics and not about science. Basically he's taken himself out of the argument."

But, Botstein clarifies carefully, he is not saying Lewontin is wrong. Not at all. "Lewontin has an intellectual business, a brand of genetics that he practices that is a serious intellectual endeavor that is quite apart from the political debate. He does serious science: population genetics. He was the leader in trying to test some of the assumptions and basic results of population genetics in Drosophila by using molecular tools. But with Lewontin and sexual orientation, it is not a science question. It is a use question." He sighs. "Listen, I'm not a psychologist, I don't know what makes him tick, but I think that like many people of my generation, Richard was deeply traumatized by what happened with genetics in the '50s and '60s and the civil rights abuses which came to light then when we were young—the ways genetics were used to pass laws about race and other things. There was a whole pseudoscientific rationale behind this stuff. And it became conventional wisdom among people who were, I think appropriately, horrified by this use of genetic information that even if we could find out whether there were racial or intelligence or moral genetic differences, we wouldn't want to find out, because the knowledge would only be abused. And that view is not rare."

Botstein adds, "I must say that as a practical matter, I am not in a hurry to figure out the genetic basis of intelligence, although I believe it's useless anyway. The genetic contribution I can see is trivial, and if there's a big contribution of genes to intelligence, I can't figure it out because it's too complicated. They used to argue about Jews being more intelligent. The political problems that caused were enormous, but there was this evidence because the kids getting the prizes and top spots were Jews. But today we know it's not genetic, it's cultural, and whatever argument there may have been in the past for Jewish intelligence genes is completely vitiated by a glance at the Bronx High School of Science, Harvard, MIT, and various places today as opposed to twenty years ago. You know what you will find? That there are 'new Jews.' Their names are Hwang and Chun and Nguyen, and they're the kids getting the Westinghouses, and Shwartz and Berg and Botstein are down on the list. So now you're going to make the argument that the Chinese and Koreans and Vietnamese are genetically smarter? Come on.

"I don't know, frankly—if I were Hamer, I wouldn't be doing this. I don't think any good will come of it. I agree with Lewontin about that. I don't think he should be censured for doing it, I just don't understand why he does it. What good can come from the result? It seems to me either way it's a no-win, no-lose for everybody, but there's a potential for serious mischief there, and I don't understand what good it's doing. I don't know."

Now, he says, taking a different tack for a moment, consider the search for the gene for violence. "I think there's more scientifically to that one, a greater likelihood of finding it, more than IQ. But it's *completely* unacceptable at the moment. You can't even talk about it. Go to any university, research center, no one—*no one*—will talk to you about this. Why? Simple. Because of the fear that there'll be a racial correlation. And there could be. And *how* would we be better off? *How?* It would, Lewontin is afraid, and I have some sympathy for this fear, mean that any scientific evidence linking some undesirable trait with black people will be used as an excuse for explicit or implicit genocide. Okay? That fear is not totally irrational. No one can dismiss that fear out of hand. Not in this century. Not looking at the Serbs and the Bosnians.

"But this is politics, not science. Look, both Lewontin and I have the same perspective on Hamer's stuff. There is nothing obviously wrong with that study, there was no stone he left unturned or bet he missed. Manic depression has been mapped three times, and they were all wrong. Alcoholism has never been seriously mapped, there was just this brain RNA study which is bullshit on its face and should never have been published. Hamer's is not like those. His is a completely reasonable, solid study. Hamer is good. But I don't think Lewontin's concern is unwarranted. I think he's in fact right. The concern is political, that there may be a class of results, whether they have to do with gays or blacks or Jews or women or whatever, for which the American public is simply not prepared."

All scientists working in this area are conscious of the fear it inspires. Sometimes they talk about it directly, sometimes obliquely. Sometimes they sympathize, sometimes they get frustrated and reject it out of hand. Mostly they think about it. In her lab in Bethesda, Angela Pattatucci, Hamer's research partner, said, "Some people find this work frightening because they think it challenges notions of free will—I don't believe it does at all—and it shatters the myth that human beings are above animals. We seem to forget that we *are* animals. We delude ourselves into believing that the distance between us and other animals is so great it's like comparing a grain of sand with a diamond. It's not. And this is something that has alarmed people since Darwin.

"There are already molecules that have been identified with behavior, like serotonin, which is associated with depression. This is what Prozac acts on, it's why Prozac works, and you would have to have been living under a rock for decades not to be aware of this. Having said that, there is definitely one thing that we have not done our homework well on—me, Dean, Simon, et cetera. We can all be taken to task for not making the disclaimer that we in no way are attempting to reduce an individual's experiences to a molecule, and I think that a lot of people think that's what we're trying to do. We're actually doing the opposite: We're trying to ask how a molecule contributes to a person's experience. There are hundreds of questions to be answered, but because we don't answer all of them immediately, people accuse of us ignoring these things. We're not. We're a small group of people, and we're doing the best we can."

The politics crowd her, push their way into her protocols and interviews and the charts of chemicals tracking her molecular progress, in part because Pattatucci is herself homosexual. "The media is always insinuating that because I'm a lesbian I'm leaning toward the genetic explanation. There's always an assumption I have a vested interest in things turning out a certain way. I find this amusing. I don't. I'm not convinced." She looks only slightly amused. "From a political standpoint, I'm not sure that even if a strong genetic component for sexual orientation is demonstrated beyond a shadow of a doubt that that will make things any better for us. Pass nondiscrimination legislation on those grounds, and you not only limit coverage to people whose sexual orientation has a genetic basis, but by that standard someone who is mulatto, has black blood but doesn't look black, would not be covered because he or she is insufficiently biologically black. You don't make laws against discrimination based on biology. You make them based on principle, regardless of the origin.

"My vested interest is very simple and that's having acceptance for gay people, and I don't care whether that happens because the public decides to accept the fact that we exist because there's a genetic component or if the public just decides that they shouldn't discriminate against people who aren't hurting others for any reason."

Pattatucci added, "In the twentieth century, those against homosexuals have cried 'It isn't biological.' In the nineteenth century, the same people said 'It *is* biological,' and gays were not allowed to breed because it was believed they would give birth to hideous creatures. Homosexuality was related to teratology, the study of monsters." She smiled faintly, a lesbian molecular geneticist in the middle of the late-twentieth-century scientific search, peculiarly aware of the irony. "Homosexuals were assumed biologically unfit, and if two of us bred," she said evenly, "we'd give rise to monsters."

If there's a gene, it will, as Lewontin would observe, be tested for. But this is another Big Question: who exactly is going to test for it? Is it ethical to give out the information? On a cloudy winter morning in Waltham, Massachusetts, Phil Reilly, head of the Shriver Center for Mental Retardation, talks about genetic testing. Reilly is a clinical geneticist with both an M.D. and a J.D. In his bearded, soft-spoken but intense way, he discussed the matter in terms of the media.

"Actually," he said, clearing the verbal air, "I'd like to argue that it doesn't matter what you or I think. For genetic testing, the only thing that matters is what the *women* who bear children think." He remembered something. "You may have seen this on *60 Minutes,* Mike

Wallace was here for hours for a segment on genetic testing. Well, all he wanted to do was talk about genes and homosexuality and abortion. I was a little miffed. This is something you and I might disagree on, but this is my fundamental position: I believe absolutely in a woman's right to control her pregnancy, including the right to terminate it."

He raised, then lowered, his eyebrows, demarcating his premise. "That being the case, and knowing that 99 percent of all pregnancy terminations are because women don't want to be pregnant, and 1 percent for medical reasons, I believe it is better to allow the woman to terminate her pregnancy for reasons we do not agree with than to control what she should do. If a woman says to me 'I don't want to have a child with Down's syndrome,' I believe that's her business. If she says, 'I don't want to have a boy,' I believe that is her business, and if she says, 'I don't want a homosexual child,' I don't believe it is the right of the state to tell her what she may or may not do. That is where I come down. Do I believe there's anything wrong with being gay? No. But if you're going to give a woman the right to abort if she has a girl, or a boy, or a kid with cystic fibrosis, how can you say she can't abort for any other reason? It's not a gay issue, it's a woman's-right-to-choose issue."

Reilly turned to *60 Minutes,* still somewhat miffed. "Now Mike Wallace took that comment and, you know how he's always on counterpoint"—He rolled his eyes—"and made it sound like I was coming out of Nazi Germany the way it was put. But basically the question is should a woman have the right to obtain information about whether her fetus has a homosexual orientation? I think she does. Does a physician have an ethical duty to supply it? *Yes.* Yes. It's her information."

Hamer has said publicly that "of course" no geneticist would test for a homosexuality gene, that no geneticist would ever give out the information. *"I* would," Reilly said. "I would—reluctantly—because I believe she has the right to choose." He looked very faintly cynical. "Look, as recently as ten years ago most geneticists said under no circumstances will we sex a fetus because we don't want people aborting on that basis. *Huh.* Now they routinely give out the information. Not comfortably, some of them, but most of them give it out." He drew a hand thoughtfully down his beard. "How in the world am I to know what this information means to this woman? I see her once in her life for ten minutes. She goes back and lives with herself. It's my duty to tell her don't take cocaine when you're pregnant, but it's not my duty to chain her to the bed to make sure she doesn't. To the extent that I become a gatekeeper withholding information, I think I run the risk of substituting my values for hers, and that's not my role."

That afternoon, at a Cambridge Chinese restaurant, which has been turned into a fortress by the wall of frozen brown sludge from the street that surrounds it—"It's amazing anyone can get in here!" Geller said as they negotiate the iced entrance—Dean Hamer, Lisa Geller, and Evan Balaban debate testing.

"Would you give out that information?" asks Geller.

"Would *you?* " Hamer counters.

"Can we order?" asks Balaban, grinning. He looks amused. "I'm not a geneticist, I'm not going to be doing this." But he's still responsible, as a scientist and all. "Oh, sure," says Balaban. Hamer orders, for everyone, in Chinese. The waiter looks mildly disturbed by this.

"We're going to need ethical protections," says Geller, "because there are already people in the army who are collecting blood samples that can be converted to DNA. They have 50,000 blood samples—and *we* give them the gay gene? You go into the army, you sign away all your rights. If I were interested in IQ, I'd say to the army give me those samples and I'd measure repeats of the Fragile X thing."

Balaban raises his eyebrows. Geller sighs. "I can't believe we're practicing phrenology again. Race and IQ."

Hamer snags another waiter to ask for chopsticks: *"Qing ni gei wo quai zu."* The waiter nods and goes off to the kitchen to recount this. Hamer says, "Military and insurance companies are the two guys I'd be worried about." They talk about ELSI, the Ethical, Legal, and Social Implications part of the Human Genome Project, which was established to deal with just these issues. They debate the medical advances. They complain about their critics, all the publicity, the media. Hamer glowers, munching moo shu, and says, "Critics always put more in your mouth than you said. That's their technique."

They debate Simon LeVay's belief that this research would benefit gays politically. They debate technology. Geller says, "National health care will obliterate these tests because caps will be put on genetic testing."

"Hunh. Who knows when *that's* going to happen," says Balaban.

"No one will do this testing," says Geller firmly.

"No one *who?* " asks Balaban.

"Maybe no one *you* know. . . ."

One scientist involved in this research would not allow herself to be identified, but quietly and firmly she said: "I would like us to approach this as a society with a childlike curiosity without involving the politics. I disagree with Simon that this work is good for gay people politically, and I think all of us working in this field have delusions of grandeur in thinking that we will have any control. If we find a gene linked to sexual orientation, there *will be* a test for it. That is a fact. Despite all the noise to the contrary. It will only be a matter of time, and someone will come up with one, and we fool ourselves if we think that we're going to be able to control things so the assholes won't be able to abuse it. We're not going to be able to control anything. And the problem is that the differences between people deemed by society to be important—you can translate that directly to 'biological questions that get funded as opposed to those that don't'—are always directly correlated to differences in power. When Dean asks, 'Where is the gene for being gay?' he is asking 'Why are they gay?' But in fact he's also asking 'Why are they not like the majority?' If there is a power difference—and there is—then knowledge of that difference can and will be abused."

As always, it cuts both ways. Christian leader Pat Robertson has remarked smugly that any moment now the gay movement will become pro-life; gay activist Larry Kramer has replied smugly that any moment now, the Christian movement will turn violently pro-choice. Waiting for Hamer's credit card receipt, the green tea grown cold, Geller's chopsticks were lying on her plate. She looked out at the freezing Cambridge sky. "How do you stop it?" Geller asks. "You could make an international agreement saying 'No medical test will be made available for this.' But how do you *enforce* it? The Nuremburg vehicles are not effective for this."

Hamer looks tired from the genes and the publicity and the responsibilities and the questions. He remarks to no one, "There was a guy at Cold Spring Harbor Labs, Alfred Hershey, got a Nobel Prize for figuring out that genes are made of DNA. He was asked, 'What's your idea of heaven?' He said, 'Finding the perfect experiment and doing it over and over and over again.' "

Hamer smiles.

Back in California, Botstein carefully searches for a conclusion to draw. Do they stop the research?

"*No.*" Botstein brings a solid, meaty hand down on his desk. "I am unequivocally on the side of *not* stopping the research. *However,* I could understand if scientists in all good conscience avoided certain problems, and they *are* avoiding certain problems. Certain problems are not being studied. They're not being studied because nobody knows what to do with the results, and they're afraid they'll do more harm than good. The genetics of violence is one of those problems."

Why are people so afraid of behavioral genetics? He takes a breath. Taps his fingers. Lets the breath out.

"I think many thoughtful people take the Lewontin point of view, that the knowledge will be abused. *I* take the view that the bigots are going to make up the facts, as you see with everyone from the white supremacists to Louis Farrakhan to Jerry Falwell. They just make them up. So it seems to *me* that we might as well have the real facts."

He is almost done. He muses about wisdom, then about intelligence. "Intelligence as a genetics problem is a waste of time," Botstein posits emphatically. "It's so complex it's almost impossible to design a study that will tell you anything. Intelligence is impervious to scientific assault." (Jasper Rine, asked about this, smiles, thinking of the Old Testament way Botstein must have delivered this opinion. Rine says bluntly of Botstein, "I love the guy. But I really disagree. Lemme back up for a sec on the technology here: Given the technological progress in genetics in the past five years, it is no longer reasonable to question whether we can understand anything. *Everything* is now within range. I mean *everything*. Some things will be easier, some frightfully difficult, but everything is now in range. Including intelligence, absolutely. There's no way of knowing without trying.")

To Botstein, in his bright office far from the cold aluminum gray of the Massachusetts winter, the question is posed: What will happen when we have the technology to get at intelligence and violence? He merely makes a dark face. *"Don't make problems,"* he growls. "We'll face those when we come to it. We've come to it already with homosexuality."

⁓

Heading south from San Francisco and down the peninsula that lounges beside the chilly waters of the Pacific Ocean, Interstate 280 uncurls itself, a wonderfully smooth tongue of cement and metal running through a stunningly beautiful valley of steep green mountains that trail the San Andreas fault. A strand of long, thin lakes—actually man-made reservoirs—line the bottom of the valley like a necklace of opalescent cobalt pearls lying in a crevice of dark green velvet, with a sheenlike trickle from a vat of metallic blue pottery glaze. The water that fills them comes from the far-off Yosemite valley. The pearls, visible from the passenger side of cars driving south on the highway, mark the spinal column of the fault, which runs north-south beneath the mountains, beside the highway, and up through the city. It waits to rip apart the peninsula and drown half of it in the Pacific. It is almost inconceivable, given the serene beauty of the place, that a destructive force would lie sleeping below, but the stunningly beautiful necklace of reservoirs is proof that it does. The fault created the valley in which they lie.

Just below Palo Alto, where the green lushness of the hills yields to more arid land, just east of Interstate 280, whose exit ramps feed workers and raw materials to ultra-clean factories and state-of-the-art design centers, is Silicon Valley. From 280, the Santa Clara exit leads to the Central Expressway, and from there to an industrial park, rows and rows of sleek buildings with glass polarized against the blazing sun. Its black asphalt parking lots rim the buildings with their artificial ponds, fountains, and grass whose thickness and greenness, contrasting with the tanned veldt along the expressway, testifies to an extensive irrigation system.

Affymetrix occupies space in one wing of this industrial park, and it is one of the new breed of companies in Silicon Valley. Much of the computer industry has moved on and outward, but venture capital and bright young minds are still flowing in here, drawn these days to a new industry: biotechnology. Where the raw knowledge of our genes will come mostly from academic research centers like Johns Hopkins and MIT and nearby Stanford, just a few minutes away up 280, the tools and equipment that will apply this knowledge, altering the genes in one way or another, changing their outcome, will come from companies like Affymetrix. The biotech industry is not ignoring its predecessor in the valley. It is carefully picking and choosing pieces of the computer and semiconductor industries nearby to incorporate into its own growth. In the spotless oasis of this Santa Clara industrial park, the sixty-some people at Affymetrix are using technology to design and build a semiconductor chip made of silicon and human DNA.

Steve Fodor is scientific director and chief technical officer of Affymetrix. He is a Ph.D. in physical chemistry and heads the science side of the business. Fodor was doing a postdoc at Berkeley when his mentor went to a biotech company on a one-year sabbatical. Fodor followed him. That company, a biotech start-up, succeeded wildly and spun off several start-ups, each with a different mission. Affymetrix is one of these, and Fodor was sent (across the parking lot outside his window) to head it up. David Singer, an MBA, is in charge of the business and marketing of the machines Fodor is working to build.

The chip Affymetrix is building, the first version of which it plans to release in 1996—its trademarked name is the GeneChip—is a tool for doing genetic diagnostics. "Say you go to your doctor," Fodor proposes. He has sandy brown hair and a friendly, direct manner. "And you ask her if you're a carrier of the version of the dystrophin gene that gives you cystic fibrosis. With DNA diagnostic tools like the GeneChip, your doctor will be able to give you the answer in her office.

"It's sort of funny," he adds, "because it's a kind of medicine we've really never practiced before. You're diagnosing a disease that you don't have. The disease hasn't appeared yet. Only the genetic blueprint for the disease exists. Or the blueprint for the normal gene. You want to find out which blueprint is there."

What is revolutionary about Fodor's project is the incredible speed and accessibility it offers. Today, we can obtain the same information the GeneChip provides but the current method of DNA sequencing is highly labor intensive, awkward, time consuming, and costly. The DNA has to be broken into pieces in laboratories and run out on electrophoretic gels, auto-raded, and so on. Francis Collins and the Human Genome Project are producing mountains of raw genetic information—which genes in which sequence create which traits—but it's just data. Fodor wants to construct the Sony Discman of genetic information, a tool that will harness, organize, and present the raw data as stories about ourselves that can be accessed with the ease of dialing the weather.

What you want to know about a gene, explains Fodor, is what version you've got. To pose the question using Jan Witkowski's analogy, is the blueprint that designed your carburetor a good one, producing a part of the engine that will last, or is it a faulty design indicating that, at some point, the carburetor is going to break down?

Consider the cystic fibrosis transmembrane regulator gene (its name is CFTR). It sits on the long arm of Chromosome 7, spread over 250,000 base pairs of DNA, and directs the manufacture of a protein that mediates chloride ion flow across membranes. This protein is exactly 1,480 amino acids in length. Its 508th amino acid, in most of us, is phenylalanine. But about one out of thirty-five Caucasians in the United States carries an allele version of the CFTR gene in which the three base pairs responsible for constructing this 508th amino acid are missing, deleted from the DNA instructions. This defective CFTR puts out a defective protein that cannot transport the chloride ions as well, and those who carry it suffer lungs clogged with thick mucus and pancreatic malfunction. Today, over 170 different versions of CFTR have been discovered, and certain symptoms correlate with certain versions. Fodor wants to be able to tell you exactly which version you have. "Rather than taking a big chunk of DNA, chopping it up, and separating the fragments in a bed of gelatin," he says, "which is pretty tedious and, frankly, unworkable for a large number of people, you could just shoot the DNA into a machine, and it would give you your answer. We wanted something so precise it would tell you, look here, the bases encoding your 508th amino acid are deleted."

They found a way to do it with probes.

Probes are short sequences of DNA that act like keys, providing information in the way that a key does: if it opens the lock, its teeth are aligned in the correct combination. Probes indicate a DNA sequence either by complementing and fitting with the DNA to which they're paired, or not. When DNA finds a sequence that complements it, it always wants to stick to that sequence. If it doesn't match, it won't stick. Say Fodor wants to know what the 146th base is in the sickle cell gene. The Human Genome Project will give him the sequence of bases in the basic model gene as well as all its popular allelic variations, so he will know that the bases on either side of 146 are, say, both T. As the DNA alphabet contains only A, C, G, or T, Fodor has only four possibilities for the unknown 146th spot. He makes four very short probes, each one three acids long, to give him his four possible keys: TAT, TCT, TGT, and TTT. ("Making the probes is easy," he explains. "You take a synthesis resin, a little pot with a lot of beads in it, and each bead has a bunch of chemical linkers on it. So you just bring in the next bead you want and stick in on the linker, and so on. Solid-phase nucleoside chemistry, a couple Nobels given for this.") He lays his TAT probe next to the 145, 146, and 147 spots in the gene and if it matches, he knows the 146th base: A. If it doesn't, he tries the other probes, TCT, TGT, and TTT, until he finds the one that sticks.

From here, it is just a mathematical step up to reading every base in the gene. Do the operation over and over down the gene's length from start to finish, probe after probe, four possibilities for each base, and the sequence will emerge. Fodor usually works with probes eight to twenty bases in length. "They're very sensitive and work well at discriminating single-base mutations." The problem, of course, is time; it would take an incredible amount to try every single key all the way down a gene, let alone the 3-billion-letter length of the whole genome.

"If I have a gene," says Fodor, "that's one kilobase long, a thousand bases, I have to have 4,000 probes—four possibilities for each base—to sequence that gene. And there's actually no human gene as short as 1,000 base pairs. Even a tiny gene, like the ovalbumin gene, is 7,700 base pairs long. The largest gene in our genome, the dystrophin gene, is 2.5 million base pairs. But be careful to understand the real question here. The question is not conceptual. We've got the concept. It's not even experimental. We've proven you can sequence genes this way. And making the probes is a breeze, I just make a probe and then photocopy it zillions of times with PCR—a lab technique you can use like a Xerox machine to make copies of genetic material—and I've got tons of 'em lying around. The question here is a technological one, and that is, how do I use them? How do I build a machine that will make this technology cheap and fast and put it in your doctor's lap?"

The technology answer was under their noses, quintessentially Silicon Valley: semiconductor manufacturing. "With a semiconductor," explains Fodor, "you take a piece of flat silicon and cover it with a photographic emulsion. You shine light through a pattern, the light transfers the pattern to the surface, you use a solvent to wash away the emulsion exposed to the light, and you've etched your chip. Now deposit metal on the chip and it sticks only to the pattern. That forms microcircuitry that works with electrical impulses. All you're left with is the metal pattern, which was initially a light pattern, and you can do it in extreme detail because it's just light that is painting these lines on the surface of the silicon. Zillions of lines.

"Repeat that process, and you can build the circuits. It looks two dimensional but it's actually three dimensional. They put up to twenty layers on top of one another.

"So what we do is basically the same thing, except that we're not using metal. We're using human DNA. And *our* circuits aren't electrical, they're chemical."

What Fodor wanted to do was create a chip that would allow a one-step procedure: The DNA of a patient would be placed on these probes *once*, and it would read back to you the whole gene, entirely sequenced, every answer for every base. To do this for, for example, the tiny ovalbumin gene, he would need 30,800 probes (7,700 bases in the gene times four possible bases). How to get all 30,800 of them on one chip? Silicon and light. "I take this glass chip. There's nothing on it, right? I illuminate one-quarter of it and paint on A. Then I illuminate another quarter and paint on C. Then G, then T."

Fodor draws it in marker on the white board opposite his desk.

"So you say, Well, that's really inefficient because it's taking you a step for each base. You've got four probes, and each of these guys is of one base each, one A, one C, one G, and one T. And it's taken you four steps.

"Ah. But now you go to the second round."

"What I do is illuminate across here." Fodor swipes the marker across his ink chip.

"I splash T across the entire surface, and in one step I've made four probes of two bases each: AT, CT, GT, and TT. So on the first round I made one probe per step, by the second round I'm making four probes per step.

"Third round. Illuminate across these guys here"—the pen slashes over the board—"add T, and by round three, with just one step I'm making sixteen probes, each with three bases: TTA, TTC, TTG, TTT, and so on. But the magic is that while the number of steps you're performing mounts linearly—4, 5, 6—the number of probes you're making goes up exponentially—16, 64, 256. It's nonlinear behavior. It rockets. And as you make more and more probes, the efficiency goes up and up. At step 28, I make over 16,000 probes. It used to be that if you wanted to make an array of 10,000 probes, it took 10,000 steps. Now it takes 30.

"And as long as you're going through all these steps, there's no reason that you can't make a whole array of arrays on the same big sheet of glass simultaneously. This is wafer-based manufacturing, and it's the technological reason that semiconductors are now so cheap. This is the manufacturing strategy we've adapted to making the DNA chips. You make sixteen arrays at a time, and when you're done you cut up the glass into chips. So we've figured out a way to make very high density arrays so that inside a chip of 1 cm by 1 cm, about the size of a dime, we can fit a million probes. And then mass manufacture the chip on assembly lines.

"And here it is."

Fodor reaches across his desk for a small rectangular piece of clear plastic with a green center and holds it up. It's an inch and a half long at most. "This is the GeneChip."

What it is, actually, is the chip inside its protective casing. The GeneChip itself is the tiny square of green glass inside the plastic case. The casing has an inlet and an outlet, small holes leading to the chip so DNA can be put in. On the surface of one side of the chip are the millions of probes, each fifteen bases long, tiny bits of human genetic sequences. "The chip is a microscope glass that's had an array of boxes etched on its surface," says Fodor, "although they've been cut so fine you can't see them. If you were one millimeter tall and were standing on the edge of the chip, it would look like an expanse of farm fields stretching out before you. Each field is growing a slightly different probe."

It is over this expanse of fields that scientists will spread DNA. Attached to a fluorescent tag and sent into the chip through the tiny In hole, DNA will wash over the squares of probes like a flash flood covering farmland, sticking wherever it meets a matching sequence. "We've created a matrix of probes designed against a particular gene of interest. Say we want to see what your chances are for developing sickle cell anemia. You've got a gene, the beta globin gene, and you've got a sickle cell chip for it. We design a chip that is complementary to that gene. So if every site is perfect in that gene, then every site should have a probe binding to it. I put in the DNA, and everyone'll light up, 'cause everyone's accounted for. And if there's a defect in that gene, there'll be probes with the defect sequence binding there. Then we take the chip and scan it with a Reader. The Reader is a laser-based instrument which focuses its beam on the surface of the chip and scans it like a compact disc. The machine reads the chip: 'probe 3,251,' 'check,' 'probe 3,252,' 'check,' and so on. It maps the fluorescent bright spots and gives you the sequence. And we'll know. We'll know that gene, the beta globin gene, its sequence, what it's going to create. And then we throw the chip away and go on to the next gene. Whichever one you're interested in."

Affymetrix could make a dystrophin gene chip for muscular dystrophy. Or a chip for the gay gene. "That's one of the prime ethical questions that arises from this technology," Fodor says. "You could do a lot of damage—or, I suppose, depending on your point of view, a lot of good. The point is, it's a lot of power." He pauses. "It's something we all talk about and wrestle with all the time."

The company's research rooms are large white spaces filled with machines busily working on slides of glass, computers monitoring the machines, and humans checking the computers. Walking through them, David Singer discusses the business side of the company, the products that actually will make profits. Tan, energetic, and young, Singer radiates a Californian entrepreneurial enthusiasm. "We see there being basically three products," he says, "for now. There's the DNA chips, the Readers, and the software with which the Reader reads the genetic information provided by the chip.

"There are four uses for the GeneChip. There's industrial markets—say someone is making a pharmaceutical and they've got 2,000-liter vats of *E. coli* bacteria producing human growth hormone. They have to make sure there are no spontaneous mutations. So they use the DNA chip as a means of quality control. Second is scientific research. Third is clinical diagnostics, high throughput screenings of blood samples anywhere from hospitals to your doctor's office. That's what Steve was talking about with the sickle cell anemia gene. There's always a market for the cystic fibrosis gene, the Tay-Sachs gene, and things like that, so we put a medically important amount of information on the GeneChip and mass produce them. We're aiming at a couple to a couple hundred kilobases of DNA per chip, enough to screen a hefty-sized gene. This is the most commercially viable area and the one we're aiming at first. And it's going to be more important the further along the Human Genome Project goes and the more genes we know.

"And last is genomics. Genomics is sort of the catch-all word that describes the things you'll be able to do with the chip that we don't even know what they are now, frankly. You approach it from the point of view of, now you have access to all this information, what are you going to do with it? Well, if the GeneChip can monitor the expression of different cells that turn off some genes, then this is the source of an activity that could help you discover new drugs: biopharmaceuticals."

The GeneChip may have its most intriguing uses in the area of disease screening, and Affymetrix may make most of its money on infectious disease testing, because doctors monitor patients constantly and so would constantly be using the chips. Strangely enough, however, perhaps the most intriguing of such uses will be a chip for monitoring *non*human genetic material. In fact, it isn't even in the form of DNA. The HIV GeneChip would sequence viral genetic material for the virus that causes AIDS.

"We think the chip can help us beat the trick the AIDS virus has of dodging the drugs we use against it," says Fodor. "It's called resistance."

Resistance is a phenomenon that has long frustrated virologists and doctors battling viruses like HIV. Viruses use an enzyme called reverse transcriptase to replicate themselves with the RNA they carry, multiplying into an army inside the body. Antiviral drugs attack viruses, but HIV presents a different challenge as it metamorphoses constantly, slipping from genetic identity to identity and evading its antiviral pursuers. In ten years, the genes of the AIDS virus change as much as ours do in 10 million years.

"The catch with HIV," Fodor explains, "is that the enzyme it uses to copy itself is kind of wobbly, and it has a high error frequency in making copies. It sounds bad, but it's actually good for the virus, because it introduces errors into each generation. Well, another name for errors is mutations. So what you get pretty soon in your bloodstream is a whole population of different viruses, basically the same type—HIV—but all slightly different. The errors make the virus a moving target. If you've got ten versions of the virus and you hit it with a drug like AZT, maybe you knock out versions one through eight, but the two viruses that have some mutation that AZT can't get at are now resistant to that antiviral, and they continue to multiply. There's a lot of debate over this, but generally, if you look at what happens when you give a patient AZT, the level of virus falls but then it rebounds, and if you hit it with AZT again, it's not going to do anything. Except make the patient sicker than hell.

"But there *is* something you can do. You can spy on the virus's RNA sequence during treatment, and we're designing GeneChips that do that. You can keep taking the viruses out and putting their RNA on the chip, just like you would the DNA from a gene, and the chip tells you okay, now it's mutated to that sequence, and so you make a drug aimed at that sequence. It's not that straightforward yet, and there are people saying we still don't understand what the variation in the sequence means with respect to drug resistance. But that's the general idea. The virus is a moving target, but the chip, we hope, will allow you to track it under all its aliases and nail it as it goes. And then you monitor what happens under various therapeutic dosages and so on. Theoretically you could be infected with HIV and with the chip you just keep one step ahead of it. You don't cure it. But you keep it constantly on the run, constantly trying to escape from you. And it's running so hard," says Fodor, "it never has the chance to harm you."

In one of the research rooms, shrouded in layers of jet-black curtains and dominated by a vast control board that governs the laser at its center, Singer worries about the commercial concerns that govern making the chip. He gives a friendly wave to the very casual recent physics grads from California Institute of Technology, Berkeley, and Stanford who lounge about, working on the board. He says, "We keep a tether on these guys. They'd go crazy scouting the outermost limits of the molecularly possible if you let them." They nod back, relaxed. "That's right," says one.

"But," says Singer, "our investors have to pay their bills. So we've gotta focus on paying the rent and other things. The chip is a series of probes, and the more probes these guys stick on, and the more ingeniously they do it, the more the chips can do. And they like that. The problem is that then it's more expensive for us to produce and so harder for our marketing people to sell. So we're searching for that happy medium. What's commercially viable."

Back in Singer's office, the question arises again: If Hamer were to find a gay gene, would Affymetrix mass produce the gay gene GeneChip? Would it debate the ethics of such a product? Singer hesitates for an instant, then looks faintly puzzled. For Affymetrix, "commercial viability" means, at least for the moment, manufacturing one chip for each gene of interest. "But that's only for the moment." Singer pauses, then asks if Fodor has talked about fitting not just a gene but the whole human genome on the GeneChip.

He has. In his office, Fodor had pointed out that it is certainly within the realm of the possible for modern semiconductor technology to place three times ten to the ninth bases on a single chip, which happens to be the size of the human genome. The entire human genome on the size of a dime. "Right now," Fodor had said, "we can only fit about a megabase of DNA info on a chip, but if we decided we didn't have to make money as a company, if we just wanted to put the human genome on a chip purely as a scientific achievement, I imagine we could do it in a couple years."

But what about the gay gene chip? Singer appears to be considering the ethical questions, but instead his answer is technical. "If," he says gently, "Affymetrix can make all the probes that it would take to sequence the entire genome, and if you can fit them all on a single chip, which is available on the open market to anyone who wants to buy it, then what we've created—what those guys in the lab want to create because it's a feat of technological wizardry, what Steve wants to create because it's an amazing step forward in biotechnology, and what I and our investors want to make because it's a great product—is the Universal GeneChip."

He sits back and waits for a moment, an eyebrow slightly raised. The meaning comes clear. At that point, it's not a question of whether to manufacture the chip to find the gay gene or not. The chip has already been made.

◡~

Thane Kreiner is one of Affymetrix's "hybrids"—Kreiner has both an MBA and a Ph.D. in molecular neuroscience from Stanford. Kreiner is gay and has, unsurprisingly, passed a number of hours with Singer and Fodor in discussion over the possible gay gene chip.

"On the ethical obligations of Affymetrix and other biotech companies," says Kreiner, "there are two arguments you can make. One is that we're not ethically responsible for the use and social and political outcome of our product. If Pfizer or Rhone-Poulenc makes a drug and people abuse it, are they responsible? Well, no, they'd say, people are responsible. If we make a chip that tests for a gay gene, we can say, look, we as a company don't think you should use this to abort gay fetuses. But what people do with the chip is not our responsibility, it's theirs." Kreiner added, "I think it's compelling to point out that if we don't make the chip, someone else will.

"Or we could take a stand and say we have an ethical responsibility never to make this chip. We have to think about this for a whole bunch of questions. For example, do we make the Huntington's chip before gene therapy for Huntington's is a real option, when we make the chip and sell it and all it's going to tell you right now is 'Oh, you're going to die and there's nothing they can do?' There's no question there's an ethical impact to developing the cystic fibrosis chip, which we're doing now in the transmembrane regulator gene. The chip will tell you if you have none, one, or two bad copies of the gene. We're developing the CF chip because there's a treatment. We're *not* doing Huntington's partly because there's no treatment. But then we have the question of whether we're erasing from our genes something that is actually beneficial, like the sickle cell version which confers resistance to malaria."

Kreiner and Affymetrix have considered making sectors of the chips "dark," cordoning off certain areas of the genome. Data comes off the chip as a fluorescent image that must be analyzed with a software program, and the data must then be compared with another program to sequences in a gene library of normal and abnormal alleles. Both programs are sold by Affymetrix. Say Affymetrix decided it did not want the army to be able to test for homosexuality, Kreiner proposes. The company's engineers could design software that would lock off information about GAY-1 or any target gene. "Your screen comes up black," says Kreiner. "Access denied. Sounds fine in theory, right? In reality it's extremely difficult."

The difficulties are technological. "Anyone who knows the chip," Kreiner says, "could break GAY-1 into pieces and put them on the chip and outsmart the software, which would have no way of knowing that it was reading the gay gene. And it would give the army the answer in pieces, which they could then reassemble to get the answer. So software is not going to work. The other way then is not to synthesize the probes that will allow people to read the gay gene. So we say okay, we're going to black out that region of the chip—but a universal chip is sequencing *everything* you put on it. If you can't sequence the gay gene, you probably can't sequence other genes that have positive utility. So it wouldn't really be a universal chip, and it's a Catch-22 as to what we're supposed to do. It's certainly easy to argue that we have no ethical obligation and that the people who use the chip have the responsibility. And there's a strong argument that we have an ethical obligation to make the universal chip for the good it will do."

There are, in fact, hundreds of positive uses for the chips, from reducing the cost of pharmaceuticals and the suffering of patients who use them to improvements in health. The chips can be used to ensure the purity of meat and to protect water sources. Someone with a lung infection could be screened for a hundred organisms at once and very quickly given the most effective treatment. The chips will allow us to look into mitochondrial DNA both for disease and as a virtually foolproof system of identity testing. Mitochondrial DNA has its own genome and is thus a perfect and eternal fingerprint, one that can be tested for with a tiny piece of skin or hair. "You could identify MIAs this way," Kreiner notes, "and the army is very interested in that."

He paused, then adds, "One of our scientists who's working on the mitochondrial DNA chip is morally opposed to the military and to killing, and he has very serious problems with our working with the army. Does he want to help them? And as for me, do I want to give them all this genetic information, this power?" Kreiner laughs, then says pointedly, "I guess I'm emphasizing that people in the company other than just me as a gay person think about ethical issues.

"Ultimately we'll make a universal chip. And we won't have any control over what people do with it. But the chip itself is ethically neutral. The chip is just knowledge. That's all it is—information. You have to get someone's DNA, target DNA, put it on the chip, and read the sequence. Who makes the target? We don't. We don't make the target. The customer does. We're just the people producing the universal chip. Some people say that absolves us of any ethical responsibility for what we do. But that's like my company makes hydrogen bombs but we don't tell our customers what to do with them. There are a lot of people who think hydrogen bombs are terrible, and a lot who think they're wonderful and have saved us from a Third World War. It depends whom you ask. There are a lot of things of controversial use. You could use our chips to genetically fingerprint everyone because we all have our own chip—which is our unique DNA sequence—and the police could keep them on a national file. It would help the police nail you if you're ever accused of a crime, and it would be your defense if you were ever wrongly accused, if there were blood in evidence on the scene. They would take your blood and compare it on your chip. Is this good or bad? Again, it depends. But to do this, you need the universal chip.

"Technology is always going to move forward. That's a certainty. We can come out and say we think people should not use them if it's not a positive utility application. But let's be realistic, that's up to everyone to decide for themself. Which, I would tend to argue, is the way it should be. Who does this cut in favor of, right or left? I don't know. You can quote the National Rifle Association, 'Guns don't kill people, people kill people.' And you can quote the National Abortion and Reproductive Rights Action League: 'Who decides?' It recalls fetal tissue research: Use the tissue of thousands of aborted fetuses and help save millions and millions of lives."

∽

Perhaps it is understandable, but the common perception that pivotal scientific advances arrive clearly announced ignores the history of the advent of scientific knowledge. If someday we are able to perform genetic surgery on gay people to change their orientation from homosexual to heterosexual, the procedure most likely will be created in exactly the place we are not looking and slip into existence unnoticed in the form, maybe, of a new technology for treating the most dreaded diseases, or a stunning medical advance in the battle against some ancient ill—possibly, for instance, in the form of a cure for cancer.

Chuck Link had only recently flown in from Iowa when he delivered a lecture on gene therapy to a group of attentive doctors gathered at Cold Spring Harbor Labs. He looks thin and blond like a college kid. "I come from a small research group," Link begins, standing at the front of the room in the Banbury Center and holding the plastic control for the slide projector. "It's called the Human Gene Therapy Research Institute and was set up about ten months ago in Des Moines, a group of about forty people. We're a small nonprofit research institute. I guess we consider ourselves sort of the nth stage of the application of molecular biology—our whole purpose is to try to take gene therapy into clinical trials, putting molecular biology right into people, basically by changing their genes."

Link is a molecular biologist. He jokes often but is slightly brittle, and he speaks very quickly. He and others like him are engaged in creating a new world of medicine based on a single superdrug, human DNA, probably the most powerful drug we will ever know. He is looking for a way to kill cancer with genes.

"I should tell you," Link says to his audience, "that I'm a medical oncologist and internist and I've given a *lot,* a *lot* of chemotherapy to cancer patients, making them terribly sick and creating all kinds of side effects, on the principle that you give them chemo, this poison, and hope that the poison kills the cancer before it kills the patient. And often you fail at that. So I don't know what you guys dream about at night, but what I dream about is this."

He takes a breath and looks at the doctors. "A woman is sitting in my office. I have just determined that she has late-stage breast cancer which has metastasized, broken up into cancerous bits which have spread the disease to her bones and liver. In my dream, I have a little syringe in my hand. Imagine, if you will for a moment, that there's a solution in this syringe, and it's a solution filled with viruses, the things that invade our bodies and target our cells and give us colds and the flu and hepatitis and polio and AIDS. But these viruses are different. I have built them precisely to my specifications, I have engineered them to target one kind of cell: the breast cancer cells in the patient who is sitting in front of me. And I have loaded each virus with a gene called the TK gene. This gene kills targeted cancer cells.

"I stick the needle into my patient's arm and inject her with millions of my gene-carrying viruses. They spill into her bloodstream, homing in on and binding to the breast cancer cells, taking over her body like thousands of tiny space probes touching down on thousands of asteroids. Each time a virus lands on one of these cancer cells, it injects its cargo, the TK gene, into the cancer cell. I then treat her with one dose of gancyclovir—a common antibiotic, easy to get, cheap—I give her one dose of gancyclovir in the office and say, 'I'll see you in six months.' And I know that when I see her again in six months, her breast cancer cells will be gone, no chemo, no mastectomy, and she'll be a healthy person. That's what I dream about. That's where we're trying to get with this technology, and I think within twenty years you'll see breast cancer cured this way."

What Link and the Human Gene Therapy Research Institute are doing to make this dream a reality is figuring out a way to get the cancer cells, which are engaged in uncontrolled growth, to kill themselves in a sort of mass suicide. The trick to doing this is the completely ordinary TK gene. Link borrows the TK gene from the herpes simplex virus. The gene codes for an enzyme called thymadine kinase (thus TK, the gene's name). Viruses use this enzyme to construct more of their own DNA. Link, on the other hand, wants to harness it for, in a sense, the opposite purpose: the destruction of cancer cells.

As Jasper Rine pointed out, the same gene that makes tubulin in yeast exists in a human version, which has a comparable function. The same is true of the TK gene. We have our own version of the TK gene, and our cells need it to survive, but while the herpes' TK churns out large amounts of the enzyme, the human version of the gene makes only tiny amounts.

But human cells are perfectly willing to accept a copy of the turbo herpes' TK gene—so willing, in fact, that Link could load every cancerous cell in his theoretical patient's body with a TK gene with no ill effect. The cancer cells would continue to multiply, murdering Link's patient. There is, however, just one catch. The catch is that once the cancer cells have unwittingly welcomed in a copy of the herpes TK gene (and once the herpes gene has gotten to work, churning out large amounts of thimadine kinase enzyme, dumping it absentmindedly into the woman's cells like so much forgotten exhaust), Link can turn this gene into his weapon. He makes an appointment with his patient, takes out a needle, and gives her an injection of gancyclovir, the common antibiotic. As she puts on her coat in Link's waiting room and prepares to go out to her car, the gancyclovir passes silently into her blood and slowly, steadily disperses itself through her system. It washes through the cancerous cells in her bones and liver, the cancer cells into which Link has loaded the herpes TK gene.

And then the changes begin. When the gancyclovir comes in contact with the herpes TK gene, something very simple takes place.

It happens that the thimadine kinase enzyme made by the herpes TK gene is exceedingly efficient at doing one particular job: converting gancyclovir into a substance called gancyclovir monophosphate, a noxious poison. As if a chemical were added to a swimming pool filled with swimmers and suddenly converted the chlorine into a deadly acid bath, the thimadine kinase enzyme converts the gancyclovir Link has injected into a lethal toxin that bathes the cells. (Our own thimadine kinase enzyme converts gancyclovir to gancyclovir monophosphate, but it does so very, very inefficiently, rarely making enough poison to kill the cell; the herpes TK gene that Link has loaded into the cancer cell, however, converts large quantities of gan-cyclovir into poison.) The TK gene becomes a suicide gene, a tiny cyanide pill delivered with precision to every cancer cell in the woman's body. The cancer, with its new gene, poisons itself and dies.

That is, at least, the working model.

The idea of using the body as its own pharmacological factory and surgeon is not new. The ancient Roman cure for asthma was tying the asthmatic to a wild horse and then letting it run. After the ride, the person was scared half to death and breathing freely. This is, bio-chemically, identical to modern practice for asthmatics in hospital emergency rooms: They are given a shot of adrenaline, which dilates the bronchi in their lungs and opens the breathing passages. In both cases, adrenaline is doing the work, although in the Roman version, the body manufactures in its cells what Pfizer or Eli Lilly make today in steel vats in laboratories. We may be returning to the days of the Romans; the pharmaceuticals manufacturers are currently spending millions of research dollars in search of the next generation of cel-lularly manufactured drugs, which could prove cheaper, more pow-erful, more tolerable, and more suitable than anything now made.

Gene therapy is a way of domesticating the body's most powerful molecule—DNA—and one of the most deadly predators preying on the inhabitants of the microscopic biological world—viruses—to make them work to change our bodies. The central problem with this treatment has always been delivery. The most perplexing question now facing Link and his colleagues (whose profession has been in ex-istence less than two decades) is how to get the gene into the cell.

In their attempts to kill each other, humans have long worked at improving not only their weapons but their ways of delivering them. Bombs were first hand-delivered, a slow, inefficient, and frequently inaccurate method that limited the effectiveness of even the best ex-plosives. Now they are sent via pinpoint-accurate ICBMs. The attempts to deliver genes to cells parallels the development of nuclear war-heads. "Now, gene delivery," Link says, taking a quick breath and plunging ahead, "has gone through a lot of different stages. We couldn't be where we are and we couldn't be talking about the ge-netic makeup of people had not a ton of groundwork gone on before us, like any other advanced medicine, and we're just now getting to the point where this is possible. The very first experiments started out with chemical methods of transferring genes."

Chemical methods, popular in the late 1970s, were tried on bac-teria. "What you did," Link explains, as if describing the first small, crude chemical rockets scientists shot off from their backyards, "was basically throw in a little calcium, throw in a little phosphate, add some DNA. Like salt crystals forming in water, the calcium phosphate would crystalize around the DNA, tiny rocks with genes inside them. You spread these around on cells—bacteria in this case—and the cells absorbed the genes. Which was great. Except that it worked with only around one cell in a million, much less than 1 percent efficiency." This was the first crude gene delivery system.

During the early 1980s, scientists developed air guns that shot tiny gold beads coated with DNA into cells and droplets of fatty molecules with genes inside them that bound to cells. These methods were somewhat more accurate—the air guns occasionally transferred their genetic payloads, depending on the cell tissue, and the fatty molecules were even better—but then biologists turned to viruses.

Viruses are useful to doctors like Link because they are, frequently, such efficient cell-invading-and-killing machines. Viruses are not alive in any way familiar to us. We have few clues as to where they came from, or why, back in the mists of the primeval origins of things. Across the earth, viruses are an immense, planetwide mixing bowl stirring and blending genetic material with trillions of microscopic blades. Sagan and Druyan write, "Viruses are looking more and more as if they are peripatetic genes that cause disease only incidentally. . . . Perhaps sex started out as an infection, becoming later institutionalized by the infecting and infected cells." In other words, a baby is conceived when a man successfully infects a woman's egg with his genes. The best way to think of viruses is as robot war drones, built solely for infection. A virus has one mission: to make more copies of itself and then deploy them to make still more copies. The predicament of the virus is that it needs a factory to build more of itself, but there are billions of factories all around it, in crowded subways and taxis and elevators and hockey rinks. The factories are human cells.

Cells manufacture thousands of biochemical products—enzymes, proteins—that make the body function. The cells in the retina look like long, rectangular planks of wood stacked on top of one another in towering walls, a fantastic microscopic lumberyard, while the blood cells, which perform a completely different function and make a different product, look like fat, floating Frisbees. We don't yet know how cells, which at the very beginning of our lives all start out exactly the same, differentiate into this stunning diversity of retinal cells and lung cells and heart cells. It is an amazing feat, similar to building an airplane factory, a brickyard, and an oil refinery from three identical blobs of raw material—and, say, a bulldozer and a motorcycle, because a cell's morphology is also determined by its function. What each cell manufactures or does depends on the production orders in its nucleus, its control room. The orders are written out in DNA. What controls the DNA controls the factory.

Viruses are invasive terrorists. (One of the research titans of the twentieth century had spent decades working on viruses but had never seen one. When the electron microscope was discovered and the very first pictures were taken of viruses—bacteriophages, as it happens, viruses that prey on bacteria—they rushed one to him. The great scientist stared at it for a moment and said, "My God! They have *legs!*") Viruses operate as parts of units in silent raids on the control rooms of cells. If undetected by antibodies and other internal security guards, a virus docks on a cell's surface, unlocks an entry portal (the AIDS virus conveniently has its own key, a molecule called GP-120, which fits a lock on the T-cell's surface), and sedately invades. Once it shuts the airlock behind it, the virus is safe from antibodies. It then turns its attention to the cell's control room, which it commandeers by neatly splicing the DNA or RNA it has smuggled inside into the genetic program that runs the cell's machinery. The program goes into override as the machinery is reconfigured, and then the cellular factory begins unwittingly churning out not the proteins and enzymes it was and should be producing but copies of the virus that has hotwired it.

The final result varies with viruses and cells. Sometimes the new virus copies trickle delicately out of the cell where they were made and into the body for raids on other cells; HIV has the efficient if brutal modus operandi of ordering the T-helper cell to make so many copies of itself that the bulging, engorged factory eventually explodes with virus, deploying a newly forged army of robots, each on its own cell reconnaissance mission. The robots leave the T-cell a crippled wreck of molecular machinery, its sides and structure blown apart, floating biological debris in the bloodstream.

Another striking aspect of viruses is that they kill people. (And they kill indiscriminately and with little apparent logic. Only one in one hundred people infected with the polio virus get the disease. On the other hand, almost one hundred out of one hundred people who are infected with HIV get AIDS. We do not know why.) Despite this, Link realized that if you wanted to get a gene inside of a cell to do genetic therapy, a virus would be the perfect vehicle. If their deadly nature could be modified somehow, viruses would serve as excellent transport devices.

In 1985 the ICBMs of gene delivery were developed. "Now we're dealing mostly with viral systems back in our laboratory," Link tells the audience at Cold Spring Harbor. "You select your delivery virus for the type of tissue or location where you want to deliver the gene." Link outlined the method's features. "Some viruses you basically gut and load up from the bottom. You modify them to carry only the genes you put in with few or none of the viruses' own genes left." He swipes at the illustration hovering on the screen above him. "Some viruses infect only specific types of body tissue, so you can aim them with great precision. For example, some delivery systems are effective in taking genes into the brain. And," he adds, "you can also size-match your viral vehicle to the size of the gene it's delivering."

Size-matching further fine-tunes the system. Genes are measured by the number of base pairs that constitute them. They range from around 3,000 base pairs for the smallest to the longest gene known at 2.5 million. A retrovirus (a virus whose genetic material is in the form of RNA) is tiny, about 1/100th the size of the cell it invades, so it can carry only a small payload—about 7,000 base pairs of DNA. Adenoviruses, on the other hand (a type of DNA virus), can carry larger payloads. "With the first generation of adenoviruses," says Link, "you could jam in about 7,000 base pairs, same as the retrovirus. But by the second generation, we will be able to fit in 10 to 12,000 base pairs. And if you use a plasmid, a circular DNA molecule"—Link clicks the slide prompter and an image of a plasmid jumps to the screen—"attach it to a chemical like polylysine and just inject the naked DNA and have it taken up by cells, your size constraint is almost limitless. You can put in huge genes that way. The problem is that the efficiency of your gene transfer falls a lot, so we're still working on that. Adenoviruses and retroviruses are by far the most efficient. Most of our work uses retroviruses. If you load one hundred infectious retroviruses with your gene and put them near a bunch of cells, you can get a *very* high gene transfer efficiency."

He clicks another slide onto the screen behind his head and turns to glance up at it. "This is a field of glial sarcoma cells we've taken out of the body and are growing in culture. We've exposed these cells to retroviruses into which we'd put the lacZ gene that we took out of the bacteria *E. coli*. The lacZ converts a chemical into a blue stain, and you can see how efficiently the viruses are transferring their genes by looking for the blue cells. The blue color indicates the cell has a new gene in it which is doing a new function." The doctors in the room gaze at the cells, which are coated in blue. Link adds, unnecessarily,

"You can see that, at least in cells we've taken outside the body, almost every cell in the field has been transferred into by the retrovirus."

He raises the lights and faces the audience. "So we've gone from successful gene delivery rates of around one in a million a few years ago to 70 to 80 percent. If you expose the cells repeatedly to the retroviruses, you reach 100 percent transduction."

The retrovirus was delivering its genetic warhead to the cancer with precision, speed, and reliability. Yet while the warhead would destroy the cancer, the viral machine that was delivering it might infect and destroy the patient's body. The next step was to make the vehicle safe.

Link dims the lights again and clicks on another slide, a photograph of two very surprised looking mice, one normal, the other startlingly and completely naked. "One of these is called a nude mouse, and the other," Link says dryly, "is not. These are siblings. This nude mouse has two genetic defects. One is in hair production. The other is in the operation of the thymus: It doesn't generate T-cells which fight off foreign invaders like bacteria and viruses and things. So if you take some human cancer tissue and put it in the nude mouse, his body can't fight it off. The cancer will grow. We use these nude mice to grow human cancers and test their treatments without testing on humans."

At the next slide, a few in the audience laugh, and Link says, "No, the mouse is not chewing tobacco here." The nude mouse, staring uncomfortably at the camera, looks as if it has a walnut in its cheek. "This is actually a human carcinoma tumor injected into the cheek pouch of the animal. It's growing now, and in another couple of weeks it'll kill the animal. You take the mouse, inject the TK gene into the tumor, give him gancyclovir for two weeks and—"

Next slide. The tumor is completely gone. The mouse gazes at the camera—"you see a much happier-looking mouse." Link cocks his head at the image on the wall. "It's just a little scarred where the tumor was." One of the doctors in the front row leans over to a colleague, and they confer earnestly, making references to the screen.

"Now the next trial we're going to be conducting," he continues, "is a trial for ovarian cancer. It's the same sort of idea. Ovarian cancer kills about 13,000 women a year in the United States. This is a tumor that is incurable. When I first talked to Dr. Ken Culver, with whom I worked to develop this trial, he told me this couldn't be done, that it technically wasn't possible. We didn't even know if the virus vector producer cells would be able to migrate inside the abdominal cavity. Well, it turns out they can, at least in animals." He turns back to the screen.

The next slide shows a mouse cut open for autopsy, its small carcass spread neatly to reveal a textbook internal anatomy. "This is a control animal, which had human ovarian cancer cells injected into its abdominal cavity." Link uses a pointer, a red dot that bounces across the screen indicating ugly yellow blobs resembling bits of rancid chicken fat infesting various parts of the animal's organs. "There's a large tumor here, another large tumor here, one here." The dot jumps between them. "Large multiple masses of human cancer that grow up and kill the animal. You inject this animal into its belly with the TK gene, add gancyclovir, and you get an animal like this—"

Another autopsied mouse fills the screen—and a hushed, collective gasp rises around the room, voices murmuring: The ugly yellow masses are completely gone. There are no tumors to be seen, nothing but strong, healthy pink tissue. "You've completely eradicated the cancer," says Link, although it is quite unnecessary to state it. The doctors are staring at the screen, frowning almost in disbelief or exclaiming quietly among themselves, and for a moment Link lets them.

"Now you might ask yourself a question here," he says, picking up again. "If I told you the best gene delivery I could do in the animal model was about 20 percent, a logical question would be, what about the other 80 percent of the tumor? Turns out that we got a gift from nature on this one. And as with all good science, blind luck also helps. Dr. Hiro Ischii, working with Dr. Bluese, tried a little experiment. They put a bunch of tumor cells in culture. Half of them had the TK gene and half didn't. Remember, these are all cancer cells, stuff you want to kill. They added the gancyclovir, this prepoison, and obviously they figured okay, the cancer cells with the TK gene will turn it into poison and die and the other cancer cells—because they don't have a TK—would, unfortunately, survive. Right? When they looked at the result, they found—surprise—about 80 percent of the cells were dead. These cells had no TK gene. They were innocent bystanders. How did they get killed?

"Well, we call it the bystander effect. It turns out that cancer cells talk to each other a lot. And they talk a lot more with each other than they do with normal tissues. But when they talk, they pass on the gancyclovir monophosphate poison, and it kills them. The more they talk, the more of them die, the better the bystander effect.

"We sort of liken this to horseshoes and hand grenades. Close counts. And that's good for us."

After he finishes and the lights come back up, members of the audience press him for details. As Link gathers his slides and papers together on the table—and Jan Witkowski, director of the Banbury Center, reminds him that the airport limo is waiting to take him to Kennedy—he talks about the strategy of getting the immune system to destroy cancer cells by giving it a genetic boost, delivering a gene called IL-2 (Interleukin-2) that makes T-cells—the cells that fight disease—grow. "Essentially a genetic vaccine to immunize the patient against cancer," Link describes it. "And if you can't *teach* the body to reject the cancer," he adds, "maybe you can *fool* it into doing it," and he talks about genetically tricking the body into rejecting the cancer cells.

You could correct the defective genes that make cancer cells multiply, he proposes, such as the RAS gene, whose job is to tell the cell when to grow. If RAS somehow gets turned on accidentally, the cells grow uncontrollably, which is cancer. Or, says Link, stuffing the last of his papers into a briefcase, you could add a gene that tells cells not to divide, such as p53. "p53 says to its cell, 'Okay, look, you're a liver cell. So don't divide, don't grow.' And the liver cell says, 'Fine, I'm a liver cell. I'm not supposed to divide or do anything fancy during this human's lifetime. I'm just supposed to sit here and clean out his blood regularly.' " Link adds, "There's actually a whole class of genes like this that block cell growth called anti-oncogenes, anticancer genes."

Witkowski hurries him out the door and into the limo, which turns, its tires spinning a bit on the gravel, and heads for the Long Island Expressway. Inside the lecture room, a dozen doctors continue to murmur to each other.

Using this technology to alter sexual orientation is speculation at this point. Dozens of obstacles present themselves. The Recombinant DNA Advisory Board, the committee that has to approve all human gene therapy trials in the United States, has tremendous power to block any experiments. The field is new and undefined. The RAC's members have stated fairly categorically that they will not consider anything for humans that is considered "enhancement" of human beings. But what is enhancement?

To get the RAC to approve a trial on humans, generally one has to show convincing experimental evidence in the animals. But what animal model do you present to the RAC for sexual orientation?

The mutant herpes viruses the geneticists are building would, one gene therapist noted ruefully, have to be injected into someone's brain, because that's where sexual orientation is. He compared it to catching a Great White Shark, an efficient but thoroughly deadly animal, giving the shark a lobotomy, which—it is assumed—tames it, and then using it to perform a service in deep water next to a group of divers. "You sort of have to be wondering," said the scientist, "gee, when it gets down there, will it somehow bypass the lobotomy and eat the divers? Herpes viruses have a natural tropism, an ability to target and infect neural tissue that's very helpful to us since we can harness them to deliver genes to the brain. But they are very large compared to our other delivery viruses—like the mouse retrovirus, which *won't* target brain cells as well for us—and so they have many, many gene functions in them, more than 170 genes. That's a lot of genes to dump into someone's head. We don't yet know exactly what the herpes virus is going do in the brain, where you've purposely injected it, *after* it delivers your gene."

Link notes that the targeting problem is severe. "That's why the theoretical example of modifying a gay gene is so complex," he says. "Say it's a gene that affects some receptor molecule, like MIH. First, we have to find out whether the MIH receptor even exists in the brain. With time this will evolve, but if you want to target a specific group of cells, one nucleus or neuron as opposed to another, there's no delivery system that can do that right now." Although many receptors are made by a single gene, Link points out that many are made by groups of genes working together. "T-cells get most of their receptors from a bunch of genes getting together to manufacture them," he says. "So is this theoretical gay gene working alone? And maybe the gay person already has the receptor but there's not enough of it, or there's a linking molecule that hooks up the receptor on the surface to a protein inside the cell, and that protein is abnormal. And maybe it's the protein deeper inside after *that,* spiraling down toward the core of this cell. It could be anywhere down along this pathway."

There is the "cascade effect," changing one gene, which changes a potentially immense number of other genes down the line. Even in simple metabolic pathways the chain reaction can be terribly difficult to slow down once it is set off. "Say we've got this gene responsible for sexual orientation," Link says. "What's going to be your cascade effect? You have to assume any gene you disturb anywhere has an effect on another gene, exceedingly complex consequences that would be very difficult to sort out.

"You know," he says cautiously, slightly wary, "I'm not a neurobiologist. It's much easier for me to think about killing this one cancer cell than all the ramifications of changing the mechanics of a neuron and altering floods of information pouring into thousands of brain cells. We may very well be able to change sexual orientation through gene therapy some day, but I'll sound like an ecologist and say that the delicate natural balance inside us is not something we want to disturb frivolously." With that, he stops.

Of course, we can speculate . . .

Imagine that Dean Hamer finds GAY-1, a gene situated, as he expected, in Xq28. He sequences it, and, building on endocrinological work done by Dr. Michael Baum at Boston University and a conceptual genetic model provided by Dr. Richard Pillard, he first deciphers what protein the gene makes and then tracks down how that protein operates molecularly on cells. It turns out that the gene's job is simple: It helps make a receptor molecule in the brain, one that studs the outside of a small nucleus (only recently discovered) in the hypothalamus of the brain called INAH-3. The receptor's function is to let in Mullerian Inhibiting Hormone, which defeminizes the part of the male brain that controls sexual orientation.

A gay man flies into Des Moines to see Dr. Link at the Human Gene Therapy Institute, formerly a small nonprofit research group but now a large medical complex that has had phenomenal success in curing human cancers. He has a gay brother, he tells Link, and from the time they were five years old, each knew his orientation was homosexual. He is interested in becoming heterosexual, although his brother isn't. Why. Why? Well . . . maybe there has been disappointment in love, a man who proved fickle, or impatient, or weak, and late at night when they are in bed together he lies awake thinking, "He is not for me. . . ." Maybe there is no man, and he is very, very lonely. Or perhaps there were parents who accepted, but with great difficulty, and they were pained (although his friends understand and accept him). Perhaps there are religious beliefs, or prejudice from a key superior in a certain career, or maybe he is just in need of a change— just a change. His brother says it will be a mistake, that he won't be the same person. But he disagrees.

Link nods and ushers the man in to speak with the institute's counselors, who are attentive, laying out options.

That afternoon Link takes some blood, extracts the man's DNA, and puts it on a recently developed device called a Universal GeneChip, a genetic diagnostic made by a fast-growing company called Affymetrix in Silicon Valley. (In fundamentally changing access both to personal genetic information and to genes that can be used medically, Affymetrix has caused a minor upheaval in the pharmaceuticals industry, and several congressional committees with jurisdiction over legislation concerning health care and insurance have invited David Singer and Steve Fodor to testify at an upcoming hearing in Washington.)

Link slips the chip under a laser reader in his office, and a computer scrolls down to the Xq28 region, then narrows in on the GAY-1 gene. The software program in the computer provides complete access to all genes in the human genome, from those causing cystic fibrosis and Tay-Sachs to those responsible for deafness and eye color, and the computer rapidly reads the sequence determining sexual orientation. And there it is: The man's GAY-1 gene is in the form of a now well-known allele that, perhaps because it is defective, perhaps because for millions of years nature has been carefully selecting for it (evolutionary biologists are divided on the question), has rendered the MIH receptor molecule in his brain nonfunctioning. The man was exposed to MIH, and since the receptors in his genitals worked perfectly, the hormone defeminized his internal organs, making his body male. But in his brain, the GAY-1 allele generated MIH receptors that did not work. The MIH molecules were unable to dock on his hypothalamus, and so the neuron in the third interstitial nucleus of the anterior hypothalamus—INAH-3—did not grow. Some of the centers of his brain controlling the gender to which he is sexually attracted were thus not defeminized, which gave him his homosexual orientation.

As he examines the sequence, Link casually mentions that this gene on the X chromosome accounts for only around 35 percent of male homosexuality. A completely separate gene on chromosome 16 has very recently been discovered by researchers at Johns Hopkins. This gene accounts for an estimated 40 percent of male homosexuality. The finding has surprised no one in genetics. A research team at Rice is now looking for another gay gene at a locus on chromosome 7.

Link and the man sit to discuss the risks of genetic surgery and the cost. The procedure, which is still relatively new although there has been an amniocentesis test available for several years, was adamantly blocked for a time in the United States by the Recombinant DNA Advisory Committee, supported vigorously by gay groups and liberal interests (the major pharmaceuticals manufacturers carefully maintained neutrality). But in Europe and Japan, testing and development proceeded, internal pressure from both the medical community and conservative groups rose intensely, and, similar to the approval of the RU-486 abortion procedure, American regulators ultimately bowed to the inevitable. (Parallels to the abortion pill's entry into the American market have been viewed with much amused irony by antiabortion activists.) Recently the man's insurance company announced that genetic therapy of this kind would be covered—the direction in which most of the insurance industry is moving—and so, having thought it through, relatively quickly he says fine, let's do it.

Link holds a little syringe in his hand. It contains a solution filled with engineered viruses, each one loaded with the dominant and by far most common allele of GAY-1, an allele that makes a functional MIH receptor in the brain. Link sticks the needle into the man's arm and injects him with millions of gene-carrying viruses. They spill into his bloodstream, making their way through his system. As they drift past the hypothalamus in the current of blood and fluid, they brush up against the nonfunctioning MIH receptors, prepare their stealthy incursion, bind to the cells, and invade. The viruses deliver their genes and then, as programmed, break apart and self-destruct.

Dr. Link shakes his patient's hand and says, "I'll see you in six months," and he knows that when the man returns, the gene will have reconfigured a tiny portion of his cellular machinery. After a few days it will begin pumping out a protein that will make the MIH receptor functional, allowing Mullerian Inhibiting Hormone into the brain cells for the first time, changing its neural architecture after a few months and coincidentally making it grow by almost two times over the next half year. And gradually there will be a change in the man's internal erotic and emotional responses. From as early as he could remember, he was attracted to other boys, involuntarily and instinctively, but now, almost without noticing it, he finds himself getting out of a taxi or stepping into a room and noticing women in a way he never has before. Involuntarily he realizes he is looking, and for the first time in his life he is sexually aroused. He wonders, after a season passes, at the fact that he was ever attracted to men, an attraction that seems not offensive now but simply alien to him, simply unfathomable, the memory of a vivid dream. All his instinctive reactions and attractions are toward women. Regardless of what he does sexually, it is women who arouse him, women about whom he fantasizes when he masturbates, who occupy his thoughts and elicit his tenderest and most romantic feelings.

His gay brother will ask him, suspiciously, maybe with a bit of hostility, maybe not, Are you the same person as a heterosexual? *How do you feel?*

And he will think about this change in his sexual orientation, not his behavior but his internal erotic and emotional being. He will think about the fact that he is now with a woman and every day marvels at the beauty of her body and her stirring femaleness, or he isn't with a woman and wants to be and is very, very lonely. Or maybe he lies in bed at night awake next to her thinking "She is not the one for me. . . ." His parents are very quietly thrilled (he hears this from his brother), although there are friends who do not understand. Things are easier, and they are not. And, shaking his head, he will say with quiet satisfaction, or perhaps confusion, or regret, or in simple acknowledgment: I feel very much different. And very much the same.

∼

All knowledge is double-edged. That certain gene therapy procedures are still beyond us today only means that certain arguments will be put off until tomorrow.

This is not to say that all knowledge has equal political impact. It does not. If or when we have the ability to make homosexuals heterosexual, it will clearly cut more in favor of those who dislike homosexuals. (That the procedure will work in the other direction as well will not, however, be merely a rhetorical curiosity. Homosexuals already have begun voicing their desire to ensure genetically that their children also be homosexual. To some, this desire resembles the reaction of deaf parents who cheer upon discovering that their child is congenitally deaf; to others, it is a medically guided manifestation of human diversity, somewhere on the spectrum near sex selection.)

To the degree to which we learn to alter sexual orientation, we will learn, axiomatically, to alter lupus, ovarian cancer, sickle cell anemia, cystic fibrosis, left-handedness, hair loss, height, and on and on. It will mean the capacity for, if not the actual performance of, fetal sex selection and prenatal surgery, alteration of weight and eye color, the eradication of deafness, and, more slowly, the transformation of physical race, enhancement of intelligence, and the coordinated biochemical management of violence. Selective growth of knowledge is impossible.

There are now more than twenty companies in the United States that are doing gene therapy, and their market capitalization has recently passed $1 billion. "There's a tremendous amount of capital flowing in," says Link, "and of course if you put a lot of capital into a technology, it moves it forward.

"We're going through the same sort of things that you saw in the computer industry. Every time I read a scientific journal, people are making positive steps forward, new viruses being adapted for delivery. The Sendai virus is one, it's a type of ilonza virus that can bind to a cell surface receptor, and you can get genes in that way. Computers began as huge machines with vacuum tubes. That's where we were when Stan Cohen and Herb Boyer founded the field of molecular biology in the late '70s, based entirely on very simple bacteria a millionfold less complex than what happens in a human cell. And now we're already starting to attack problems in human cells, and that's actually pretty impressive. It's been a real acceleration in the pace of discovery and technology."

On the prospect of conducting gene therapy on sexual orientation, Link has little opinion. "It's not going to be for a long time," he says, more or less dismissing it.

What is a long time?

"Oh," he says with a shrug towards a far distant horizon, "years and years from now."

How many years? Thirty years?

He starts, looks slightly offended, and laughs very, very briefly. His eyebrows go up. "Well, *no,*" he says, "not *that* long."

Chapter Eleven

THE KNOWLEDGE OF GOOD AND EVIL

Science only advances by renouncing its past.
—*Niels Bohr*

IN 1609 a middle-aged Italian inventor and scientist, Galileo Galilei,
looked up at the night sky through his new invention, a metal tube
with two glass lenses at either end, and gazed at four tiny points
of light. They were the moons of Jupiter, and Galileo observed that
they were revolving around that huge planet. If those bodies were governed by the gravitational force of a larger body, so, he reasoned, must
the earth's moon be by the earth, and the earth by the sun.

"It moves," observed Galileo of the earth. It was the beginning of
the scientific age and, to most men and women at that time in Western Civilization, the end of the world.

The assault of science on religion defines the modern era. The current late twentieth-century political and social battle over the question
"What is sexual orientation?," a battle that appears to us so desperately
important, is merely the latest clash between received wisdom and contradictory new knowledge. Since the publication in 1543 of Nicholas
Copernicus' *De Revolutionibus,* scientists have been bringing about the
end of the world with a certain tenacious regularity. The Vatican
placed Copernicus' work firmly on its Index Librorum Prohibitorum,
the Index of Banned Books. (Friar Paulo Sarpi, Galileo's mentor and
a man possessed of an independent Renaissance mind, acidly, if privately, remarked that the Index was "the first secret device religion ever
invented to make men stupid.") What Copernicus had written was a
mathematically complex—in fact almost unreadable—work of astronomical theory indicating that the sun was at the center of the solar
system.

The Catholic Church, by contrast, had for centuries maintained that the earth was stationary, that the sun and planets revolved around *it,* and that the stars were on a huge winch system called the primum mobile, which was cranked around the earth every twenty-four hours by an angel. The sixteenth and seventeenth centuries were a time of religious upheaval. From 1546 to 1563, the Council of Trent created and launched the Counter-Reformation in response to the massive Protestant incursions, both theological and geographical, of Luther and Calvin. Protestantism cited Joshua 10:12, "The sun stood still and the moon did not move until the nation [of Israel] had conquered its enemies." If the sun stood still, this proved that the *sun* moved, not the earth. It was the Protestant Churches that launched the first, and some of the most virulent, assaults against the new theory of heliocentrism in what would soon become a full-scale theological war.

It is difficult to overestimate the Christian fury directed against the idea that the earth would revolve around the sun. Those who believed in it lived in fear. Johannes Kepler, a young astronomer at the Protestant University in Tubingen, privately sent Galileo a copy of the manuscript of his book, which revealed that he too was a Copernican. The older Galileo replied enthusiastically though circumspectly, grateful for an ally: "It really is pitiful that so few seek the truth, but this is not the place to mourn over the miseries of our times. I shall read your book with special pleasure, because I have been an adherent of the Copernican system for many years. It explains to me the causes of many appearances of nature which are quite unintelligible within the commonly accepted hypothesis." Galileo further confided that although he had formulated many arguments, if not proofs, that had convinced him heliocentrism was true, "I do not publish them for fear of sharing the fate of our master, Copernicus." He ended wistfully to the young man he had never met, "If there were more people like you, I would publish my speculations. This not being the case, I refrain."

And he did, in the end, refrain. In 1633 the Church brought him before Rome's Holy Inquisition, which found him guilty of having treated the earth's motion as real. Galileo recanted. For this, he has been much criticized; Brecht's portrayal of him as a coward is particularly savage. But then, Brecht had the luxury of living centuries later; standing shakily before his accusers in Rome, the by-then elderly astronomer was well aware of the recent fate of the astronomer and mystic Giordano Bruno. For refusing to recant his belief in the Copernican system and his opinion that the universe was infinite (which would make it as great as God), Bruno had in 1600 been led to Campo dei Fiori square in Rome, his head encased in a metal bridle that sent a spike snaking into his mouth and up to pierce his palate and another down onto his tongue, and was burnt alive at the stake.

The execution slowed pursuit of heliocentrism but naturally had no effect on astronomical facts. Science possesses a certain almost disquieting inertia. Answers have a habit of appearing, welcome or not, and amid the chaos of debate is the virtual certainty that the biological origins of sexual orientation will become known to us. In his large, sunny, carpeted office in the Stanford biology department, David Botstein was sitting back in his chair, gesturing at the ceiling in the middle of a discourse on the genetics of race—"No one has bothered to try to map the genes for being black or Asian because, look, why the hell bother?"—when he remarked, almost in passing, "Besides, it'll fall out sometime in the future from some other research." We will find the answers, sooner or later, whether or not we look for them. Someday scientists will have gone so far in so many directions that they will turn around and find answers that have surfaced, unnoticed, behind them. Some will be answers for questions we might rather not have posed, questions regarding intelligence, violence, and sexual orientation. But knowledge is the inevitable by-product of research, and research will always go on.

But the origin of sexual orientation is a fascinating biological question. "No one is going to study why your tongue does or doesn't roll," Jane Gitschier commented, "because it's not interesting. But the genetics of homosexuality is *very* interesting. Why? Because sexual orientation is very common and tied to who knows what other interesting things." She thought for an instant. "Take cell cycle regulation, how cells decide to divide and synthesize DNA and stuff. It's *the* hot area at UCSF these days. The most powerful people here are currently working on cell cycle. They're not doing sexual orientation directly. But it would be surprising if in doing *that* they didn't find out something about *this.*"

Another researcher noted enthusiastically, "The fascinating question now emerging from the observed profile of this trait, one which we're forced to pose very seriously, is this: are male and female homosexuality genetically independent of each other? If they are, and if we can figure out the genetics of how sexual orientation develops in both sexes, it'll have immense scientific implications. It's not just a sexual orientation problem. It's a fundamental part of the study of biology."

∼

And we will certainly be afraid of what we find. We always have been. In 1859, the year that Charles Darwin published *On the Origin of Species,* the theory of evolution was born into an era of political turmoil, split between an adamant Right and an inflexible Left. The nineteenth century was a brutal age ravaged by the forces of industrialization, colonialism, and wrenching class struggle. Positivists clashed with Republicanists who warred with Secularists, Materialists, Monarchists, and Free Traders. Just twenty years before in the summer of 1839, Leftist Chartist mobs and radical workers, after losing a Reform Bill, had begun arming themselves bitterly. The Socialists and the Lamarckians were on the move, rioting had broken out across the country, and pamphlets were being distributed in the streets denouncing marriage, property, and the government, and part of the cause (everyone knew) was Darwinism.

It is inevitably the Right that we remember when we think of Darwin: obdurate, reactionary, cloaked in religious rhetoric, the churches uniting for pitched battle. Talk of the day was of the "physiological division of labour," the roles naturally occupied by the various races. In the Potato Famine of 1845, 700,000 had died, and the impoverished questioned their impoverishment, their class, and the class of those above them, and most of all what they had always been told was their "natural" place in life. Could it be they were not, in fact, made by God to live this way in perpetuity? Natural selection threatened the bizarre concept of "human nature" held dear by the Right, yes. What is usually forgotten about Darwinism (and, today, about the research of sexual orientation), however, is that it threatened just as much the equally bizarre view of "human nature" held by the Left.

The Left, propounding an idealistic, utopian view of a benign, perfectible mankind, feared and despised Darwin's theories. Athiests like the publisher George Holyoake derided true Darwinism and a human nature echoing the cruelty of unstable species competing selfishly for ephemeral life, driven (scientists would discover a century later) by tiny mindless acid molecules blindly pursuing self-perpetuation. The unpalatable aspects of selection were repressed by the Left in favor of a politically correct evolution in which men and women, naturally equal, worked together harmoniously and selflessly for the greater good. The weak did not die, the strong did not rule, and human nature was infinitely, wonderfully malleable, correctable.

Standing under the deep blue sky on top of the dry, hot, dusty California hills near his hyenas, squinting in the sunlight, Laurence Frank said in his forthright way: "There are some really basic biological reasons why males would be expected to be more promiscuous and less selective than females. It's hard to think of an animal example where they aren't, and certainly in some aspects of sex behavior—let's take baboons—you have a series of aggressive behaviors that are typical of males and others that are typical of females. There has been a lot of interest in baboon aggression because ecologically they're a lot like humans and phylogenetically they're not very far, and males have dramatic weapons—their teeth—and dramatic aggressive behavior. Females are pretty nasty among themselves, too, but not anything like what the males do with their dagger-like canines. And it has profound effects on many aspects of their biology. And *yes,* there are thus many specific male behaviors and specific female behaviors in relation to dominance and aggression, and they're quite different in the two sexes. There are also behaviors that are pretty much identical, foraging behavior, grooming, or resting—you know, where they go to sleep at night—where there is markedly less real sex difference.

"I am not at all threatened by biology's influence on behavior. It seems to me that what's dangerous is to deny those influences or say the potential for political abuse of this knowledge is so high that we can't afford to learn this stuff. And a lot of people will tell you that. They'll say, 'Look at what the Nazis did with fake science.' Well, *sure*—but, what, we'd be better off not learning how bacteria work and using the knowledge to fight disease?"

In Berkeley's classrooms, Frank has not gone over well with his liberal students, who have often insisted that, among other things, there are no differences between male and female. "I can't even call myself a liberal anymore," said Frank, suddenly glowering. "Anybody who has looked at animals and humans critically for five consecutive minutes—I mean, I got up in front of a class on behavioral endocrinology a little while ago and was talking about androgens and aggression, and I gave a number of examples. It's been known for millennia that castrated males, animals and humans, are less aggressive than intact males, for instance, and the universal observation is that in any human culture you look at, young males are going to be a lot more aggressive and violent than any other component of the population. And afterward, this young lady comes up, and she was shocked! *shocked* that I could say such a thing! Why, didn't I *know* about—and then she mentioned two cultures comprising maybe ten individuals or something, one in Asia and one in South America, which were matriarchal societies and the males were less aggressive."

He paused. "I don't believe it. That the males were less aggressive,

I mean. And frankly even if they are, two societies out of thousands demonstrates, if it demonstrates anything, the virtual universality of the observations I'd been talking about.

"Look, acknowledging that we can expect certain behaviors from, for example, young males, wouldn't it be better to channel it into Little League or whatever? Wouldn't it be better to deal with biological reality? And if you find a person has a genetic locus that predisposes them to some really nasty form of violence, I think it's worth discussing whether you might offer him a pill he could take every day so he doesn't end up in jail or hurting somebody else. In Europe, these pills exist; they're drugs known as anti-androgens. They are well known to depress the influence of androgens, and Europeans use them quite successfully for violent sex offenders, a process known as chemical castration. It's very well demonstrated that if you've got some guy who is a violent rapist or sexually aggressive toward children, you can knock out the behavior with these pills, essentially by knocking out his libido. In Scandinavia, in particular, they're much less likely to put sex offenders in jail than to say, okay, if you take these pills every day, you don't have to go to jail. And it works in a significant portion of cases. In this country, this is unthinkable, people go ballistic and say 'How could you even *think* of such a thing!' Well, wait a minute—think of *what*? Allowing these guys to avoid hurting other people, stay out of jail, and lead relatively normal and productive lives? The question is not biological, because biologically we can do this right now, today. The question is purely political. Frankly I've never sympathized with the opponents. I think they're cruel. But then, prison is not rehabilitation. It's retribution. If someone blew away a member of my family, I wouldn't be satisfied to have them just waltzing around taking a little pill every day. I'd want them to pay.

"The observation that behaviors are biologically directed is scary to liberals because that means people aren't infinitely malleable. It means you can't pass laws and do social engineering to change the nasty parts about people, and liberals—and Marxists in a more extreme sense—are completely and totally committed to the notion that we can change *anything*. All we need is good will. I think it's two things. I think one major liberal-conservative difference is that liberals have great faith in the inherent goodness of people and conservatives in the badness of people, and if you really want to believe that humans *can* be good if only they're given the right opportunities, then you simply don't want to hear about a billion years of evolution which say that males should be more aggressive and more promiscuous than females. You just don't want to deal with it. And I find it intellectual dishonesty on a really disturbing scale. It is simply being unwilling to want to deal with the fact that some things are bred into us, and there are reasons for that, and we have to take the world as it is.

"You know, trying to deal with human behavior and the nasty parts of human behavior like criminality or domestic violence or aggressiveness or racism while pretending that evolutionary and biological reasons for them don't exist is like attempting to cure a disease without dealing with bacteria and viruses. It's voodoo, medicine based on spirits and vapors, which is tantamount to trying to deal with human social behavior without taking into account the biology of social behavior. I think one of the defining aspects of humans, to a more extreme degree than any other mammals, is the us-and-them business, the human ability to set whoever is us apart from whoever is them and find reasons to hate each other on the basis of that. That has deep biological roots in our social evolution, and to attempt to deal with phenomena in cultures like war and nationalism and violence, to attempt to deal with these aspects of people without attempting to understand why people behave this way—where are we ever going to get?"

Frank gazed across the tan hill through the fence at Mouse and Owl, who stared back, keeping their eyes on him. "I guess I tend to be a bit blunter about this than most people, but it seems to me just—extravagant stupidity to pretend devoutly that humans are totally cultural and environmental creatures."

∼

James Fallows, the Washington editor of the *Atlantic Monthly* and a Democrat, has thought seriously about the political implications of the research of sexual orientation on the Left versus the Right. His conclusion is surprising. We reflexively think of the evidence that some of us are homosexual as threatening the Right. What is forgotten, Fallows points out, is the more profound threat the research poses to liberalism and the liberal view of human nature.

"Boil down conservative policies and liberal policies to their simplest, bumper-sticker versions," Fallows proposes, "and you get, obviously, two very different approaches to any issue you can name, from economics to education to health care to crime. Take one example: homelessness. When a Conservative and a Liberal look at a homeless man, the Conservative will basically say, 'The reason this guy is homeless is because he's lazy and irresponsible, he made bad choices, and he failed to seize the many opportunities that our dynamic society presents since we know anyone can make it in America if he really tries. Something is wrong with him. So we simply need policies that are going to make it easy for him to work if he wants to and not coddle him if he doesn't.'

"The Liberal will look at the guy and say, in essence, 'Oh, the reason this guy is homeless is because of racism, sexism, homophobia, ableism, Reagan policies, and deindustrialization that denied low-skill workers an income, and if there'd just been a government-funded Head Start program when he was a kid, he'd be a nuclear physicist today. So we need to change all these external factors, we need programs that will alter his environment.' "

These two political views raise the question: *why?* Why, in particular, do liberals emphasize environment and the malleability of outcomes? "At the bottom of all political differences," answers Fallows, "is the way people think about human nature. Let's give the very essence of what we could call the Conservative View of Reality and the Liberal View of Reality: to the Conservative, the crucial factors are all internal, the guy's innate motivation, discipline, and intelligence, but the crucial forces at work as far as the Liberal is concerned are all external. And if you assume, as the Liberal does, that options are only closed off to people by external factors, anything from prejudice to lack of opportunity to hostile circumstances, then any and all possibilities can be open to any and all persons—the world can be a utopia." Look at liberal programs, says Fallows: at their heart, their aim is to give more people more opportunities by changing their external circumstance; the classic twentieth-century example is the post–World War II G.I. bill. "It took in large numbers of people who would otherwise never have gone to college, it gave them an education, and it changed these people's lives," says Fallows. "Conservatives would have said, 'These people aren't college material.' The liberals said they only needed a chance."

But liberalism faces the research of sexual orientation uneasily. Liberalism's most crucial tenet is that people are "equal" in all the ways that matter, says Fallows. Obviously people are different in many ways: size, sex, race, handedness, and so on. But the basic belief of liberalism is that these differences don't matter in any sense of morals or merit or human standing because they don't affect the ways that really count: intelligence and ability. "And what, according to liberals, determines how that intelligence and ability can be expressed? Environment."

Fallows thinks for a moment, then says, "I'll put it another way.

Liberals believe, though they don't often come out and say it, that any traits that can't be changed—like whether your body happens to be male or female, for example—will matter less and less as society evolves, and traits that do matter, like ability and behavior, can be changed with the right programs.

"So it's a real problem for liberal thought when it is confronted with the idea that things that matter might not be changeable. That different genders might carry different facilities, that some behaviors and temperaments are given to us. And," he added warningly, "the research of sexual orientation and what it tells us is only the tip of the iceberg." The plain fact is that a scientific look at sexual orientation and the behaviors and traits that correlate with it—and an objective look at gender and many behaviors correlating with *it*—inherently support the conservative's view of reality and contradict the liberal's view. Why? Because they suggest that people are different in basic, unchangeable, and important ways. "The news media have usually presented this research as damaging to conservatives—'You can't claim that gay people are willfully violating God's law if they are born that way.' And in the *short*-run, the media is right on this. Liberals will benefit from an accurate understanding of homosexuality, which is that it is basically like left-handedness. And yes, the research does negate the crudest kind of right-wing stereotypes, and the notion that gay people choose to be gay."

But the long-run outcome is quite different. "This stereotype has always forced Conservatism, a philosophy holding that the environment has little to do with outcomes—and that Liberal programs meant to alter it are a waste of money—to make an inconvenient exception on homosexuality and argue, contradictorily, that young people can be pushed one way or another into profound aspects of their personalities by education and society. Which is *exactly* what Liberals have wanted them to admit forever." Being forced to shed an internally illogical position (not to mention one that is empirically false) will only make the Conservative position stronger. In the long run, says Fallows, researching sexual orientation threatens Liberalism, not Conservatism, and it threatens it in a fundamental way. It is a time bomb, ticking away unnoticed. The problem for liberals is that research into homosexuality is, as Gitschier and Botstein point out, inextricably part of a vaster body of clinical and biological research that tells us that nature controls important aspects of our personalities, behaviors, and intellectual capacities. "This is truly repellent to the liberal mind," Fallows says. "The implications are profound. Liberalism, which has for the past four hundred years ridden to triumph on science, is now all at once at odds with science, which is showing deeper remnants of our animal past than Liberals are comfortable with.

"The trouble for Conservatives is merely temporary. Sure, the Republicans are having trouble accepting differences in people—it's happened to them before. The religious fundamentalists in the party will probably continue to argue that homosexuality is immoral. They can do this because by definition religion is not subject to empirical disproof. But Conservatives have over time come to accept non-whites, non-males, and non-Christians to build their political majority, and it is almost certain that in order to maintain basic electability Republicans will come to accept non-straights and decide not to become a fundamentalist sect. They will bow to the inevitable, adding sexual orientation to race and gender and religion on the list of traits that don't justify discrimination." Once they have repositioned themselves tactically, Fallows predicts Conservatives will find their basic philosophy—that the individual is responsible for his or her own fate—stronger. Liberals will find themselves weaker. "If Liberalism is to survive," concludes Fallows, "it must come to terms with this body of research. It must adjust its view that people are naturally alike and that everything important in human beings is malleable by social force. It is always hard to change a basic philosophy; Conservatives are lucky not to have to do that here. But if Liberals don't adjust to the growing mountain of evidence, they're going to end up like the Creationists in the face of Darwin."

And there is yet another hidden benefit for Conservatives in the research. One researcher said with an impatient roll of his eyes, "As a biologist, I've never understood why the homophobes align themselves with the environmental position. It's really dumb, frankly. They want to believe homosexuality is a disease, so they look for environmental factors." He laughed. "They could just as easily look for biological factors! Because their whole argument comes down not to the question 'What causes homosexuality?' but 'Is homosexuality a disease or not?' What's the answer? The answer is that we've known for a long time that it fits the definition of disease no better and no worse than does left-handedness. Not that that's stopped them. But the conservative position surprises me—I think that it shows a lack of intelligence on their part—because while they can't win the disease argument, and they can't win the choice argument, admitting that it's biological and unchosen is their trump card. It means they've won on an argument that is still only on the horizon today: changing it. A medical trump card, like the abortion pill. Their people can immediately adopt a strategy of 'Okay, let's find the gene or neuron and figure out how to fix it.' And the abortion debate has taught us that no matter how much others dislike their doing this, no one is going to take away their freedom to do it."

Fallows's observation about the conflict between what is observed and what is merely believed is one that scientists have long noticed. The distribution curve for sexual orientation, which contradicts a number of cherished beliefs, is an example of this. In the evenings at Cold Spring Harbor Laboratory, geneticists and biochemists shut down their microscopes and turn out the fluorescent lights in their clean, glassy labs. As the harbor silently turns the colors of the setting sun and, across the Long Island Sound, Connecticut glows dimly, they filter down or up the quiet hills to the bar in the basement of Blackford Hall. They laugh about botched experiments and discuss the latest issue of *Cell* magazine and listen to the music in what resembles (a little bit) a college social in the basement of a dorm. Over a beer, Jan Witkowski was talking about the curious ways in which evidence seems not to matter in the face of a desire to believe otherwise.

Witkowski recalled one far-right conservative who firmly believed that "we're all really born heterosexual," a claim that dictates a very specific population distribution for the trait sexual orientation. "He said, 'It's common sense,' " said Witkowski. "When I asked him why it was 'common sense,' he said, 'Because sex was naturally made for procreation.' " He drank from his beer. "I did not point out that although on one level this was true, since human beings are extremely complex animals it was evolutionarily and biologically empty logic."

I in turn recounted a conversation I'd had with staunchly liberal newspaper columnist Ellen Goodman. "I've always felt," she had said earnestly, "that we're all really born naturally bisexual, and it's only social pressure that forces us to be gay or straight." She added that to her this was "just common sense. Look at all the other different ways human beings have of living." Again, this view mandates a distribution curve for the trait.

Witkowski joked that it seemed as if you should be able to graph this. "Belief in bisexuality correlates positively with liberalism," he said with a grin, "belief in homosexuality negatively with conservatism." And so, as he finished his bottle of beer, I stole his paper napkin, slightly stained with peanut oil and salt, picked up a ballpoint, and graphed on the back of the napkin the empirical, Conservative

and Liberal Distribution Curves for Sexual Orientation. The empirical curve, derived from research, looks, of course, like this:

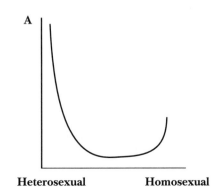

The Conservative Curve looks like this:

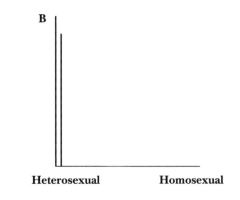

And the Liberal Curve looks like this:

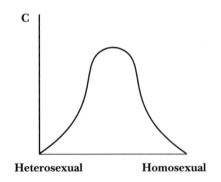

⮑

It is easier to obfuscate, prevaricate, and lie about a trait when science is still working out some of its parts. Research is the margin in which those who will say whatever it takes to forestall the inevitable can operate. And they do.

The most thoroughly exploited still-incompletely researched biological trait is probably not sexual orientation, but another trait, one also much debated in the twentieth century.

In the two decades between 1910 and 1930, American production of cigarettes rose from an annual 10 billion to 123 billion. Much of this growth was due to World War I; the compact cylinders of tobacco were easier for soldiers to carry into battle than their pipes and cigars. Demand became such that in 1917, three years after the war began, General Pershing cabled Washington: "Tobacco is as indispensable as the daily ration. We must have thousands of tons of it without delay."

In 1919, the chairman of the department of medicine at Barnes teaching hospital in St. Louis, Missouri, invited his students to observe an autopsy he was to perform. It was an opportunity, the chairman informed his students, for them to get a good look at a vanishingly rare disease, since they would most probably never see another case of lung cancer again.

One of the students present at that autopsy, however, did see it again, although not for seventeen years. In 1936, Dr. Alton Ochsner was presented with nine cases of lung cancer. It was astounding. What in the world could be the cause? By asking his patients questions—as Hamer and Pattatucci and all clinical researchers ask their subjects questions—he discovered that all the patients smoked heavily and had done so since World War I. "I had the temerity," Ochsner would write later, someone sardonically, "to postulate that the probable cause of this new epidemic was cigarette use."

In 1932, the *American Journal of Cancer* linked tobacco tars to cancer. In 1950 the *Journal of the American Medical Association* found that practically all lung cancer correlated with longtime, heavy cigarette use. In 1953, Sloan-Kettering announced that tobacco tars painted on the skins of mice produced cancers.

Today, the tobacco industry maintains that science has not yet established "ultimate causation" between tobacco use and lung cancer, throat cancer, heart disease, and emphysema. The tobacco executives are, of course, absolutely correct. The biological evidence that tobacco causes cancer stops at about the same point as the biological evidence for the nature of sexual orientation and handedness: empirical observation. We are only just beginning to understand tobacco smoke's biochemical effect on tissue at a molecular level, the level of mutagen and cell, just as we are only beginning to understand sexual orientation and handedness on the molecular level of hormone and neuron and gene. And it was precisely at this stage at which astronomy was poised at the end of Galileo's life in 1642. At his trial, it was noted that he had not supplied "proof." His evidence was the mere observation of a phenomenon: the earth rotated around the sun. He did not, and could not, prove what *caused* what he saw. Indeed, it was difficult enough a feat in 1609 merely to make the observation. Gravity, the force that drove the phenomenon Galileo observed, was a Black Box, its exterior dimensions measured but its origins and interior mechanics unknown. And it still is today—physicists cannot prove on the level of the particle that gravity even exists—though the existence of gravity no longer inspires theological opposition.

Thomas Kuhn, the historian of science, has written that a few of Galileo's more fanatical opponents refused even to look through the telescope on the grounds that if God had meant man to gain knowledge with such a machine, he would have given him telescopic eyes. Those who looked denied that what they saw *proved* Galileo's contentions. "In this, of course," notes Kuhn with scientific precision, "they were quite right. Though the telescope argued much, it proved nothing." Galileo's evidence, as the evidence of the brilliant astronomer Tycho Brahe, whose measurements of arc and parallax helped Kepler refine the sun-centered model, was observation. And nothing more. As a conclusion, it is the astronomical equivalent to the epidemiologist's conclusion that tobacco use correlates with cancer. It is the equivalent of the clinical researcher's conclusion that sexual orientation is a bimodal orientation distributed among the population in approximately the same ratio as handedness, that it runs in families, does so with a characteristic "maternal" pattern, is detectable in early childhood, and is not chosen, changeable, or pathological. Such observations, as endocrinologist Paul Licht of the Berkeley hyena team would quickly point out, prove nothing on a molecular level. Why, then, are they enough to start such great controversies? Because they predict, with coldest, remotest oracular disdain for the emotional wreckage to come, what science will in the end find. Kuhn astutely observes that fear of Galileo and Brahe's observations "derived from . . . a subconscious reluctance to assent in the destruction of a cosmology that for centuries had been the basis of everyday practical and spiritual life. The conceptual reorientation that . . . meant economy to [Galileo] frequently meant a loss of conceptual coherence to men like Donne and Milton whose primary concerns were in other fields."

Donne and Milton were not the only ones to fear. Science has at times frightened even the scientists; it is often forgotten that Tycho Brahe himself was an *opponent* of heliocentrism who proposed his own wretched model of an earth-centered solar system, one with most of the planets badly off center and an ungainly imaginary physical mechanism to account for their motions. What is even more often forgotten is that Copernicus himself, as a devout Catholic, never reconciled what his mathematical senses showed him was reality with what his religion told him. The author of the "Copernican system" did propose a fixed sun, but his work was infested with archaic residue, and written in fear, and he never had the courage to wander too far from the Ptolmeic system adhered to by the Church, a system he knew was incorrect. It was left to Kepler, who had the luxury of living somewhat later, to overcome that ancient inheritance.

Today the old religious and political pressures on scientists are joined by new ones (the media, mostly, and tenure considerations), often intense, but none, at least, that threaten burning at the stake. If there are among the Hamers and Pattatuccis, the Pillards and Allens, patiently hunting sexual orientation in their labs, no examples on Brahe's scale of withholding upsetting results (and I have come across none), there is some delicate downplaying before young, liberal students who are shocked by findings that men and women are different, or before Christians appalled by the existence of homosexual people. There are some careful reassurances to those afraid of biotechnology that the difficult answers are "still, oh, years away." The tacit understanding among the labs involved decrees that this particular body of research will be discussed with a certain tact.

In the twentieth century, a century of political turmoil, nuclear warfare, revolution, and epidemics, science has frightened scientists more than ever before. Thane Kriener of the biotech company Affymetrix stated with evident concern, "I think it would be a huge mistake for the human race to make itself genetically homogenous." It would, he explained, be like replacing a giant natural forest of a thousand varieties of trees and their varied wildlife with a single-tree industrial plantation. In the forestry industry, the result of such botanical standardizing has been the death of millions of trees in a single sweep of a single disease against which they had no resistance, a bio-logical disaster made possible by genetic uniformity. "Do we risk eras-ing from our population something that is sometimes harmful but sometimes beneficial, like the sickle cell version or, quite possibly, the gay gene? What will this do to us as a species?"

In her large, imposing office atop the Carnegie Institution over-looking Washington, D.C.'s Dupont Circle neighborhood, molecular geneticist Maxine Singer leans forward, frank and a bit grim. "Listen, let me tell you," she says firmly. "I really don't think sexual orienta-tion is going to be the Big Thing, either biologically or politically. When people learn that a lot of their behavior in a lot of things is in-fluenced by their genes—notice I don't say 'caused' because it's more complex, and 'nature-nurture' is a *stupid* discussion—but when they learn this, they won't like it. But it's going to be true. When they learn there are genes that influence their intelligence, they won't like it. But it's going to be true. And that biological knowledge is going to affect a lot more people than what we find out about sexual orienta-tion. A *lot* more. When people learn that there are alleles of genes that make you more likely to suffer from mental disease, they won't like it. But it's going to be true. There's no question about it. And we're going to find out. The NIH is spending several billion dollars a year to work out the technological problems, and we learn more all the time.

"Right now, we are all upturned about sexual orientation," Singer says, sitting back with a shrug. "Eventually, we won't be. There'll be too much else."

～

Galileo's most famous words were written in 1623: "Philosophy is writ-ten in this grand book, the universe which stands continually open to our gaze. But the book cannot be understood unless one first learns to comprehend the language and read the letters in which it is composed. It is written in the language of mathematics, and its char-acters are triangles, circles, and other geometric figures. Without a knowledge of them, it is humanly impossible to understand a single word of it. Without these, one wanders about in a dark and obscure labyrinth."

The Church begged to differ. Galileo's view, the epitome of the scientific view of things, has never been universally shared. Only in 1822 did the Church allow books to be printed treating the earth's motion as real. (In 1991, the year of the first major neuroanatomical study of homosexuality and 350 years after the Inquisition found him guilty, the Catholic Church pardoned Galileo while avoiding any admission of error. It is the reaction of an institution under siege cutting its losses in a losing battle; that same year it joined in earnest the attack on the research of sexual orientation.)

The Church's 1822 decision was made during a time of political chaos in a century of political chaos, terror, and repression. In nineteenth-century Europe, revolts by the underclasses against their semifeudal rulers were undergirded by what was, to the traditional monarchies, the more alarming growth of revolutionary ideas. The earth's motion was just another incomprehensible facet of an incomprehensible century. In February 1831, when the new and ultraconservative Pope Pius VIII, the "apogee of reaction," was met with revolts in Modena, Parma, and the Papal States inspired by the July 1830 Revolution in Paris, he brutally suppressed them with the aid of Austrian troops. When the ambassadors of the powers demanded much-needed reforms in the Papal States, Pius contented himself with an amnesty and a few minor concessions in administration. In March, fresh insurrections aimed at freeing Italy and giving her a republican constitution broke out. Again the Austrian soldiers forcibly imposed order.

But in 1846, it seemed that everything would change. Giovanni Cardinal Mastai-Ferretti was elected pope. He took the name Pius IX and was hailed by Liberals at his election as "the pope of progress." Pius was democratically minded, relaxing censorship laws and proclaiming an amnesty for political prisoners, but during the 1848 Italian revolution against their Austrian masters, the pope proclaimed his neutrality. For this, the Liberals turned against him and forced him to flee. It was after the brutal two-year war that Pius returned, an embittered man now hostile to Liberalism in all its forms.

Historian William L. Langer described the result. Pius IX "flung down the gauntlet of defiance to the new social and political order in the encyclical *Quanta cura,* with the appended *Syllabus errorum.* The pope censured the 'errors' of pantheism, naturalism, nationalism, indifferentism, socialism, communism, freemasonry, and various other nineteenth-century views. He claimed for the Church the control of all culture and science, and of the whole educational system; rejected the liberty of conscience and worship enjoyed by other creeds, and the idea of tolerance; claimed the complete independence of the Church from state control; upheld the necessity of a continuance of the temporal power of the Roman See, and declared that

'it is an error to believe that the Roman Pontiff can and ought to rec-
oncile himself to, and agree with, progress, liberalism, and contemporary civilization.' "

Pius IX also issued the Dogma of the Immaculate Conception,
which in 1869 established papal infallibility in faith and morals. Darwin's heretical theory had been published ten years earlier, and
Catholics and Protestants used biblical evidence to counter it, "And
the Lord God formed man of the dust of the ground, and breathed
into his nostrils the breath of life; and man became a living soul," from
Genesis 2:7.

And yet it is Genesis that contains the wisest and most unerring
commentary on the nature of science. The words—in Genesis 3:5—
speak to the nature of scientific knowledge itself, the cause of the great
furor. They read: "And ye shall be as gods, knowing good and evil."
They say that knowledge is always dual in nature, and they are un-
consciously repeated, in various ways, by molecular geneticists Singer
and Pattatucci and Hamer, by gene therapist Link and neuroanatomist Byne and endocrinologists and virtually every scientist
conscious of the power of their work. They are, of course, the words
of the Serpent.

If many in the Church disputed Galileo's view of the necessity of
knowledge, others did not. It was Friar Sarpi, Galileo's mentor and a
believer, who commented that true religion dealt with heaven, and
politics and the things of this life dealt with the world, and the two
were and should be separate. Four centuries later, the friar's opinion
is still hotly debated. The revulsion that met Galileo's discovery is no
longer surprising. It is, rather, the very essence of the history of science. And on it goes. The conflict continues, as it always has and always will.

NOTES

Chapter One The Black Box

1. JOHN D'EMILIO, *Sexual Politics, Sexual Communities: The Making of a Homosexual Minority in the United States, 1940–1970* (Chicago: University of Chicago Press, 1983). See also: Alfred Kinsey et al., *Sexual Behavior in the Human Male* (Philadelphia: W.B. Saunders Company, 1948).

2. RICHARD ISAY, *Being Homosexual: Gay Men and Their Development* (New York: Avon, 1989), 111. Isay continues dryly with a concise anecdote from this dismal period: "In a study of 106 gay men, [psychoanalyst] Irving Bieber and his associates claimed that 19 percent of those who had been exclusively homosexual switched to heterosexuality as a result of psychoanalytic treatment. (Irving Bieber et al., *Homosexuality*, Basic Books, New York, 1962, 275–302.) Wardell Pomeroy, a co-author of the Kinsey Report, has maintained a standing offer to administer the Kinsey research questionnaires to any of the patients who were reportedly cured. Bieber acknowledged to Pomeroy that he had only one case that would qualify, but he "was on such bad terms with the patient that he could not call on him." The last remaining warrior for the psychoanalytic model, Charles Socarides, claims in his book *A Freedom Too Far* to have cured or altered or influenced a number of homosexuals in the direction of or toward a mimesis of heterosexuality—it is difficult to ascertain what, exactly, Socarides claims. In any case, despite a similar invitation from other psychiatrists to confirm claims of sexual orientation alteration, Socarides has never produced an example. It is widely felt that such claims have contributed to the decline of the psychoanalytic profession.

Chapter Two Definitive Proof that Homosexuality Is Biological

1. SIMON LEVAY, "A Difference in Hypothalamic Structure Between Heterosexual and Homosexual Men," *Science* 253 (August 30, 1991): 1034–7.

2. C. A. BARRACLOUGH and ROGER GORSKI, "Evidence that the Hypothalamus Is Responsible for Androgen-induced Sterility in the Female Rat," *Endocrinology* 68 (1961): 68–79.

3. G. W. HARRIS, "Sex Hormones, Brain Development, and Brain Function," *Endocrinology* 75 (1965): 627–648.

4. Roger Gorski et al., "Evidence for a Morphological Sex Difference Within the Medial Preoptic Area of the Rat Brain," *Brain Research* 148 (1978): 333–346.

5. Christine de Lacoste and Ralph Holloway, "Sexual Dimorphism in the Human Corpus Callosum," *Science* 216 (June 25, 1982): 1431–2.

6. Another of these communications cables, called the anterior commissure, runs between the temporal lobes of the brain and specializes in relaying sensory (olfactory, auditory, visual) information as well as other sorts of information still unknown. These bundles of axons that act as data freeways are actually extensions of brain cells. A brain cell on the right side sends an axon snaking out like a phone line looking for information, and the axon shoots through one of these bundles and hooks the brain cell up with a brain cell on the left side. Like almost everything in the human brain, these cables are rather mysterious. If, for example, you cut the anterior commissure of a monkey, the monkey is able to distinguish between left and right *better* than if it had an intact anterior commissure. And studies in humans have found that men who are missing a third cable, the massa intermedia, score higher on certain cognitive tests than men whose massa intermedia are present and functioning. About all that is certain is that these bundles carry information, and the corpus callosum is chief among them, the main interstate highway connecting the right and left halves of the brain.

7. D. F. Swaab and M. A. Hofman, "An Enlarged Suprachiasmatic Nucleus in Homosexual Men," *Brain Research* 537 (1990): 141–148. This finding didn't get as much press coverage as LeVay's, probably in part because it came out of Europe, in part because it was not published in a prestigious scientific journal like *Science.*

8. Laura Allen, Melissa Hines, James Shryne, and Roger Gorski, "Two Sexually Dimorphic Cell Groups in the Human Brain," *Journal of Neuroscience* 9 (2) (1989): 497–506.

9. Kulbir Gill, "A Mutation Causing Abnormal Mating Behavior," "*Drosophila* Information Service," 1963.

10. J. Michael Bailey and Richard Pillard, "A Genetic Study of Male Sexual Orientation," *Archives of General Psychiatry* 48 (December 1991): 1089–96.
    This is the best known of the studies for concordance rates in twins. There are others. Frederick L. Whitam, Milton Diamond, and James Martin, in "Homosexual Orientation in Twins: A Report on 61 Pairs and Three Triplet Sets," *Archives of Sexual Behavior* 22 (3) (1993), arrive at a 65.8 percent concordance for homosexuality for MZ twins and 30.4 percent for DZ twins. (MZ pairs were 34 male, 4 female.) There were also three sets of triplets; in two of the sets, there was an MZ pair concordant for homosexuality and one DZ sibling, who was heterosexual, while the MZ triplet pair were

all three concordant for homosexuality. The figures slightly differ from those found by Bailey and Pillard, but the percentage ratios, which is the fundamentally important point, are almost identical. More extensive work will be needed to establish precise percentages for MZ and DZ concordance.

Arriving at a concordance figure for MZ twins for any trait presents difficulties, primarily due to the differences between pair-wise versus proband-wise concordance measurements. Proband-wise concordance always inflates the true concordance, so a researcher must know the population base rate of the trait. (Obviously, the lower the population frequency, the more robust a particular rate appears.) Also, in proband-wise concordance, a twin and a co-twin can, due to the logic of the method, be counted as two concordant pairs. However, proband-wise concordance carries some statistical advantages.

11. J. MICHAEL BAILEY and RICHARD PILLARD, et al., "Heritable Factors Influence Sexual Orientation in Women," *Archives of General Psychiatry* 50 (March 1993): 217–223.

Chapter Three Definitive Proof that Homosexuality Is Not Biological

1. Most of the brain's vast communication network works on electrical impulses, but at the tips of the cables brain cells use to talk to one another, the system works with neurotransmitter "juices." It is an odd hybrid, as if electrical impulses transmitted data over 99 percent of the AT&T system, but, for some reason, at one small connection spot between all these cables the data was transmitted with squirts of liquid. The gaps between the ends of the billions of cables are called synapses, and the cables themselves are called axons and dendrites. Axons and dendrites act, in a sense, as the arteries and veins of neurons, carrying neurotransmitters respectively out from and in to neurons. The neurotransmitter liquid waits around in the little vesicles at the bulbous end of an axon until an electrical impulse comes down the axon to it, telling the neurotransmitter to release itself into the synapse gap. The neurotransmitter leaps across the synapse, bridging the axon-dendrite gap, and the end of the dendrite accepts it, then signaling an electrical impulse to fly up to the neighboring neuron. The system is electrical everywhere but the terminus, where it is chemical.

Chapter Four How to Look at a Brain

1. WILLIAM BYNE and BRUCE PARSONS, "Human Sexual Orientation: the Biologic Theories Reappraised," in *Archives of General Psychiatry* 50, March 1993: 228–239.

2. LAURA ALLEN, "Sex Differences in the Corpus Callosum of the Living Human Being," *Journal of Neuroscience* 11 (4) (April 1991): 933–942.

3. EVELYN HOOKER, "The Adjustment of the Male Overt Homosexual," *Journal of Projective Techniques* 21 (1957): 18. The theory of a connection between sexual molestation and homosexual orientation was studied in the 1950s–1970s and, no correlation being found, was dispensed with. Data has, in fact, established that heterosexuals are more likely to molest children than homosexuals. See work of Charlotte Patterson, University of Virginia.

Chapter Five Biological Archaeology

1. Dr. Angela Pattatucci, a molecular geneticst at NIH, suggests that the way to think of this hormonal process of creating a human being, which is mostly completed through the nine months of gestation, is to compare it to a single day of producing a daily newspaper. The editors direct production, make decisions, etc., acting in effect like hormones. If the paper closes at 11:00 P.M. then at 8:00 A.M. it is extremely malleable; it could be almost anything. By definition it's going to have a basic structure—front page, a sports section, advertising—just as humans will have a brain, two eyes, a spine, one heart, and so on, but what goes into a particular section is both dependent on what happens during the day in world events outside and on the decisions the editors make internally on how to put these events in the paper. As the hours tick by, the edition begins to take form; editors decide which items will definitely be published while leaving other things in the air and dropping others. By the 11:00 P.M. deadline, the edition is finalized, the production day closes, and the paper is printed.

2. GUNTHER DORNER, "Hormonal Induction and Prevention of Female Homosexuality," *Journal of Endocrinology* 42 (1968): 163–164. See also, for example, Gunther Dorner et al., "A Neuroendocrine Predisposition for Homosexuality in Men," *Archives of Sexual Behavior* 4 (1975): 1–8.

3. HEINO MEYER-BAHLBURG, "Psychoendocrine Research on Sexual Orientation. Current Status and Future Options," *Progress in Brain Research* 61 (1984): 375–398.

4. R. C. FRIEDMAN, F. WOLLESEN, and R. TENDLER, "Psychological Development and Blood Levels of Sex Steroids in Male Identical Twins of Divergent Sexual Orientation," *Journal of Nervous Mental Disorders* 163 (1976): 282–288.

5. JOHN MONEY, M. SCHWARTZ, and V. G. LEWIS, "Adult Erotosexual Status and Fetal Hormonal Masculinization and Demasculinization: 46XX Congenital Virilizing Adrenal Hyperplasia (CVAH) and 46XY Androgen Insensitivity Syndrome (AIS) Compared," *Psychoneuroendocrinology* 9 (1984): 405–415.

6. J. MONEY and A. EHRHARDT (1972). *Man, Woman, Boy, Girl.* Johns Hopkins University Press, Baltimore, London.

7. R. W. GOY and J.A. RESKO, "Gonadal Hormones and Behavior of Normal and Pseudohermaphroditic Nonhuman Female Primates," *Recent Progress in Hormonal Research* 28 (1972): 707–733. Of particular significance is that using hormones, Goy has also succeeded in masculinizing behavior in female rhesus monkeys *without* masculinizing their bodies. Critics had argued that the correct interpretation of the early work with androgenized rhesus was that their masculine behavior was due simply to the fact that they had been made hormonally to look male; therefore their peers treated them as though they were male; therefore they acted male (a social explanation). This anthropomorphic argument is notably disproven by Goy's feat: R. W. Goy, F. B. Bercovitch, and M. C. McBriar, "Behavioral Masculinization in Independent of Genital Masculinization in Prenatally Androgenized Female Rhesus Macaques," *Hormonal Behavior* 22 (1988): 552–571.

8. RICHARD GREEN, *The "Sissy Boy Syndrome" and the Development of Homosexuality.* New Haven, CT: Yale University Press, 1987.

9. J. MICHAEL BAILEY and KENNETH ZUCKER, "Childhood Sex-Typed Behavior and Sexual Orientation: a Conceptual Analysis and Quantitative Review," *Developmental Psychology* 31 (1) (1995): 43–55. Finding: about 75 percent of boys who exhibit "extremely feminine behavior" are gay as adults and about 10 percent of girls who exhibit "extremely masculine" behavior are lesbians as adults. For boys and girls who, respectively, display "feminine" and "masculine" behavior, in adulthood 51 percent and 6 percent are homosexual.

10. FREDERICK L. WHITAM and ROBIN M. MATHY, *Male Homosexuality in Four Societies: Brazil, Guatemala, the Philippines, and the United States.* New York: Praeger, 1986. And Frederick L. Whitam and Christopher T. Daskalos, "Male Sexual Orientation, Femininity, and Childhood Cross-Gender Behavior in Four Societies: Brazil, Peru, Thailand, and the United States," presented at the Society for the Scientific Study of Sexuality, Miami, November 1994.

Whitam estimates that roughly 25 percent of homosexual men as children display highly gender atypical behavior and are most likely to engage in cross dressing as adults; 50 percent show marked gender atypical behavior; 15 percent are less athletic than heterosexual boys but display no gender atypical behavior and are traditionally masculine in other respects; and 10 percent are heterosexual-male typical macho in all respects.

11. FREDERICK L. WHITAM, "Culturally Invariable Properties of Male Homosexuality: Tentative Conclusions from Cross-Cultural Research," *Archives of Sexual Behavior* 12 (3) (1983).

12. These experiments are detailed in the National Holocaust Museum in Washington, D.C. See also Richard Plant, *The Pink Triangle: The Nazi War*

Against Homosexuals. New York: Henry Holt, 1986. In an interview on October 19, 1994, Dr. Klaus Müller, a scholar at the Holocaust Museum, detailed some of the medical experiments:

"Castration was done in the camps as an experiment to change sexual orientation. In a draft law in 1943, it was described, when one was sentenced to the camps for homosexuality, as a medical solution. The other [solution] that was debated was liquidation. Castration became more prominent as an easy medical solution. It didn't have any effect, which, as we know from much medical history, doesn't matter.

"The experiments carried out in Buchenwald were set up as medical experiments to see if one could change sexual orientation. Dr. Carl Vaernet, who was Danish and a member of the SS, performed operations designed to convert homosexual men into heterosexuals by implanting capsules in their legs which released testosterone. He came back to his patients and asked his patients if they were still attracted to men, and because they were eager to help, they said no. So he reported to Himmler that he was very optimistic about finding a surgical solution to homosexuality. Two people died from the surgery, but all the others were ready to testify that they were heterosexual. They weren't, of course. The other path was castration, and we have documentation from the late 1930s of those who were either forced to undergo castration or were exported to concentration camps.

"The legal situation, in 1936, said officially that castration was a voluntary act and shouldn't be done under pressure, so victims were made to sign a statement saying they were undergoing it voluntarily. In 1939 Himmler dissolved this law. From that time, if a homosexual in the camps were to undergo the operation, he would be released. There is a medical case in 1944 about the long-term effects on castration on homosexuals in which close to 700 men were researched. Only a few were released. If you were released, you became an object of medical curiosity and research on the long-term effects of castration on sexual orientation. Vaernet experimented on about 20 people. We don't know how many people in all were experimented on in these ways, but we talk about several hundred in total."

Chapter Six Genetic Grammar 101: A Crash Course

1. The reason all the As, for example, don't wind up in the same place in the A lane of the gel pool is that each letter is attached to all the letters that come before it. If the sequence is, from the first letter, TGGACCA, the first "A" in the sequence is actually put in the gel as TGGA (4 letters into the sequence), and the second A is TGGACCA (7 letters into the sequence). So when Hamer looks at the autorad, the 4th letter position of the sequence, which is really occupied by a chunk of DNA 4 letters, shows up as A, the last

letter of the chunk. The 5th position (in which is really sitting a chunk 5 letters long, TGGAC) shows up as its last letter, C. And so on.

2. The author of the "Map of the Y Chromosome" is molecular geneticist Jane Gitschier of the University of California, San Francisco.

3. In defense of those doing the work, however, no DNA fingerprinter would check merely one gene. Many, many alleles are checked, and the statistical likelihood of a person having one allele is multiplied by the likelihood of having the next, which is multiplied by the next, and the next, and on and on until reaching a certainty factor of, frequently, 99.9 percent. This is the reason O. J. Simpson's attorneys never contested the DNA evidence; statistically, the likelihood of the blood at the scene of his ex-wife's murder being his was virtually 100 percent.

4. Byne's warning about "different species, different rules" fractures not only in the astounding interchangeability of genes but in other ways as well. Neural systems such as the dopamine system in the human brain exist, in just as sophisticatedly a form, in rats. In fact, the dopamine systems of rats are so well developed that scientists are able, quite effectively, to test schizophrenia drugs meant for humans on them—these drugs work on dopamine—*without* saying "the rats are schizophrenic." This implies that specific molecular aspects of the biology of sexual orientation can be researched in rats—or in yeast—without saying the rats are homosexual.

5. Actually, neither women nor men have a functioning backup X in a given cell; both sexes have to make do with only one allele for each gene on the X per cell, men because they have only one X, women, because one or the other of the two X chromosomes in each cell becomes an inert little mass called a Barr body and is nonfunctional. The reason women effectively have a "backup" is not that they have two active Xs per cell but that, in different cells, it isn't always the same X that becomes the Barr body but, usually, the X with the non-functioning alleles.

6. The role/ use/ utility of women as subjects in experiments is touchy. More genetic work on sexual orientation has been done on men than women, and women often raise the issue of why men and not women are in general more often accepted as research subjects. Maxine Singer, a geneticist at the Carnegie Institution in Washington, D.C., commented on this in her matter-of-fact manner. "Sometimes an experiment just works with men and not women. Many of these things on the X are examples. Having said that, there is a long, long history of not using women in all kinds of clinical investigations, probably for the singularly lousy reason that men were doing the study and made the argument that they would have fewer biological variables like hormonal surges and such to deal with if they kept it to men. So they just did men, and their argument was both true in that narrow, short-term way

and absolutely unconscionable. Inexcusable. And one day someone woke up and realized well, wait a minute, you test this drug only on men and then give it to women and you're surprised that women sometimes react very differently. But *women* aren't surprised! Believe me. They are in some cases harmed."

She tapped her fingers on the arms of her chair, looking annoyed, and said briskly, "You know, for years I was on a committee that was talking about redoing science curricula, and one of our tasks was to review a document about what kids should learn about biology. We were all reading this document, and there was a whole section on human development, from birth to aging to death. It covered puberty and sexual maturation, which marks biologically the passage to adulthood. And all of sudden there's a sentence I see: 'There are no physiological events which mark the passage toward old age.' " Singer held the invisible document in her hand and stared at it. "And I look at this sentence. And I think well, this is *strange.* . . . So I go to the next meeting, and I say 'I would like to discuss this sentence because it seems to me that menopause fits your definition, and this sentence is wrong.' "

She paused. "There was *dead silence.* Nobody wanted to hear the word 'menopause.' Not the men. Not the women. Nobody wanted to talk about it. And you have to force them to recognize that there is in one sense a very clear demarcation later in life as clear as the onset of puberty. Ah, but this one only exists in women." Singer lifted her eyebrows. *"Hm."*

Chapter Seven The Gay Gene

1. As usual, nature provides several exceptions to the usual female/male roles. Sagan and Druyan, noting Stephen Emlen's work with Jacana birds, recount that among Jacanas, "males do all the parenting, and the females compete vigorously for something like a harem of males. Those females who don't possess a harem don't reproduce, so the dominant females are often challenged by lower-ranking females. When a takeover attempt succeeds, the incoming female routinely destroys the eggs and kills the chicks. She then sexually solicits the males, who now have no young to distract them—and so are able to attend to propagating the genetic sequences of the incoming female. . . . There are species in which the female is eager to mate with many males and there are species in which the male plays a major, even primary, role in raising the young. Over 90 percent of the known species of birds are 'monogamous'; so are 12 percent of the monkeys and apes. . . . [but] monogamous doesn't mean sexually exclusive; in many species . . . the male . . . is also sneaking out for a little sex on the side; and she is often receptive to other males." DNA fingerprinting reveals that as many as 40 percent of the young reared by "monogamous" bird pairs are sired by extramural encounters, and numbers almost as large may apply to heterosexual humans.

2. Were the basic plan of the human genome available to Hamer, his work would be simplified exponentially. With the complete genomic map, he explained, every A, C, G, and T would be known, and Hamer could simply extract DNA from the families and ask, "What specific genetic sequences do the gay people in this family share that the heterosexuals don't?" He would then use a very powerful computer to compare all three times ten to the ninth bases, which would today take a matter of weeks, "In the end," Hamer says, "it would simply tell you, 'Okay, the gay ones have the same sequences here, here, and here.' But till we have this map, forget it!" In fact, assembling this basic genetic map is precisely the object of the Human Genome Project, which has run consistently ahead of schedule. Hamer should have to wait only until 2005, and perhaps not that long.

3. The order in which geneticists home in on genes is first finding the chromosome, then narrowing the search down to the locus, then finding the gene, and finally deciphering the letters that spell out the triplets in the gene (like finding the chapter, the paragraph, the sentence, and last, the letters that spell out the words in the sentence). The steps are called (1) *mapping,* getting a fix on the general location of the gene with linkage analysis, (2) *cloning,* the word geneticists use to mean "isolating" the gene, identifying exactly where it starts and stops. (Cloning uses a technique called polymerase chain reaction (PCR) that multiplies copies of the gene as a photocopier multiplies copies of an original document), and, finally, (3) *sequencing,* picking apart the gene to get the spelling of the bases—ACTTGCTGACG, etc.—which is done with electrophoresis gels.

4. This kind of luck is not at all unknown in genetics. When researchers began their search for the Huntington's gene in the early 1980s, they were faced with the grim task of surveying the entire human genome. It could have been sixty or seventy thousand tries before they found a marker gene close to the Huntington's gene; it turned out to be the twelfth. Nancy Wexler, a scientist who led that search, has written, "It was as though, without the map of the United States, we had looked for a killer by chance in Red Lodge, Montana, and found the neighborhood where he was living."

5. However ingenious the linkage analysis method may be, Hamer does make a qualification: "In general, the robustness of a scientific finding is inversely proportionate to the amount of statistics needed to demonstrate its hypothesis. In other words, if you have to do a whole lot of stat work to get a significant result, it indicates you don't know what the hell is going on. Usually the simpler things are, the more believable they are."

6. DEAN H. HAMER, STELLA HU, VICTORIA MAGNUSON, NAN HU, ANGELA M.L. PATTATUCCI, "A Linkage Between DNA Markers on the X Chromosome and Male Sexual Orientation," *Science* 261 (July 16, 1993): 321–327.

7. DEAN H. HAMER and ANGELA PATTATUCCI et al., "Linkage Between Sexual Orientation and Chromosome Xq28 in Males But Not in Females," *Nature Genetics* 11 (November 1995).

Chapter Eight What Does "Genetic" Mean?

1. As noted at the end of Chapter 7, Hamer has since looked at the straight brothers, and they, in fact, do inherit the same X locus at a drastically lower rate than their gay brothers. What this indicates is that the penetrance of the gay allele is, in fact, not all that low; it seems to be a reasonably powerful gene in the population Hamer studied.

2. Biologist Norman Carlin has noted that a high-penetrance gay gene is more consistent with a disease model. The various sociobiological hypotheses about the gene's non-disease adaptive advantage selected for evolutionarily would be bolstered by low penetrance.

3. There is yet another scenario, favored by liberal biologists, that blacks could possess a genetic endowment that, all things being equal, makes them intellectually superior to whites, enabling them to perform as well as they do despite nutritional deprivation. Of course, this premise requires an assumption of even more extreme deprivation, but this is not outside the possible.

4. MAOI's—monoamine-oxidase inhibitors—an early and crude group of psychopharmaceuticals, were "dirty" drugs that, due to their high tyrosinase content that dangerously raised blood pressure, could cause all sorts of unpleasant side effects. Users had to limit their diets strictly. A marvelous, though vanishingly subtle, joke playing off this fact was slipped into the Thomas Harris novel and the movie *The Silence of the Lambs.* In one scene, Hannibal Lecter, the mad cannibal genius and serial killer, grins as he tells Clarice, a young FBI agent, that he murdered one of his victims "and I ate his liver with some fava beans and a nice Chianti." Lecter was not taking his medication; patients obediently taking their MAOIs are specifically forbidden to eat fava beans, Chianti, and liver.

Chapter Nine How the Gay Gene Might Work

1. Cassandra Smith's point about sexual orientation being merely one smaller component of the greater rubric Sexuality—a common inheritance comprising many aspects—can be given diagrammatically:

SEXUALITY

Gender identification

Tastes in race of partner

Erotic response

Physical build

Sexual aggression

Sexual drive (frequency desired)

Sexual drive (number of partners desired)

Sexual orientation

Tastes in height, weight, and race of partner

Sexual fantasy

Sexual attraction

Self-identification of sexual orientation

Physical gender

Chromosomal gender

Morality and religious views

Age

Cultural norms

2. In various labs around the world, the search for the biological origins of transsexualism is proceeding parallel to and (given the intriguing similarities) in connection with the search for the biological origins of sexual orientation. Simon LeVay's 1991 discovery of a difference in the hypothalamus between heterosexuals and homosexuals has its counterpart in the transsexualism research, a finding by a team of Dutch researchers led by Dick Swaab that the bed nucleus of the stria terminalis in the brains of male-to-female transsexuals is the same size as that of non-transsexual women, consistently smaller than that of non-transsexual men. *Nature* (1995)

The four distinct traits, 1) sexual orientation 2) physical gender 3) gender identity, and 4) sexual behavior, combine differently—and, the important point, independently—in different individuals. Take straight men and gay men, for example. The traits, gender identity, physical gender, and sexual behavior, are both male-typical in gay and straight men alike, but the trait sexual orientation is reversed (in gay men, it is female-typical). The breakdown is as follows:

Straight Men	Physical Gender:	male
	Internal Gender:	male-typical
	Sexual Behavior:	male-typical
	Sexual Orientation:	male-typical
Gay Men	Physical Gender:	male
	Internal Gender:	male-typical
	Sexual Behavior:	male-typical
	Sexual Orientation:	female-typical
Transsexual Men		
(Pre-operative)	Physical Gender:	male
	Internal Gender:	female-typical
	Sexual Behavior:	female-typical
	Sexual Orientation:	male-typical

Straight Women	Physical Gender:	female
	Internal Gender:	female-typical
	Sexual Behavior:	female-typical
	Sexual Orientation:	female-typical
Lesbians	Physical Gender:	female
	Internal Gender:	female-typical
	Sexual Behavior:	female-typical
	Sexual Orientation:	male-typical
Transsexual Women		
(Pre-operative)	Physical Gender:	female
	Internal Gender:	male-typical
	Sexual Behavior:	male-typical
	Sexual Orientation:	female-typical

3. This example is not an isolated quirk of nature. In parts of Asia where malaria is endemic, including Thailand and Myanmar, several deadly alleles are genetically selected for. They can occur in any of the globin genes and create diseases called thalacemias by interfering with red blood cell function, but they confer, again, a huge advantage in resistance to malaria.

4. In a sense, mitochondria act as biochemical maquiadoras, the refining and assembly plants operated just south of the U.S.-Mexican border. Chemicals are imported into them from the cell and processed into the precursors of steroids. Some manufacturing and refining takes place, mostly not understood. "It's like they're little factories in some free trade zone for which we don't have travel papers yet," one scientist said. "All we know for sure is there's lots of stuff constantly moving in and out." The materials are then exported back out to the cytoplasm of the cell to be distributed to various points where more refining and finishing takes place and the final chemical product is shipped to its destination.

5. Biologists are not certain about how every aspect of this rather peculiar means of reproduction operates, but in this species of wasp, males are made male by being haploid (having only one copy of each gene) whereas if a wasp is diploid, with two copies, it is female. Unfertilized eggs, which are naturally haploid, become males while fertilized ones become females. The bacterium takes charge of this process and changes the outcome. If an egg is laid unfertilized (and, thereby, destined to be male), the bacterium somehow ingeniously causes the first mitotic division of the egg cell to fail, the first two haploid nuclei fuse together to make the egg diploid, and the resulting wasp is female. It can then transmit the bacterium.

6. Wilkins was coincidentally the source of many research subjects for the well-known sex researcher John Money. Parents with intersex children

brought them to Wilkins to see what he could do, and Money began his valuable psychobiological studies with these children.

7. Michael Crichton chose an MIH/testosterone-like mechanism as one of the security controls in his *Jurassic Park*. On an island where dinosaurs have been re-created with dinosaur blood from the bellies of ancient mosquitoes, the main characters in the movie version, among them a Dr. Ian Malcolm, played by Jeff Goldblum in the movie, watch dinosaurs hatching. "Surely these aren't the ones that breed in the wild," says the Goldblum character. A geneticist, played by B. D. Wong, responds breezily, "They don't breed in the wild. There's no unauthorized breeding in Jurassic Park." Goldblum: "And how do we know they don't breed?" Wong: "Because all the animals in Jurassic Park are female. We've engineered them that way. It's really not that difficult. All dinosaur embryos are inherently female anyway. They require an extra hormone given at the right developmental stage to make them male. We simply deny them that."

8. Once again, nature has provided us with an exception. Only recently the first person to lack completely all estrogen receptors was discovered.

GLOSSARY

A,C,G,T: The four bases that make up genetic material.

ALLELES: Versions of a gene. Different alleles create different versions of traits. (adj.: allelic.)

ANDROGENS: From the Greek *andros*, for "man," male hormones.

AUTOSOME: Any chromosome that is not an X or Y sex chromosome; humans have twenty-two pairs of autosomes and only one pair of sex chromosomes.

BASE: Also called a nucleotide, the four bases make up DNA: adenine (A), cytosine (C), guanine (G), and thymine (T).

CHROMOSOME: The double helix in a cell's nucleus made up of genetic material; the structure that carries genes.

CONTROL GENES: Genes that profoundly affect the functioning of genes adjacent to them.

CLINICAL RESEARCH: Research carried out on the phenotype (the expression of genes known as traits) rather than on the genes themselves; research done by looking at human traits in an orderly, controlled way.

DOMINANT: An allele (gene version) sufficiently powerful to express itself when paired against most other alleles.

EPIGENETIC: A trait created by a combination of a genetic input and an environmental input.

EPISTATIC: A number of genes working together to create a trait.

EXPRESSION: Genetic action that creates a trait.

EXPRESSIVITY: The degree to which a gene is expressed.

GAMETES: Eggs and sperm. Each gamete carries only one-half of a gene pair.

GENE: A complete strand of ACGT bases that, when read, is the recipe for a protein.

GENOTYPE: An individual's complete genetic map; the full complement of a person's genes. (See also *phenotype*.)

HETEROZYGOUS: A given gene pair in which the two copies are two different alleles (gene versions).

HOMOZYGOUS: A given gene pair in which both copies are the same allele.

INAH 1-4: Interstitial Nuclei of the Anterior Hypothalamus. Four nuclei in the hypothalamus discovered by Laura Allen. Simon LeVay found INAH-3 to be dimorphic regarding sexual orientation.

LINKAGE: Used in genetic analysis: "When two gene pairs sit close together on the same chromosome, they do not show independent assortment and are thus said to be linked." More simply, two genes near to each other on a parent's chromosome are more likely to be passed on to off-spring together than two genes far apart.

MULLERIAN DUCT: The duct present in the embryonic female containing the female internal reproductive organs—internal vagina, uterus, fallopian tubes. (See also *Wolffian duct*.)

MUTATION: A change in the spelling—and thus the effect—of a gene.

MUTAGEN: Anything that creates mutations in genes, such as X rays and gamma rays that alter the spelling and function of genes in the tissues through which they pass.

PEDIGREE ANALYSIS: Tracing a trait through several generations of a family using a family tree.

PHENOTYPE: From the Greek for "the form that is shown"; the physical expression of genes; the trait(s) created by gene expression.

PENETRANCE: The probability that a given gene will be expressed as a trait.

PLEIOTROPY: Genetic side effects; the phenomenon whereby genes create more than one trait, often a primary trait selected for by evolution due to an advantage it carries and a secondary trait that is incidentally created.

PROBABILITY: The number of times an event is expected to happen divided by the number of chances it has to happen (the number of trials). The probability (p) of rolling a two on a six-sided dice is $p = 1/6$, or .166.

RECESSIVE: An allele (gene version) that is weaker than other alleles and that, when paired with them, is not expressed.

RIBOSOME: A particle that reads genes and then constructs amino acids from the genetic instructions, assembling them into proteins.

SEXUAL DIFFERENTIATION: The process whereby the fetus is made internally and externally male or female.

SEXUALLY DIMORPHIC: Any organ or body part that is of a different size or shape according to sex, one size in men and another in women.

SEXUAL-ORIENTATIONALLY DIMORPHIC: Any organ or body part that is of a different size or shape according to sexual orientation, one size in heterosexuals and another in homosexuals.

SIGNIFICANCE: The statistical assessment of how likely it is that a scientific finding occurred merely by chance.

STOCHASTIC: Random biological processes.

TESTOSTERONE: The most common male hormone.

WOLFFIAN DUCT: The duct present in the embryonic male containing the male internal reproductive organs—vas deferens, seminal vesicle, and ejaculatory ducts. (See also *Mullerian duct.*)

X-LINKAGE: The characteristic inheritance pattern in men shown by traits created by genes on the X chromosome.

XQ28:   The twenty-eighth region of the q (long) arm of the X chromosome, where Dean Hamer discovered a genetic locus linked to homosexual orientation in some men.

INDEX